Translational Research in Environmental and Occupational Stress

Shashi Bala Singh • Nanduri R. Prabhakar
Srinivas N. Pentyala
Editors

Translational Research in Environmental and Occupational Stress

Springer

Editors
Shashi Bala Singh
Defence Institute of Physiology
and Allied Sciences (DIPAS)
New Delhi, India

Srinivas N. Pentyala
Department of Anaesthesiology
School of Medicine
State University of New York
Stony Brook, New York, USA

Nanduri R. Prabhakar
Department of Medicine
and Director for System
The University of Chicago
Chicago, USA

ISBN 978-81-322-1927-9 ISBN 978-81-322-1928-6 (eBook)
DOI 10.1007/978-81-322-1928-6
Springer New Delhi Heidelberg New York Dordrecht London

Library of Congress Control Number: 2014946730

Printed on acid-free paper

Springer is part of Springer Science+Business Media (www.springer.com)

Preface

The book on "Translational Research in Environmental and Occupational Stress" introduces the various aspects of translational research which resulted from bench to bedside and bedside to bench for the benefit of the society at large. This book includes the tangible and non-tangible products which have been developed after rigorous research in the field of Environmental and Occupational Stress. It serves as a guide as well as a reference book for practicing clinicians, scientists, and research scholars.

Translating Physiological Knowledge to Health Application: This chapter introduces the problems encountered by soldiers operating in extreme operational environments such as high altitude, cold and desert. This chapter describes how the physiological responses during high altitude acclimatization were formulated as the staging of acclimatization schedule at high altitude. Likewise, physiological effects of heat stress resulted in formulation of an ergogenic drink. The extreme cold conditions led to the development of anti-frostbite cream. The translational research led to the development of products/technologies/processes which helped in providing health solutions.

Grassroots Solutions to Overcome Abiotic and Biotic Environmental Stress in Agriculture: This chapter gives an insight into the myriad problems in agricultural farming due to environmental stress and their counter measures. The authors give an account of the grassroot solutions and innovations by the farmers to overcome abiotic and biotic stress. Moreover, they also highlight various efforts of the farmers to conserve germplasm and improve the quality of their produce.

Stress Research: Varied Paradigms: In this chapter, the authors describe different existing perspectives for the thorough understanding of the concept of stress from multiple viewpoints. Further, they focus on discussing different research paradigms. Beginning with the early physiological notions of 1920, the authors analyze genetic, biological, psychosocial, developmental, and interactional paradigms.

Therapeutic Potential of Intermittent Hypoxia: Lessons from Respiratory Motor Plasticity: In this chapter, the authors consider distinctions between the intermittent hypoxia (IH) protocols giving rise to pathology versus beneficial effects and then consider the mechanisms and implications of respiratory and non-respiratory motor plasticity elicited by modest protocols of IH. Specifically, the goals of this chapter are to: (1) differentiate between the experimental IH protocols giving rise to beneficial versus pathologic effects in multiple body systems; (2) review the impact of modest IH protocols on respiratory and non-respiratory motor function; (3) review recent advances in our understanding of mechanisms giving rise to that IH-induced motor plasticity; and (4) consider the clinical implications of IH-induced motor plasticity.

High-Altitude Research and Its Practical Clinical Application: In this chapter, the authors highlight their experiences and discoveries that change the way disease is treated in hypoxic environment of high altitude. This chapter describes life under chronic hypoxia which is as normal as sea level and diseases at high altitude, particularly in the cardio-respiratory areas which behave differently because the organism adapts to chronic hypoxia, even deviating from the normal and optimal status of good health.

Nanomaterials in Healthcare: In this chapter, authors present a summary of different nanomaterial based technologies in healthcare with emphasis on therapeutics, medical devices and diagnostics.

Nanotoxicity and Cellular Stress Response: Physical and Chemical Properties and Their Link to Translational Research: This chapter is an elaborate account on nanomaterial toxicity wherein the author emphasizes on the various physicochemical properties of nanomaterial such as size, shape, chemical composition and surface chemistry which is required to understand their interaction with the biological system.

HIF-1 and EGLN1 Under Hypobaric Hypoxia: Regulation of Master Regulator Paradigm: This chapter highlights the adaptation and acclimatization to high altitude which are driven by the alterations in the O_2-sensing pathway. The authors describe the role of HIF-1 and EGLN1, the central regulatory molecules of O_2-sensing pathway which will help in our understanding of the pathophysiological mechanisms.

Electrochemical Biosensors for Hypoxia Markers: In this chapter, the authors describe the development of a biosensor which measures the biologically important hypoxia biomarkers as it provides valuable information regarding the personnel at high altitude. This is a perfect example of translational research which led to the development of a cost effective and portable electrochemical biosensor for the measurement of various hypoxia biomarkers in volume miniaturized samples using screen printed electrodes.

Determining Nutritional Requirements of Indian Soldiers: An Outcome of Translational Research: This chapter stresses upon the adequate nutritional requirement for maintaining highest level of physical fitness under different climatic conditions and operational situations. The planning of ration for combat operations is a challenging task, as balance between nutrient requirement for optimum health and palatability needs to be ensured. Translational research has played a great role in formulating dietary recommendations and ration scales for Armed Forces and at the same time, various ready to eat products became the innovative products of military nutrition research.

Improvements in Adjuvants for New-Generation Vaccines: This chapter highlights the development of new herbal adjuvants, which will lead to the generation of both Th1 and Th2 immune response to mitigate the diseases caused by bacteria or viruses.

Rapid Acclimatization Strategies for High-Altitude Induction: This chapter describes the possible ways to prevent high altitude induced diseases. Authors claim that acclimatization to high altitude is the best strategy to prevent high altitude induced diseases, but it can be achieved by hypoxia preconditioning by the use of interventions like hypoxia mimeticsm, viz, cobalt chloride ($CoCl_2$) and sphingosine-1-phosphate (S1P) and nanocurcumin. They also discuss an alternative approach to induce acclimatization and reduce incidence of acute mountain sickness by the use of intermittent hypoxia exposure.

Noise, the Silent Killer: This chapter highlights noise induced hearing loss and its effective counter measure by carbogen breathing. This chapter describes the development of remedial measures for conserving hearing and effectively counteracting the adverse interactions of noise on the efficiency and performance of people working in noisy environments.

Yoga for Preventive, Curative, and Promotive Health and Performance: In this chapter, the authors describe about Yoga, which is an ancient Indian system of philosophy, culture, tradition and a way of maintaining better life, established in India thousands of years ago. The authors also highlight that regular yogic practices endow perfect physical and mental health to its practitioner which improves aerobic capacity, anaerobic capacity, joint flexibility and muscle strength. Authors claim that Yoga is not only a discipline to be practiced by spiritual aspirants but also has got relevance to the spirit of military activities.

Technology Translation from Heat Physiology Research: The authors describe the effect of high ambient temperature exposure on human body and its effects. Authors also highlight the translational aspects of their study, which resulted in the formulation of replenishment drink DIP-SIP and development of a Man Mounted Air Conditioning System.

Improved Habitability Under Extreme Environments at High Attitude: This chapter deals with improving habitability under extreme environmental conditions. This chapter highlights the solar shelter 'Sourja' for cold conditions, space heating device (Bukhari) and oxygen enriched shelters at high and extreme altitudes to combat hypoxia and solar snow melter.

Inhaled Nitric Oxide Therapy for Treatment of High-Altitude Pulmonary Edema: This chapter describes about the inhaled nitric oxide therapy for treatment of high altitude pulmonary edema. This chapter also highlighted the development of tailor made NO delivery system for patients of high altitude pulmonary edema.

High-Altitude Medicine: The Path from Genomic Insight to Clinical Applications: This chapter describes the role of genomics in high altitude induced diseases. Authors highlight that the ultimate goal would lie in making genomic information readily accessible for more informed and better management of high altitude environmental stress.

Hypoxia in Acute Chemical Emergencies: Toxicity, Mechanism, and Treatment: In this chapter, the authors discuss the toxicity, molecular mechanism(s) of action and treatment modalities of chemical asphyxiants. It also addresses the possible implication of organophosphorus compounds in producing chemical hypoxia. This information will be useful for medical management of hypoxia-related chemical emergencies.

Hypertension at High Altitude: This chapter describes the cause for sustained elevation of blood pressure during exposure to high altitude and the factors associated with inter-individual variation. Overall, this chapter reviews available literature on systemic blood pressure responses to high altitude.

Herbs for Mitigating High Altitude Maladies: This chapter highlights the translation of traditional ethnopharmacological wisdom into a process and product with a scientific rationale and development of products for mitigating high altitude maladies. Herbs have wide range of therapeutic effects and through systematic scientific investigations using various animal models and clinical trials, research at DIPAS has translated this knowledge into products for protection against high altitude maladies.

Lessons from a 20-Year Investigation of Intermittent Hypoxia: Principles and Practices: In this chapter, the author presents an elaborate description about intermittent hypoxia. An in depth account on the key mechanisms involved and the effect of intermittent hypoxia on the cardio-respiratory system, immune system, reactive oxygen and nitrogen species signalling, hypoxia inducible factor signalling and epigenetics is given. The author also talks about the clinical applications of intermittent hypoxia treatment being practiced in treating various disease states.

New Delhi, India — Shashi Bala Singh
Chicago, IL, USA — Nanduri R. Prabhakar
Stony Brook, NY, USA — Srinivas N. Pentyala

Acknowledgements

The book on Translational Research in Environmental and Occupational Stress is an effort to acknowledge the recent advances in the area of high altitude physiology and its translational applications that have revolutionized developments in contemporary high altitude medicine and biology. The preparation of this book required the efforts of many individuals. The Editors particularly wish to thank each of the authors and contributors. Dr. K. P. Mishra, Mr. Manmeet Singh, Dr. Som Nath Singh and Dr. Prasanna K Reddy saw that the process of receiving, editing and assembling the chapters went as smoothly as possible. Without their efforts, the completion of this book would have been immeasurably more difficult. Editors wish to gratefully acknowledge the efforts of each of the members of the editorial and production staff of Springer India who participated in the preparation of this book.

Contents

About the Editors

Dr. Shashi Bala Singh Outstanding Scientist and Director, Defence Institute of Physiology and Allied Science (DIPAS), obtained her Ph.D. degree in Human Physiology from All India Institute of Medical Sciences, New Delhi, and started her career in DRDO at DIPAS, Delhi, as scientist 'C'. She also served as Director, Defence Institute of High Altitude Research (DIHAR), Leh, before taking over as Director, DIPAS. She has immensely contributed to the understanding of high altitude physiology and pioneered the development of nutraceuticals and prophylactics for several high altitude maladies that include hypophagia and cognitive impairment. She has developed supplementation with antioxidant and cholinomimetic drugs to improve high altitude induced impairment in cognitive functions. She has been instrumental in devising technologies for improving the quality of life in extreme environmental conditions of the combatants. Some of the pertinent products include zero energy based solar shelters, Sourja, improvised space heating devices and microclimate controlled oxygen enrichment enclosure. She received "Scientist of the Year Award" by Hon'ble Prime Minister in year 2010 and is the elected Fellow of National Academy of Sciences, India (FNASc.), Indian Association of Biomedical Scientists (FIABMS) and Indian Academy of Neuroscience (FIAN). She is the President of Federation of Indian Physiological Societies (FIPS) and also holds the editorship in many prestigious journals.

Dr. Nanduri R. Prabhakar received his Ph.D. and D.Sc. in Physiology from the M.S. University, Baroda, India, and Ruhr University, Bochum, Germany, respectively. He was a fellow of the Swedish Institute at the Nobel Institute for Neurophysiology, Karolinska Institute, Stockholm. He joined the Case Western Reserve University, Cleveland, USA in 1984 as an Assistant Professor and became Professor and Vice-Chairman of the Department of Physiology (1997–2007). In 2007, he joined the University of Chicago and is currently Harold H. Hines Professor and Director, Institute for Integrative Physiology and Center for Systems Biology of O_2 Sensing at the same university. He is a leading authority on O_2 sensing mechanisms and physiological consequences of hypoxia. He is the Director of Program Project Grant funded by the National Institutes of Health, USA, and published more than 230 papers and edited 7 books.

Dr. Srinivas N. Pentyala is a faculty member of anesthesiology, urology, health sciences, physiology and biophysics at Stony Brook Medical Center, NY. He is also the Director for Translational Research in Stony Brook's Anesthesiology Department where he is working on several applied projects, including drug discovery, diagnostic markers, and devices for health care settings. His research has received funding from both public and private sources. Dr. Pentyala has several publications and patents to his credit and also received numerous honors and awards.

Translating Physiological Knowledge to Health Application

W. Selvamurthy and Shashi Bala Singh

Abstract

Physiology forms the basis for medical practice. Earlier in India, physiological research undertaken by medical teachers and researchers was primarily focused towards publications rather than translating that knowledge into a device/technology/process. The Defence Physiological Research and Development endeavours have provided a new paradigm since 1960 with the establishment of Defence Institute of Physiology and Allied Sciences (DIPAS). Most of the research efforts were aimed at finding solution to problems encountered by soldiers operating in extreme operational environments such as high-altitude, cold, desert, underwater and aerospace environments. Even the basic physiological research had an ultimate aim of application for the well-being of the soldiers thereby focusing on translational component. To cite a few examples, physiological responses during high-altitude acclimatisation studied on sojourners were translated to formulate the staging of acclimatisation schedule at high altitude. The nitric oxide and oxygen therapy for treatment of high-altitude pulmonary oedema (HAPO) resulted from the pathophysiological studies undertaken on the patients with HAPE. This resulted in devising an equipment to deliver a precise concentration of nitric oxide and oxygen to the patients while monitoring the concentration of nitrogen dioxide. Studies carried out to assess the physiological effects of heat stress resulted in formulating an ergogenic drink to keep the fluid electrolyte balance, thereby optimising the physical and mental efficiency of soldiers operating in desert environment. Development of a radio sensitiser, namely, 2-deoxy-D-glucose (2DG), came out of physiological research to find a method to ameliorate the adverse effects of radiotherapy in cancer patients. Such examples of translational research in physiology led to the development of products/technologies/processes which helped in providing health solutions which are illustrated in this chapter.

W. Selvamurthy (✉)
Former Chief Controller R&D for Life Sciences (LS) and International Cooperation (IC), Defence Research and Development Organization (DRDO), New Delhi, India
e-mail: Selvamurthy2003@yahoo.co.in

S.B. Singh (✉)
Defence Institute of Physiology and Allied Sciences (DIPAS), New Delhi, India
e-mail: drshashisingh@gmail.com

S.B. Singh et al. (eds.), *Translational Research in Environmental and Occupational Stress*, DOI 10.1007/978-81-322-1928-6_1,

Introduction

Integrated physiology has been in the forefront to facilitate the understanding of various functions of the organs and systems, which formed the basis of medical practice. Even the Nobel Prize in medicine is given as 'Physiology and Medicine' signifying the greater role of this discipline of medical sciences. Subsequently, physiology gave birth to many subset of disciplines such as molecular biology, genetics, etc. which have gained importance in the recent years. Physiological research in India in the early 1960s was primarily focused on the neurophysiology of food intake, sleep, consciousness, sensory physiology and respiratory functions [4, 34, 35, 37, 38, 43–48]. These led to the understanding of brain functions and other physiological systems. The clinical physiological research demonstrated the pathophysiological responses in different disease conditions [3, 5, 10, 25]. During the late 1960s, a new trend was set by the physiologists who focused on the applied and translational aspects of physiological research which benefited the soldiers and the society significantly. The establishment of Defence Institute of Physiology and Allied Sciences (DIPAS) as a part of Defence Research and Development Organisation (DRDO) was an important milestone to set the trend of translational research in physiology and allied sciences which enabled the soldiers to maintain optimal level of health and efficiency even in extremes of environmental and operational conditions while providing spinoff benefits to the society at large.

High-Altitude Physiology

The Indo-Chinese conflict in 1962 brought a concerted physiological research effort to understand human acclimatisation at high altitude (HA). The cardiovascular, respiratory, metabolic and neurophysiological responses during acclimatisation were studied on sojourners on acute induction and during prolonged stay at HA and compared with acclimatised low landers and high-altitude natives [22, 26, 41, 42, 49]. These endeavours ultimately resulted in the formulation of staging of acclimatisation [23]. If a soldier has to be inducted to extreme altitude (>5,000 m), he has to be taken through three levels of staging for better acclimatisation. The introduction of this staging procedure significantly reduced the mortality and morbidity and the operational efficiency of troops at HA.

High-Altitude Pulmonary Oedema

One of the serious clinical problems encountered among the sojourners at HA is high-altitude pulmonary oedema (HAPO). Increase in pulmonary hypertension, capillary permeability and decrease in surfactant were observed as the pathophysiological factors culminating in the development of HAPE [8, 33, 50]. Various hypotheses were promulgated to explain the pathophysiological mechanisms underlying the genesis of HAPE. These include transarterial leakage, exaggerated sympathetic nerve activity, J-receptor suppression, hypothalamic neurogenic mechanism, reduced chemoreceptor sensitivity and vascular and cerebrovascular water retention [21, 24, 28, 39, 53]. The introduction of diuretics resulted from studies on ADH and aldosterone leading to water retention which needs to be reduced for keeping the effective circulating blood volume at optimal level [12]. In view of hypobaric hypoxia being the important aetiological factor for the genesis of HAPO, a life-saving device, namely, HAPO bag (Fig. 1), that increases positive pressure simulating lower-altitude conditions was developed and supplied to large number of field stations over the Himalayas. This has resulted in saving a large number of lives which would have otherwise succumbed to HAPE [19]. Anand et al. demonstrated for the first time the application of nitric oxide (NO) at 15 ppm and oxygen (O_2) at 50 % concentration to be beneficial in the treatment of HAPE [1]. DIPAS scientist developed a device which can deliver precise concentration of NO and O_2 to the patients with HAPE while ensuring the safety of patients regarding elevated levels of nitrogen dioxide (NO_2) [11] (Fig. 2). A screening method

Fig. 1 HAPO bag

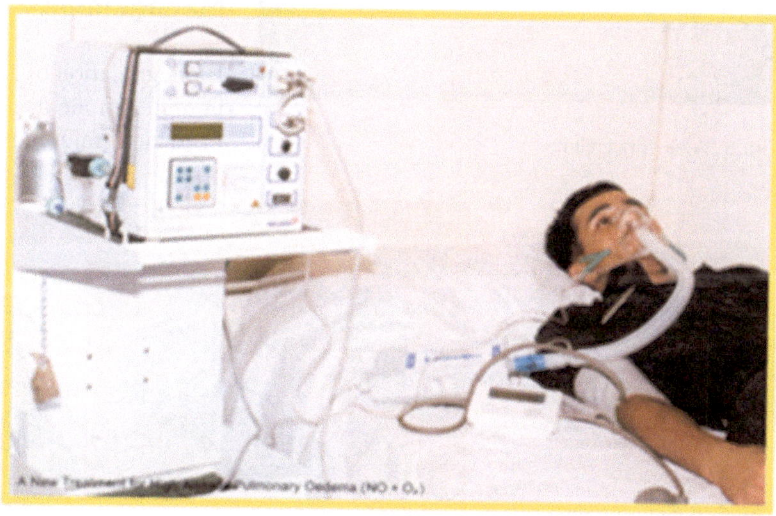

Fig. 2 Nitric oxide delivery system

to assess the susceptibility of individuals to develop HAPE was formulated using the exhaled NO measured by the chemiluminescence method [6]. This can be used to screen out susceptible individuals during rapid induction in emergency. In addition, a modified nebuliser was developed at the Institute of Nuclear Medicine and Allied Sciences (INMAS), another DRDO laboratory, to deliver pulmonary vasodilator drugs into the alveoli in nano size which has been successfully trial evaluated in HA regions [18].

Cold Injuries

Efforts were made to understand the aetiopathology and pathophysiology leading to various forms of cold injuries including frostbite which is the most severe form of cold injury. Subsequently, a new protocol for treatment using rewarming of the affected parts in tea decoction medium at temperature closed to body temperature and application of aloe vera cream was effective in the treatment of frostbite [31]. A large number of aloe vera cream jars were distributed among the soldiers for both prophylactic and therapeutic applications (Fig. 3). Battery-operated heating gloves and socks were also developed to keep the extremities warm, thereby preventing frostbite (Fig. 4).

Fig. 3 Aloe vera cream

Oxygen-Enriched Shelter

Acute mountain sickness (AMS) is a common transient clinical problem manifested by headache, nausea, vomiting, sleep disturbance, loss of appetite, etc. These symptoms normally disappear after acclimatisation for a week [36]. However, severe cases of AMS need to be treated with administration of oxygen. In order to provide oxygen treatment to a large number of AMS patients, an oxygen-enriched shelter was developed and established in a few places at HA (Fig. 5). The solar energy was harnessed to light and heat the shelter [17].

Biodigester for Human Waste Management

Biodegradation of human waste is rather difficult in high altitude due to prevailing cold conditions in which microbes are unable to effectively

Fig. 4 Battery-operated socks and gloves

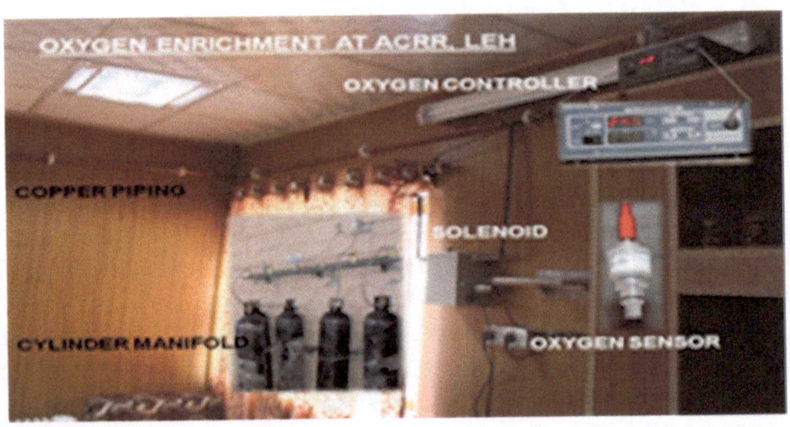

Fig. 5 Oxygen-enriched shelter

Fig. 6 Biodigester

biodegrade the constituents of the faecal matter. Scientists from Defence Research and Development Establishment (DRDE) collected a consortium of microbes from Antarctica, cultured them in the laboratory, studied their kinetics and developed a bioreactor providing a congenial medium and environment to produce a large quantity of this consortium. An underground structure to house these microbes was designed and developed which was linked to the superstructure of the toilet (Fig. 6). Since the consortium included bacteria which carried out the process of hydrolysis, acidogenesis, acetogenesis and methanogenesis and its high level of tolerance to cold and heat, these biodigesters are effective in a wide ranging environments from high altitude to desert [13]. These have already been installed in large number in Himalayan high-altitude regions, Lakshadweep Islands and in other parts of India.

Aerospace Physiology

The fighter aircrafts pose physiological challenges due to high G environment, hypoxia and explosive decompression during emergency ejections. G-induced loss of consciousness (G-LOC) and disorientation result in major air accidents. A sharp decline in cerebral perfusion due to blood pooling in lower extremities due to G force is the major factor responsible for G-LOC. Anti-G suit has been developed and introduced in Indian Air Force in large quantity. These clothing ensembles were made flame retardant [17].

Heat Acclimatisation

Environmental physiologists have evaluated the physiological effects of hot dry and hot humid environments on physical and cognitive functions [40]. This knowledge was translated to climatic mapping for various stations in hot humid and hot dry conditions prevailing across India for providing cooling and air conditioning in the zones of requirement [29]. Work–rest schedule of soldiers was formulated which minimised the incidence of heat casualties and improved work efficiency [2]. The fluid–electrolyte imbalance due to work in heat was mitigated by developing a suitable ergogenic beverage with potassium supplementation [30]. A thermoelectric cooling suit based on Peltier effect was developed and successfully demonstrated to have effective microclimatic cooling which could be worn by people working in high-temperature environments (Fig. 7).

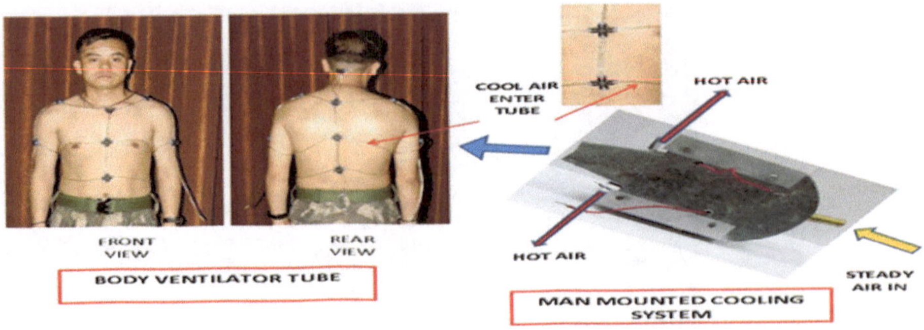

Fig. 7 Vortex cooling suit

Protection Against Noise-Induced Hearing Loss

Noise is a health hazard not only in military environments but also in industry, road traffic due to vehicular noise and even in living environments. Physiological research indicated the processes underlying temporary threshold shift (TTS) of hearing and permanent hearing loss (PHL) due to the damage of hair cells due to high-intensity noise. The conventional method of protection was by using barrier methods with the help of ear defenders such as ear plugs, ear muffs and active noise cancellation devices [9]. Scientists at DIPAS developed a novel approach using carbogen (5 % CO_2 and 95 % O_2) mixture to attenuate noise-induced TTS to protect from PHL. A mobile device was developed to facilitate carbogen breathing before and after noise exposure which resulted in significant protection against noise hazard [7] (Fig. 8).

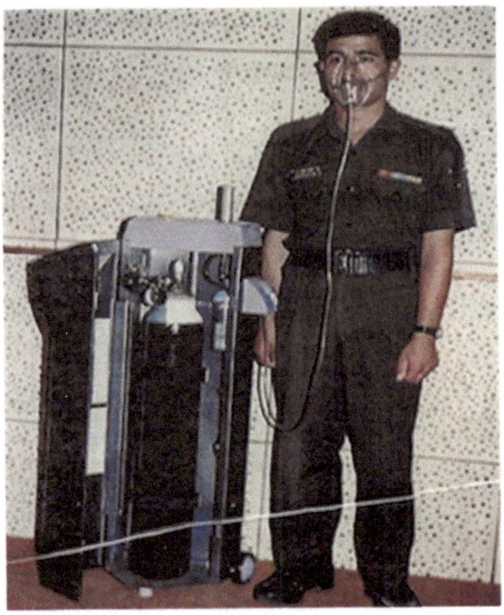

Fig. 8 Portable carbogen breathing system

Underwater Physiology

Submarines operate at various depths under the sea in hyperbaric environments. In the event of submarine accidents, the crew need to escape from the submarine and pass through a gradual ascent to avoid decompression sickness and rapture of the lungs. An indigenous submarine escape hydro-suit along with the life support system for providing oxygen, nitrogen and helium was developed by the Defence Bioengineering and Electromedical Laboratory (DEBEL) and inducted in Indian Navy [27] (Fig. 9).

CBRN Defence Technology

Chemical, biological, radiological and nuclear (CBRN) threat is a global concern. A large number of protective devices have emerged from physiological and engineering innovation. Autojet injector is one such device which is used for administering atropine sulphate and PAM chloride for individual exposed to nerve agents.

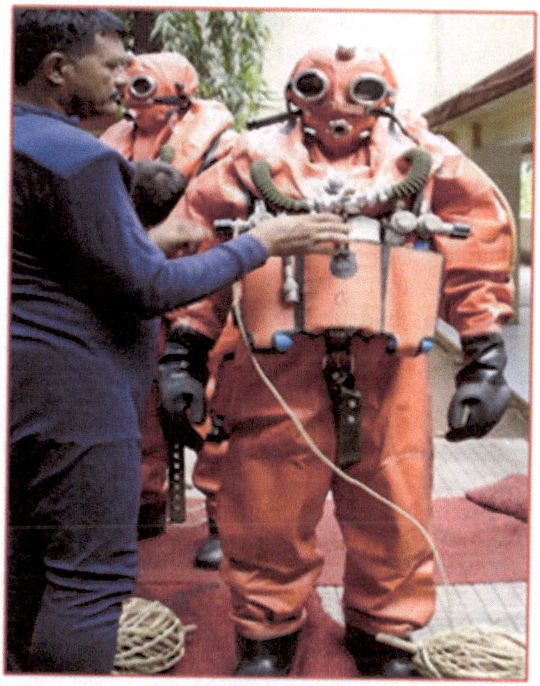

Fig. 9 Submarine escape hydro-suit

Fig. 10 Herbal mosquito repellent

These cholinesterase blocking agents protect neuromuscular transmission of impulses thereby protecting reversibly the individual exposed to nerve agents such as soman, sarine, VX and tabun [16, 52]. Efficient decorporating formulation using Prussian blue as the main ingredient has been developed and successfully trial evaluated for removing radionuclides ingested or inhaled [32]. A very innovative approach at INMAS to develop a radio sensitiser resulted in the formulation of 2-deoxy-D-glucose (2DG) which has completed phase III clinical trials successfully and ready to be launched for marketing. This can ameliorate the adverse effects of radiotherapy in cancer patients [14]. The radio-protective effect of Tulsi (Ocimum sanctum) was demonstrated and taken to phase II clinical trials.

Mosquito Repellent

Mosquitoes are the vectors for many diseases such as malaria, dengue and *chikungunya*. A tropical country like India provides congenial temperature, humidity and other environmental factors where mosquitoes can easily breed and transmit pathogens through biting either during the day or night. Mosquito repellents developed by DRDO provide long-lasting protection (8–10 h). These have been developed by undertaking the electrophysiological recording of antenogram of insects and testing the efficacy of different formulations having repellent properties [15] (Fig. 10).

Biomedical Technologies

Ambulatory physiological monitoring is required in many clinical as well as basic physiological research. However, to use such devices in difficult environments such as HA, desert, aerospace environment and Naval operational environments require adequate ruggardisation which has been accomplished by scientists at DEBEL [17]. A critical care ventilator was developed to provide optimal ventilation for patients in ICU and under critical care [51] (Fig. 11). Extensive research was carried out on a polyherbal preparation which was found to be very effective to treat leucoderma [20] (Fig. 12).

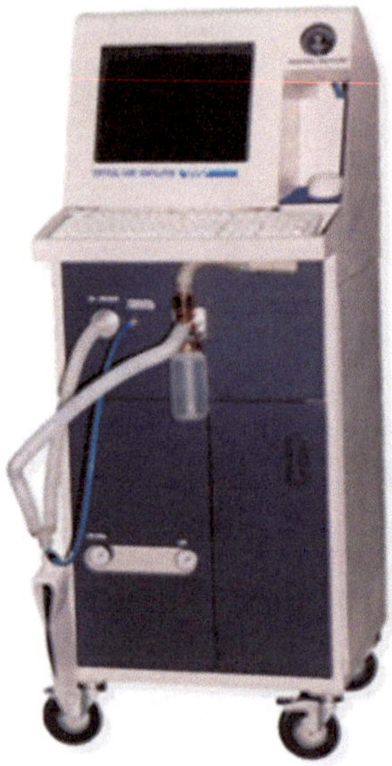

Fig. 11 Critical care ventilator

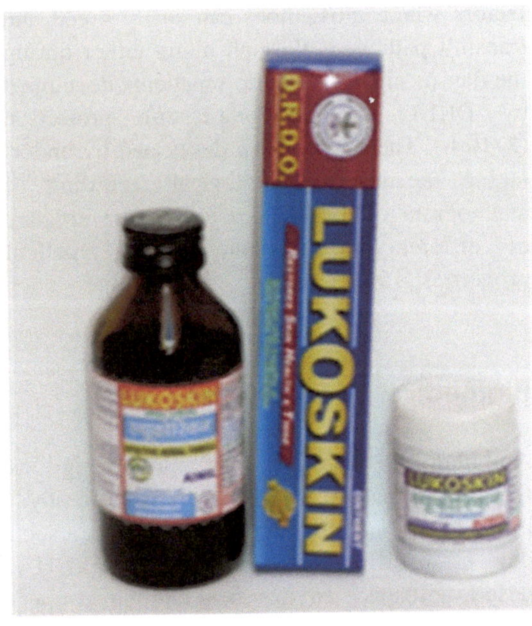

Fig. 12 Leukoskin

Conclusion

While basic research forms the foundation for seeding new technologies and generating new concepts and hypotheses, the applied research and its translations is vital to bring innovation to benefit the society at large. Translation from science to technology to product requires close interaction among three stakeholders, namely, academia, R&D institutions and industries. DRDO laboratories were able to successfully integrate these three entities which resulted in the development of a number of products, technologies, processes and solutions. Even the basic research conducted at our educational institutions and universities should be directed towards a specific goal or problem. Thus, the translational research in physiology or any other discipline will have greater impact on the welfare of the society, quality of life including health and generating wealth.

References

1. Anand IS, Prasad BAK, Chug SS, Rao KRM, Confield DN, Milla CS, Singh N, Singh S, Selvamurthy W (1998) Effects of inhaled nitric oxide and oxygen in pulmonary edema. Circulation 98:2441–2445
2. Baburaj TP (2006) Heat stress problems in desert operations and its management. DIPAS project no. DIPAS/17/2006
3. Bartsch P, Swenson E, Paul A, Julg B, Hohenhaus E (2002) Hypoxic ventilator response, ventilation, gas exchange and fluid balance in acute mountain sickness. High Alt Med Biol 3:361–376
4. Basu CK, Selvamurthy W, Bhaumik G, Gautam RK, Sawhney RC (1996) Respiratory changes during initial days of acclimatization to increasing altitudes. Aviat Space Environ Med 67:40–45
5. Brahmachri HD, Malhotra MS, Joseph S, Radhakrishnan U (1973) Glucose tolerance at high altitude in man. Indian J Med Res 61:411–415
6. Busch T, Bartsch P, Pappert D, Grunig E, Elser H, Falke KJ, Swenson ER (2001) Hypoxia decreases exhaled nitric oxide in mountaineers susceptible to high altitude pulmonary edema. Am J Respir Crit Care Med 163:368–373
7. Chaturvedi RC, Rai RM, Sharma RK (1990) Therapeutic role of carbogen in impaired hearing. Indian J Med Res 92:420–42371
8. Duplain H, Vollenweider L, Delabays A (1999) Augmented sympathetic activation during short term hypoxia and high altitude exposure in subjects susceptible to high altitude pulmonary edema. Circulation 99:1713–1718
9. Fausti SA, Wilmington DJ, Helt PV, Helt WJ, Konard-Martin D (2005) Hearing health and care: the need for improved hearing loss prevention and hearing conservation practices. J Rehabil Res Dev 42(4 suppl 2):45–62

10. Gupta KK (1960) Acclimatization at high altitude and causes of its failure. Indian J Chest Dis 4:211–218
11. Himashree G, Dass D, Banerjee PK, Selvamurthy W (2003) Nitric oxide and the respiratory system. Curr Sci 85:607–614
12. Kumar R, Pasha Q, Khan AP, Gupta V (2004) Renin angiotensin aldosterone system and ACE I/D gene polymorphism in high altitude pulmonary edema. Aviat Space Environ Med 75:981–983
13. Life Sciences Compendium (2008) Published by Directorate of Life Sciences, Defence Research & Development Organisation (DRDO), Ministry of Defence, DRDO Bhawan, New Delhi, p 26
14. Life Sciences Compendium (2008) Published by Directorate of Life Sciences, Defence Research & Development Organisation (DRDO), Ministry of Defence, DRDO Bhawan, New Delhi, p 61
15. Life Sciences Compendium (2008) Published by Directorate of Life Sciences, Defence Research & Development Organisation (DRDO), Ministry of Defence, DRDO Bhawan, New Delhi, pp 156–157
16. Life Sciences Compendium (2008) Published by Directorate of Life Sciences, Defence Research & Development Organisation (DRDO), Ministry of Defence, DRDO Bhawan, New Delhi, p 175
17. Life Sciences Compendium (2008) Published by Directorate of Life Sciences, Defence Research & Development Organisation (DRDO), Ministry of Defence, DRDO Bhawan, New Delhi
18. Life Sciences Compendium (2008) Published by Directorate of Life Sciences, Defence Research & Development Organisation (DRDO), Ministry of Defence, DRDO Bhawan, New Delhi, pp 21–22
19. Life Sciences Compendium (2008) Published by Directorate of Life Sciences, Defence Research & Development Organisation (DRDO), Ministry of Defence, DRDO Bhawan, New Delhi, p 6
20. Life Sciences Compendium (2008) Published by Directorate of Life Sciences, Defence Research & Development Organisation (DRDO), Ministry of Defence, DRDO Bhawan, New Delhi, p 62
21. Maggiorini M, Melot C, Peiere C (2001) High altitude pulmonary edema in initially caused by an increase in capillary pressure. Circulation 103:2078–2083
22. Malhotra MS, Mathew L (1974) Effect of prolonged stay at altitude (4000 m) on autonomic balance. Aerosp Med 45:869–872
23. Manual of health for the armed forces (2002) Published by Directorate General, Armed Forces Medical Services New Delhi, vol 1, pp 81–82
24. Mathew L, Gopinathan PM, Purkayastha SS, Sengupta J, Nayar HS (1983) Chemoreceptor sensitivity and maladaptation to high altitude in man. Eur J Appl Physiol 51:137–144
25. Mathew L, Purkayatha SS, Jayashankar A, Radhakrishnan U, Sengupta J, Nayar HS (1985) Responses of the autonomic nervous system in altitude adapted and high altitude pulmonary oedema subjects. Int J Biometeorol 29:131–143
26. Mathew L, Purkaystha SS, Sengupta J, Malhotra MS (1976) Body temperature and basal metabolic changes during acclimatization to altitude in man. Indian J Physiol Pharmacol 20:197–202
27. Murthy GSN, Prakash YS, Kumar A, Padaki VC (2009) Submarine escape test facilities. Def Sci J 59:441–446
28. Paintal AS (1967) Mechanism of stimulation of aortic chemoreceptors by natural stimuli and chemical substances. J Physiol 189:63–84
29. Pichan G (1981) Effect of heat on troops operating under field service conditions. AFMRC project no. RD/Y/DIPAS/65/P1/21
30. Pichan G (1987) Role of potassium supplementation on physical performance and aldosterone rennin-angiotensin system under heat stress. Project no. AFMRC-1140/1987
31. Purkayastha SS, Bhaumik G, Chauhan SK, Banerjee PK, Selvamurthy W (2002) Immediate treatment of frostbite using rapid rewarming in tea decoction followed by combined therapy of pentoxifylline, aspirin & vitamin C. Indian J Med Res 116:29–34
32. Ravishankar L (2011) Radiological and nuclear emergencies: medical management of radiation injury. Def Sci J 61:113–117
33. Schoene RB, Hackett PH, Handerson WR et al (1986) High altitude pulmonary oedema. Characteristics of lung lavage fluid. JAMA 4:63–69
34. Selvamurthy W (1984) Neurophysiological problems in snow bound high altitude areas. Def Sci J 34:397–415
35. Selvamurthy W (1987) Neurophysiological responses at high altitude. In: Proceedings of national symposium on physiology human performance. DIPAS, Delhi, pp 01–15
36. Selvamurthy W (1989) Physiology of man at high altitude-recent trends. Ann Natl Acad Med Sci (Indian) 25:175–186
37. Selvamurthy W, Raju VRK, Ranganathan S, Hegde KS, Ray US (1986) Sleep pattern at an altitude of 3500 meters. Int J Biometeorol 30:123–135
38. Selvamurthy W, Saxena RK, Krishnamurthy N, Suri ML, Malhotra MS (1978) Changes in EEG pattern during acclimatization to high altitude (3500 m) in man. Aviat Space Environ Med 49:968–971
39. Selvamurthy W, Saxena RK, Malhotra MS (1978) Pulmonary edema—central mechanism of causation. AFMJ 34:343–349
40. Sharma VK, Sridharan K, Pichan G, Panwar MR (1986) Influence of heat stress induced dehydration on mental functions. Ergonomics 29:791–799
41. Sharma VM, Malhotra MS (1976) Ethnic variations in psychological performance under altitude stress. Aviat Space Environ Med 47:248–251
42. Sharma VM, Malhotra MS (1976) Cognitive functions at high altitude. Psychologia 19:23–28
43. Singh SB, Sharma A, Sharma KN, Selvamurthy W (1996) Effect of high altitude hypoxia on feeding responses and hedonic matrix in rats. J Appl Physiol 80:1133–1137

44. Singh SB, Sharma A, Yadav DK, Verma SS, Srivastava DN, Sharma KN, Selvamurthy W (1997) High altitude effects on human taste intensity and hedonics. Aviat Space Environ Med 68:1123–1128
45. Singh SB, Chatterjee A, Panjwani U, Yadav DK, Selvamurthy W, Sharma KN (2000) Effect of high altitude on sensitivity to the taste of phenylthiocarbamide. Int J Biometeorol 44:20–23
46. Singh SB, Thakur L, Anand JP, Yadav D, Amitabh, Banerjee PK, Selvamurthy W (2004) Changes in visual evoked potentials on acute induction to high altitude. Indian J Med Res 120:472–477
47. Singh SB, Thakur L, Anand JP, Banerjee PK, Yadav D, Selvamurthy W (2004) Effect of high altitude on human auditory brainstem responses. Indian J Physiol Pharmacol 48:230–234
48. Singh SB, Thakur L, Anand JP, Panjwani U, Yadav D, Selvamurthy W (2003) Effect of high altitude on event related brain potentials. Indian J Physiol Pharmacol 47:52–58
49. Srivastava KK, Rao YBM, Bhardwaj SK, Malhotra MS (1973) Serum creatine phosphokinase activity and urinary excretion of creatine in man during acclimatization to high altitude and in high altitude natives. Aerosp Med 44:272–275
50. Stream JO, Grissom CK (2008) Update on high altitude pulmonary oedema: pathogenesis, prevention and treatment. Wilderness Environ Med 19:293–303
51. Technology Focus (2010) Published by DRDO, Critical Care Ventilator, Defence Scientific Information & Documentation Centre (DESIDOC), Metcalfe House, New Delhi, vol 18
52. Vijaya Raghavan R, Jain N, Gautam A (2007) Evaluation of the antidotal efficacy of atropine sulphate and pralidoxime chloride given by autoinjectors against nerve agent (Sarin) toxicity. J Med CBR Def 5:1–10
53. West JB, Mathieu- Costello O (1992) Stress failure of pulmonary capillaries: role in lungs and heart disease. Lancet 340:762–767

Grassroots Solutions to Overcome Abiotic and Biotic Environmental Stress in Agriculture

Pawan Kumar Singh, Vivek Kumar, Nitin Maurya, Hardev Choudhary, Vipin Kumar, and Anil K. Gupta

Abstract

Agriculture systems, the world over, are evolving at a fast pace to meet various challenges like fluctuating precipitation pattern, warmer climate, depleting ground water resources, declining soil fertility, insect and pest resistance, etc. Fighting such challenges amid uncertainty, farmers have been continuously devising ingenious solutions. These are a testimony to their experimental efforts. Mostly derived from locally available resources, such solutions are low cost, frugal and easily replicable. These include improved crop varieties, agricultural practices for plant protection and production, or eco-friendly farm practices. This contribution of farmers for coping with environmental stresses and addressing food security problems is now slowly being recognised globally by scientific researchers. It is hoped that formal scientific institutions will try to build upon the coping strategies of creative farmers further, add value and develop innovative products and practices. Such a blending of the best of formal and informal science can only ensure a sustainable future for all by mitigating present and imminent agricultural challenges.

P.K. Singh • H. Choudhary
Value Addition Research and Development Department, National Innovation Foundation-India, Satellite Complex, Premchand Nagar Road, Ahmedabad 380015, Gujarat, India

V. Kumar
Scouting and Documentation Department, National Innovation Foundation-India, Satellite Complex, Premchand Nagar Road, Ahmedabad 380015, Gujarat, India

N. Maurya
Dissemination and Social Diffusion, National Innovation Foundation-India, Satellite Complex, Premchand Nagar Road, Ahmedabad 380015, Gujarat, India

V. Kumar
National Innovation Foundation-India, Satellite Complex, Premchand Nagar Road, Ahmedabad 380015, Gujarat, India

A.K. Gupta (✉)
Centre for Management in Agriculture, Indian Institute of Management, Ahmedabad and Executive Vice Chairperson, National Innovation Foundation-India, Satellite Complex, Premchand Nagar Road, Ahmedabad 380015, Gujarat, India
e-mail: anilg@iimahd.ernet.in; anilgb@gmail.com

S.B. Singh et al. (eds.), *Translational Research in Environmental and Occupational Stress*,
DOI 10.1007/978-81-322-1928-6_2, © Springer India 2014

Introduction

Stresses caused by the increasing fluctuations in the climatic and edaphic conditions have to be dealt with by either enhancing the resilience of the agricultural ecosystems or inducing human interventions for the purpose. Many farmers and local communities have met these challenges through (a) in situ conservation of varieties developed through careful selection, (b) development of coping strategies by manipulating agronomic or general management conditions and (c) by controlling biotic stresses such as pests and diseases through herbal plant protection innovations. Honey Bee Network has been tracking and pooling such examples of farmer creativity for the last 25 years. This paper comprises specific examples where we have documented and shared the creative coping strategies with or without value addition [1]. In both formal and informal sectors, experiments for improving existing crop genetic resources and developing sustainable practices have begun to attract added attention in the recent years.

National Innovation Foundation – India (NIF), an autonomous institution of the Department of Science and Technology, Govt. of India, provides institutional support to such farmers and other grassroots innovators. NIF tries to ensure that they receive due acknowledgement for their efforts and that such innovations diffuse widely through commercial and/or non-commercial channels generating incentives for farmers/innovators and others involved in the value chain. NIF, with the help of the Honey Bee Network (HBN), has been able to document thousands of farmers' developed agricultural practices and plant varieties from different parts of the country.

Coping with Stress

The practices deal with the biotic factors like weeds, pest diseases and plant pathogens and abiotic factors such as salinity, drought and extreme cold or hot temperature. Several sociological processes also help in coping with risk such as intra-household risk adjustment strategies, inter-household strategies, common property or pool resources-based alternatives and public interventions [2–4]. However, the scope of this paper is restricted to technological innovations, and it does not provide any insight to these practices.

Innovative Agricultural Practices

Harbhajan Singh (Haryana) combats water stress by sowing cotton crop on two-foot-wide ridges which are separated by six feet. Irrigation of alternative row reduces water requirement to half without affecting yield. In addition, the pest and weed control expenses also lessen. Plants require moisture and not water [5]. Frequent or heavy irrigation causes succulence in plants which in turn renders pest attack. Numerous farmers have adopted this irrigation pattern which reduced their irrigation and seed costs while increasing the yield. Scientific studies also confirm that alternate row irrigation of cotton field has great potential in reducing the uses of water without compromising the yield of lint [6]. The use of lady's finger as a border crop around the cotton fields is another interesting example of economical grassroots experiment performed by few farmers in Maharashtra, India, for controlling pests. The lady's finger's flowers are similar to cotton but it blossoms earlier than cotton's flower. Hence, it attracts the pests and reduces the burden on the main crop [7]. Despite of the availability of such non-expensive practices, it is tragic that we still encounter increasing rate of suicides in large-scale farmers of the cotton-growing regions, as these non-monetary practices to reduce cost and increase resilience of the cropping systems were never publicised and most of the farmers are not aware of such practices. Farmers also manage and maintain some weeds, which are used for the biological control of various pests, hedge, etc. [8]. Weeds are controlled by creative communities by growing *berseem* (*Trifolium alexandrium*) or other such fodder crops in weed-infested fields. Once in 3 years or so, farmers grow such crops, harvest them intermittently to feed livestock and thus prevent weeds to grow to maturity, set seeds and increase their seed load in soil. Thus in subsequent years, there are much lesser weeds. Two lady farmers from Tamil Nadu, Ariyammal and Pushpam, developed purple-coloured *Chinnar* 20 paddy variety using selection method from ADT-46 paddy variety. The plants have purple pigmented leaves and culm; hence, differentiation between weed and crop becomes easier. Weeds can thus be removed saving time and labour cost and preventing loss of the crop [9].

Farmers have developed growth promoters and herbal pesticides by studying the properties of various weeds as well as other plants and fruits. Haribhai Narola from Bhavnagar, Gujarat, provides a basal treatment to the soil with the week-old flour of pearl millet as an alternative to chemical fertilisers. With the above treatment, improvement in the growth and yield was recorded in wheat, cotton, chilli, brinjal and other vegetable crops. In 1995, he conducted a systematic study to compare the effects of 'bajra' flour with those of chemical fertiliser, diammonium phosphate (DAP) on wheat. He obtained a yield of 250 kg/ha in the plots treated with 'bajra' flour. Haribhai also observed small patches of rust in DAP-treated plot, while the plots treated with 'bajra' remained healthy. The growth and boll setting was much better when the cotton was sown in the same plot in which wheat was grown with 'bajra' flour treatment in the previous season [10].

M A. Chinnathambi had developed a pesticide by using the aerial parts of cactus kodikaali (*Euphorbia nivulia*) or madakalli or thirugukalli (*Euphorbia tirucalli*), cut into pieces, immersed in water and fermented for 15 days. After fermentation the extract is then filtered and sprayed on the crops. The pesticide effectively controls the leaf curl disease in chilli and brinjal, the mosaic disease in lady's finger, jassids in paddy and sucking pests in cotton [11]. Pravinsinh Jadeja, Kutch, Gujarat, suggested a remedy for protection of wheat crop from termites by placing crushed 'kharsadi' (*Euphorbia tirucalli*) and 'thor' (*Euphorbia neriifolia*) plants in watering grooves within the field [12].

Germplasm Conservation

Conservation and development of high yielding and disease-resistant crop varieties is also one of the ways farmers overcome environmental stress conditions. A round chilli variety has been conserved by a community in Bihar. This variety bears fruit for 5–6 years continuously. The main features of the variety are high pungency, ability to grow under shade and tolerance to common diseases and pests [9]. Kir community from Jaipur region of Rajasthan, India, has preserved and propagated a traditional tinda (round gourd) variety for the past 35 years for its superior yield. For conserving this variety, they selected two main characters, i.e. big size of fruits and dark green colour with hairy skin. The fruits of this traditional variety of tinda are tender, delicious in taste and flat round in shape and weigh about 100–200 g at the time of harvesting. Fruits of this variety are famously known as 'Sahapur tinda' in Jaipur and surrounding areas [13].

Gangaram Kir of the same community has been continuously trying to improve a traditional variety of brinjal through successive or recurrent selection. The criteria used for selection are pests and disease-free plant with more spiny fruits per plant. The variety has features like strong spininess of calyx of fruit, larger fruits, high yield and less number of seeds per fruit as compared to other varieties, with sweet taste and better cooking quality [9].

Improved High Yielding Varieties

Farmers make selections in the available diversity produced through natural mutations, mixtures or outcrossing. Sometimes, diversity may increase due to natural stresses, which creates selection pressure. This provides opportunity for less common characters to become noticeable.

Chunni Lal from Rajasthan has developed an improved variety of ridge gourd (*Luffa acutangula*) by selecting the seeds from a traditional variety with particular attributes. The developed variety is tolerant to powdery mildew disease. The vine climbs the trees of height of 9–10 m and bears 15–20 fruits per vine with fruit size of 3–7 ft. The fruits contain very less number of seeds as compared to conventional varieties [9]. Santosh Pachar, a lady farmer from Rajasthan, has developed a high yielding variety of carrot. Sixteen years back, she had collected some carrots for consumption, which had good colour, taste and no forked roots. She adopted the root-to-seed method of planting at her farm to produce seeds from the above carrots. She collected the seeds from first year plants, sowed them again in the field and adopted the same selection method continuously for 5 years. The carrot variety is now stable in yield and has desirable traits. The length of carrot is up to 1.5 ft, with less or no forked roots; the carrot is sweet in taste and adapted to high temperature [14].

Prakash Singh Raghuvansi from Varanasi, Uttar Pradesh, has developed various improved plant varieties of wheat, paddy, mustard, pigeon pea, etc. He developed three wheat varieties, namely, Kudrat 5, Kudrat 9 and Kudrat 17 (developed from Kalyan sona and RR21 varieties), which have higher yield (65–70 q/ha), lengthy spikes, robust stem and water lodging resistance with high protein content. Paddy variety Kudrat 1 developed from HUVR-2-1 gives high yield (60–70 q/ha) with higher number of tillers (30–35/plant) and seeds (280–290 seeds/tiller) as compared to other locally popular varieties. The pigeon pea variety developed by him gives high yield (30–34 q/ha), has bold seeds, robust stem and bears more number of pods per plant. The variety is also tolerant to common pest and disease of the crop. The variety of mustard has bunchy siliquae, bears about 40 seeds per siliqua, contains higher oil content (42.3 %) and shows synchronous maturity. Farmers of many states are cultivating their varieties with commercial intention [15]. Rajkumar Rathore from Madhya Pradesh has developed an improved variety of pigeon pea *Richa* 2000, which is very famous in his area. It has low seed rate (3–4 kg/acre) and contains pods in bunch at apical region, with double harvest (December and April) and higher yield (24 q/acre) [16].

Disease-Resistant Varieties

Jagdish Prasad Pareek from Sikar, Rajasthan, developed a high yielding variety of cauliflower, which grows in all the seasons, is tolerant to hot temperature, is resistant to disease and is less susceptible to insect attacks. The fruit weighs up to 12 kg [17]. A wilt-tolerant groundnut variety, 'Dhiraj-101', was developed by Thummar Dhirajbhai Virjibhai from Gujarat [18]. The yield of this wilt-tolerant variety is higher than locally cultivated varieties (GG 20 and GG 2) with higher oil content (52–55 %). Only 0.47 % stem rot incidence has been found in Dhiraj-101 as compared to 16.37 % in the check variety (GG 20), reported in the trials conducted at Oil Seed Research Station, Junagadh Agricultural University.

Using grafting technique, R.G. Hegade from Karnataka has developed a drought-resistant black pepper variety NP 77, which has about 30 % higher yield than the prevalent varieties. It contains about 80–100 spikes per plant and 100–120 berries per spike, and the plant height reaches up to 50–60 cm. The yield of dark black-coloured berries is about 3,000 kg per acre with high dry recovery percentage, early maturity and resistance to quick wilt. It also contains heightened flavour and aroma [9]. An improved variety of paddy 'HMT' has been developed by Dadaji Ramaji Khobragade, of Maharashtra, which is now included as a standard reference for thinness by Protection of Plant Variety and Farmers' Right Authority (PPV&FRA). This variety has an average yield of 40–45 q/ha with short grains, high rice recovery (80 %), better smell and cooking quality in comparison to the parent ones. He also developed another improved paddy variety 'DRK', which gives average yield of 60–80 q/ha. The grains are short, slender, white in colour and aromatic and have good cooking quality. The most important feature of this variety is its ability to tolerate biotic and abiotic stress conditions [19].

Varieties with an Ability to Withstand Drought and Salinity

Sabu Varghese from Kerala has developed a cardamom variety, *wonder cardamom*, which is known to have drought-resistant characters and can be grown along with the rubber plantations. The variety was developed using seeds collected from a morphologically different plant. The plants were propagated by vegetative multiplication. The other specialty of this variety is its branched panicles, which have attracted interest from the scientific community due to its better adaptability at lower altitudes and lower rainfall regions. Traditionally, these areas are not known for cardamom production and generally used for the cultivation of rubber [20]. Another farmer from Kerala, KJ Baby, has developed a white flowered variety of cardamom from *Vazhuka* type of cultivars through mass selection method, which took about half a decade. The variety bears white flowers and has high productivity compared to other commonly grown cardamom varieties. It can be grown in waterlogged areas as well. It has wider adaptability to different shade conditions apart from having higher production with good quality than other locally popular *Mysore* and *Vazhuka* cultivars, viz. *Njallani*, Green-bold, *Palakkudi* and *Veeraputhara* varieties. It has sturdy plants, robust tillers and deep root system, which also makes it resistant to various biotic and abiotic stresses [21]. *Alakhpura* selection, an improved chilli variety, has been developed through mass selection method by Balwan Singh from Haryana. He started the selection in the year 1984 based on the criteria of fruit length, diameter, dark red colour, pungency, etc. The variety is known to perform extremely well under saline conditions and is also tolerant to extreme heat and humidity. The yield of the variety is 400 q (green) and 40 q (dry) per hectare and is known to fetch high market price due to its skin colour [9].

Water shortage inspired Manaram Choudhary, Rajasthan, to breed an onion variety requiring less irrigation. He succeeded in developing a highly productive, early maturing and drought-resistant variety of onion, which gradually became famous in Haryana, Delhi, and Rajasthan for its delicious taste. This onion variety, known as *Rashidpura*, can be cultivated as a winter crop due to adaptability in varying climatic conditions. It has higher yield (40,000 kg/ha) as compared to normal yield (25,000–30,000 kg/ha). It also matures early (110–115 days) with less irrigation and hence gives good result in drought conditions [22]. Drought conditions cause reduction in the yield, and also the higher frequency of irrigations may lead to excessive vegetative growth in heavy soils [23]. Sundaram Verma, also from Rajasthan, has been experimenting on the arid crops varieties. He has successfully developed and conserved several cultivars of plant varieties during his many years of farm research. He has developed a *kabuli*-type bold-seeded chickpea variety, which gives very good yield in drought conditions. The potential yield of the variety is about 28 q/ha, which is higher than other commercially cultivated varieties and fetches good market price due to the seed boldness [9].

Conclusion

Farm practices, herbal pesticides and stress-resistant crop varieties discussed illustrate the enormous potential that exists for learning from farmers and partnering with them to develop solution to deal with various stresses in the environment. Such solutions, also referred as frugal innovations or Gandhian innovations, are incredibly economical and eco-friendly and help the crops to withstand biotic or abiotic stress conditions. These grassroots innovations can revolutionise sustainable food security solutions. Being improvement in locally adapted varieties or practices, farmer innovations often tend to have a robustness and frugality which may elude in many externally developed solutions. Respect, recognition and reward for grassroots innovations, Honey Bee Network has argued, can pave the bridge between formal and informal science and technology system and increase mutual trust and reciprocity.

References

1. Gupta AK (1988) Survival under stress in South Asia: a socio-ecological perspective on farmer risk adjustment and innovations. International Conference on Plant Physiology, New Delhi
2. Gupta AK (1984) Small farmer household economy in semi-arid regions: socio ecological perspective. Center for Management in Agriculture, Indian Institute of Management, Mimeo, Ahmedabad, p 452
3. Gupta AK (1985) Socio-ecological paradigm to analyse the problem of poor in dry regions. Eco-Dev News (Paris) 32:71–75
4. Richards P (1989) Agriculture as a performer: farmer first. Intermediate Technology Pub, London, pp 39–43
5. Cotton cultivation in water stress condition. http://nif.org.in/awards/profile/Cotton-cultivation-in-water-stress-condition/150
6. Subramani et al (2008) ag.arizona.edu/pubs/crops/az1513/az1513c.pdf
7. Gupta AK (2009) http://www.sristi.org/anilg/papers/Seduce_the_scientist.pdf
8. Altieri MA, van Schoonhoven A, Doll JD (1977) The ecological roles of weeds in insect pest management systems: a review illustrated with bean (Phaseolus vulgaris L.) cropping systems. Pests Articles News Summ (PANS) 23:195–205
9. NIF database. http://www.nif.org.in/
10. Gupta AK (ed) (1996) Bajra' flour: higher yield, lesser rust. Honey Bee 7(2):14
11. Gupta AK (ed) (2002) Cactus as pesticide. Honey Bee 13(2):8
12. Gupta AK (ed) (1999) Tradition of invention: simultaneous innovations across space and time. Honey Bee 10:4
13. Conservation of traditional variety of *Tinda* round gourd by Kir-community. http://nif.org.in/awards/profile/Conservation-of-traditional-variety-of-Tinda-round-gourd-by-Kir-community/74
14. Improved variety of carrot *Laxmangarh* Selection. http://nif.org.in/awards/profile/Improved-variety-of-carrot-Laxmangarh-Selection/631
15. Improved Varieties of Wheat, Paddy, Mustard and Pigeon pea. http://nif.org.in/awards/profile/Improved-Varieties-of-Wheat-Paddy-Mustard-and-Pigeon-pea/17
16. *Richa 2000* perennial pigeon pea variety. http://nif.org.in/awards/profile/Richa-2000-perennial-pigeon-pea-variety/387
17. Cauliflower variety of 12 Kg weight. http://nif.org.in/awards/profile/Cauliflower-variety-of-12-Kg-weight/185
18. Groundnut variety *Dhiraj 101* which is resistant to wilt due to stem rot. http://nif.org.in/awards/profile/Groundnut-variety-Dhiraj-101-which-is-resistant-to-wilt-due-to-stem-rot/19
19. *HMT* an Improved Paddy Variety. http://nif.org.in/awards/profile/HMT-an-Improved-Paddy-Variety/313
20. Wonder cardamom a new variety of cardamom. http://nif.org.in/awards/profile/Wonder-cardamom-a-new-variety-of-cardamom/358
21. White Flowered Cardamom Variety. http://nif.org.in/awards/profile/White-Flowered-Cardamom-Variety/354
22. Improved Onion variety. http://nif.org.in/awards/profile/Improved-Onion-variety/360
23. Gaur PM, Tripathi S, Gowda CLL, Ranga Rao GV, Sharma HC, Pande S, Sharma M (2010) Chickpea seed production manual. International Crops Research Institute for the Semi-Arid Tropics, Hyderabad, p 28

Stress Research: Varied Paradigms

Vijay Parkash, Updesh Kumar, and Manas K. Mandal

Abstract

The term 'stress' is used to denote varied type of concepts in different disciplines. The concept of stress and has gained augmented attention of researchers during the past century, and it still continues to remain a popular research domain. Researchers and scientists of different behavioral and health sciences have attempted to look into the concept of stress from varied perspectives. Literature witnesses that numerous paradigms exist to delineate the conceptual understanding and stress research; however, there is hardly any meta-analytic synthesis that presents different perspectives and paradigms together to the reader. Present article reviews different existing perspectives for the thorough understanding of the concept of stress from multiple viewpoints. Different research paradigms, running through which the topic of stress has gained a huge momentum, have been discussed concisely. The authors present a meta-analytical review of different conceptual models and research approaches to stress. Beginning with the early physiological notions of 1920s, the genetic, epigenetic, biophysiological, psychological, sociological, developmental, environmental stress, and occupational stress paradigms have been reviewed and more future researches with combined methodological approaches tested with multivariate analytic methods have been called for to further simplify the understanding of the concept of stress from a multidisciplinary perspective.

V. Parkash, Ph.D
Scientist 'C', Mental Health Division, Defence Institute of Psychological Research, DRDO, New Delhi, India

U. Kumar, Ph.D
Scientist 'F' and Head, Mental Health Division, Defence Institute of Psychological Research, DRDO, New Delhi, India

M.K. Mandal, Ph.D (✉)
Distinguished Scientist & Director General (Life Sciences), Defence Research and Development Organization, New Delhi, India
e-mail: mandalmanask@yahoo.com

Stress Research: Varied Paradigms

Being used to convey several concepts in different disciplines, stress seems to be a nebulous term. While the origins of the concept of stress dates back to Hippocrates, the construct has been marked by broad variations in the physical, physiological, behavioral, and psychological elements used to define it. In the seventeenth

S.B. Singh et al. (eds.), *Translational Research in Environmental and Occupational Stress*, DOI 10.1007/978-81-322-1928-6_3, © Springer India 2014

century, the word *stress* was used to describe strain, hardship, adversity, or affliction. Later in the eighteenth and nineteenth centuries, the term stress referred primarily to an individual's "force, pressure, strain or strong effort" on some object [1]. It was these early definitions used in physics and engineering that began to influence the notion that stress may affect individuals, where external forces are seen to exert pressure on an individual, producing strain. In the past century, this concept of stress and the stress of daily life have gained increased attention of researchers. Although the term stress as it relates to the human condition has been in the scientific literature since 1930s, while the word did not become so much popular until the last quarter of the century. At present, interest in phenomenon of stress and coping has reached all-time high in psychology as well as other behavioral and health sciences. Lazarus [2] has rightly commented, "Never before has there been so much interest in stress world-wide, among social and biological scientists, and on the part of the general public. . . .Stress has become a household word, and we are flooded with messages about how it can be prevented, eliminated, managed, or just lived with" (p. 27). Today, stress has become inevitable part of everyone's life. "Stress is not something to be avoided. Indeed, by definition it cannot be, ... Complete freedom from stress is death" [3, p. 32].

As the time has progressed, researchers and scientists of different behavioral and health sciences have attempted to look into the concept of stress from varied perspectives. Literature witnesses that numerous paradigms exist to delineate the conceptual understanding and stress research; however, there is hardly any meta-analytic synthesis that presents different perspectives and paradigms together to the reader. In the coming sections of this article, we attempt to review different existing perspectives for the thorough understanding of the concept of stress from multiple viewpoints. Further, our efforts focus on discussing different research paradigms running through which this topic has gained a huge momentum. Beginning with the early physiological notions of 1920s, we attempt to analyze the genetic, biological, psychosocial, developmental, and interactional paradigms.

Conceptual Models of Stress

There have been three major conceptualizations of stress in the literature. It has been conceptualized as either a stimulus, a response, or as an interactional process combining both stimulus and response. Different disciplinary approaches have utilized these models, though there may be specific matching measurement approaches to a particular model. However, few critic researchers believe that "in addition to being itself, and the result of itself, stress itself is also the cause of itself" [4, as cited by 5]. There are other researchers who believe that stress is etiologically ambiguous in the sense that it can logically be both an input and an output [6]. The ambiguity in stress further expands when all three models – the stimulus model, the response model, and the interactional model – are combined uncritically such that stress is simultaneously a stimulus (the cause), response (the result), and process [5].

The stimulus model considers stress as the sum of biological and psychological disturbances (stimuli) caused by any aggression on an organism [7]. According to the stimulus model-based understanding, stress may be termed as "objective stress," and it is considered as an independent variable in terms of a causative stimulus. The important questions answered by this approach concern which particular situations are stressful, and accordingly the disturbing environment or external stressors are taken as the basis for defining or explaining stress. The response-based approach considers stress as the response of an organism to a noxious or threatening condition [7]. In the terms of this model, stress is conceptualized as a dependent variable in the form of a person's response to certain adverse stimuli. According to this approach, stress is defined as the response elicited by an individual when confronted and stimulated by a stressor [4]. Hence, this approach may also be taken as "subjective stress" model [5].

From the interactional perspective, stress is the process wherein the adaptive capacity of an organism is taxed or exceeded by environmental demands or pressures [8]. The interactional model explains that stress is a lack of fit or the degree of misfit and discrepancy between environment (environmental demands) and person (personal capabilities to meet those demands). In these terms, stress is somewhere between its antecedent factors (the stimulus) and its effects (response). Stress here can be understood as a dynamic system of interaction between person and environment that consists of an individual perceptual phenomenon shaping out of the imbalance between external (that arises externally from the environment) and internal (that arises from individual's inherent psychological and physiological needs) demands on the individual and his/her ability to cope. This approach emphasizes the role of a cognitive appraisal of a stressful situation and a person's psychological and physiological capability to cope. Researchers believe that the adoption of an interactional model of stress and thereby considering stress as a "dynamic state of imbalance" has been instrumental to avoid many of the pitfalls related to a more diminutional thinking [4]. Recent researches are attempting to explore different dimensions of this interaction. The isomorphic theory of stress that attempts to describe a one-to-one correspondence between the person and environment along the three dimensions of control, uncertainty, and personal relationship [9] is an example of a research effort in this direction.

Research Approaches to Study Stress

Literature reveals that different approaches focusing on measurement of stress primarily include life events, chronic stressors, and daily hassles [5]. Researchers believe that life events are essentially self-limiting in their nature, and these are discrete, acute, and observable events that require major behavioral readjustments within a relatively short span of time [10]. As compared to life events, chronic stressors require readjustments over prolonged periods of time as they are relatively enduring, persistent, or recurrent demands, conflicts, threats, or problems [11]. Daily hassles are mini-events, which constantly require small behavioral readjustments during the course of routine life [12]. Most of the existing research attention so far has focused on life events and chronic stressors, with a relatively lesser focus on daily hassles [13]. A small body of intervention research also exists regarding examining how to reduce stressors, alleviate stress or its effects, or a certain combination of these approaches [5].

The earliest measures of stress have been the life events, generally taken in the form of a checklist of events sampled from across various domains and hierarchies and weighted for frequency of occurrence, intensity, and/or standardized importance of each event, mostly subjectively by respondents themselves. Although a checklist normally aims to be a representative sample of the major events that occur in people's lives [8], some of the researchers believe that there is some evidence that the universe of possible "life events" has not been sampled uniformly [14], with events occurring to young adults being over-sampled, while those occurring to women, minorities, and the poor being under-sampled. There may be several different types of chronic stressors including work-overload, role conflicts, interpersonal conflicts, and other ambient stressors. Chronic stressors could include barriers in the achievement of life goals, inequity, status inconsistency, goal-striving stress, lifestyle incongruity, disjunction of economic goals and means, and social and economic hardships including poverty, crime, violence, overcrowding, noise, and chronic disability [11]. Anticipated events which may cause anxiety, whether or not they actually occur, may also act as chronic stressors [1].

Daily hassles are the frustrating, irritating, and distressing demands that characterize everyday transactions with environment to some extent [15]. Usually, daily hassles and uplifts are also measured with a checklist approach in broad domains, such as work and family. These items are also generally weighted by self-reported severity of different events. Being in the form of "recurrent micro events," daily hassles can be considered as an intermediate position between chronic stressors and life events [10]. Attempting to understand these approaches to study stress, it seems evident that life events have conventionally been equated with objective, discrete events that are not the result of the individual's psychological function (stimulus-based model). On the other hand, chronic stressors are seen as subjective and influenced by individual's emotional functioning (response-based model) [11]. Though life events and chronic stressors are conceptually distinct, these can be interrelated in the sense that they both can lead to each other and they provide meaning and context for each other [16].

Exploring the domain of stress from varied angles, some researchers [5] refer to a combination of recent and distal life events as well as chronic stressors that affect an individual at any one point in time and term it as "operant stress" [17]. Similarly, the amalgamation of current stressors with previous significant traumas that continue to be the sources of stress has been conceptualized as "cumulative stress" [17]. Researchers have further tried to study "stress sequence" which is another aspect of the interplay between types of stressor, for example, a chronic stress followed by a life event [12, 16]. "Stress proliferation" is also a related concept according to which a "primary stressor" (e.g., a specific life event) may lead to a "secondary stressors" (i.e., chronic stress) [18]. Attempts have been made to even study the "carry over" effects of stress where stress is translated between people, across role domains, and across stages of life [12]. Varying relative impact of stressors across the life course has also been studied by few of the researchers [17]. Further researches have also attempted to study if stressors may also act as mediators/moderators of the effects of other stressors, can act to mobilize coping resources or to deplete them, and can even deter the occurrence of certain other stressors ([19, p. 57]; as cited by [5]). Thoits [12] put forth that there are four major approaches in literature to operationalize the meaning of stress: (1) the process of "appraisal" that determines whether a demand is perceived as a threat or challenge, harm/loss, and/or controllable or not; (2) considering a person's biographic history, plans and goals, sequences of events, and the surrounding context; (3) "identity salience" (the relevance of the role domain in which a stressor occurs); and (4) "belief systems" of individuals. The major approach of appraisal can be either primary or secondary. Primary appraisal involves the evaluation of an environmental stimulus as stressful whereas secondary appraisal is the individual's assessment of whether or not he/she can reduce or eliminate the effects of the stressful stimulus [20].

Going beyond the operationalization of meaning, the research focusing on stress "intervention" has taken more than one approaches. Of the two major approaches, the first approach which is referred to as "ecological stress perspective" focuses on changing the environment to better accommodate and diminish stress in individuals across the dimensions of control, uncertainty, and social support. The second approach, known as the stress "adaptation" perspective, motivates the individual to adapt to the situations in order to reduce stress and focuses on generalized expectancies of control, tolerance for ambiguity, and self-reliance [9].

Stress Research Paradigms

Stress research in the domain of life sciences and behavioral sciences has been conducted with numerous varied perspectives. Enlisting each and every research perspective on all the concepts related to the stress domain is really a herculean task. In the forthcoming section, we attempt to concisely discuss different major stress research paradigms that include the genetic perspective, bio-physiological perspective, developmental perspective, environmental stress perspective, psychological perspective, and sociological perspective.

Epigenetic Perspective

The concept of stress has essentially captured the attention of diverse disciplines ranging from biology to behavioral sciences for around a century. Though different perspectives will be delineated in the forthcoming sections, we begin with the relatively newer but the most insightful perspective on stress. There have been various linkages between biological and physiological aspects and stress; in recent times, scientists have gained new insights into the molecular processes that link a stressful environment with gene expression [21]. The newly emerging field of epigenetics is transforming the conceptualization of gene–environment interactions that influence our vulnerability or resistance to stressful life events and health [22]. "Epigenetics is defined as the study of functional alterations in gene expression that do not result from alterations in the basic DNA (deoxyribonucleic acid) sequence but arise during development and from the environment" [21, p. 71]. Mathews and Janusek [23] believe that it has been demonstrated by the advances in behavioral epigenetics that stressful environments can affect gene expression by altering the epigenetic pattern of DNA methylation and/or chromatin structure. The advancements in the epigenetics are based on explorations of early-life problems which cause epigenetic alterations in relevant regions of brain that impact adults' behavior and their response to various stressors [24].

Recent evidence shows that adults also respond epigenetically to environmental cues and the epigenetic modifications may be reversible [22, 24]. By this insight, the role of environment in health has got more emphasis, and the horizon has widened for the potential newer approaches to eliminate the stressful impact of adverse environments. The researches in the domain of epigenetics first attempt to study the epigenetic modifications – DNA methylation, histone modification, and chromatin remodeling – and then explore the linkage between adverse early-life experiences and the epigenetic modifications. Recent developments in epigenetics have demonstrated the impact of early-life experiences on neurobehavioral development which is mediated through modifications in the epigenetic pattern [21]. These researches are based on the epigenetic modifications caused by the early-life experiences. This demonstrates the plasticity of the epigenome across life span and provides considerable evidence that stress exposure in adults results in epigenetic modifications in the regions of brain that regulate the neuroendocrine stress response. Adverse life experiences (stress) have been viewed as a source of mental health problems for many years. The recent researches in the epigenetic paradigm suggest that epigenetic modifications, which are possible even in adults, may mediate these poor stressful outcomes [21]. It has also been found that neonatal stress is one of the causes (via epigenetic programming) of early onset of cardiovascular disease in adults [25].

Bio-physiological Perspective

The biological perspective on stress focuses on the activation of bodily mechanisms or the physiological systems that are particularly responsive to physical and psychological demands. Prolonged or repeated activation of these bodily systems and the biological mechanisms that they undergo is thought to place an individual at risk of the development of a range of physical and psychological disorders. According to the biological paradigms, two interrelated systems that are viewed as the primary indicators of a stress response are the sympathetic–adrenal medullary system (SAM) and the hypothalamic-–pituitary adrenocortical axis (HPA) [8].

Sympathetic: Adrenal Medullary System

Interest in the study of the impact of the sympathetic–adrenal medullary system (SAM) activation on bodily reactions to emergency situations may be traced back to [26] early work on the fight or flight response. According to [26], the SAM system reacts to various emergency states with increased secretion of the hormone epinephrine. His proposition is supported by a large body of research evidence that indicates increased output of epinephrine and norepinephrine in response to a wide variety of psychosocial stressors [27]. In addition to increased secretion of epinephrine and norepinephrine, other components of the SAM that get elicited by stressors include increased blood pressure, heart rate, sweating, and constriction of peripheral blood vessels. Researchers assert that if SAM activation is excessive, persistent for longer duration, or is repeated too often, it may result in a sequence of responses that may ultimately result in development of illness or any disorder. These responses include functional disturbance in various organs and organ systems [28] and ultimately permanent structural changes of pathogenic significance [29] at least in the individuals who are predisposed. The secretion of the epinephrine hormones by the adrenal medulla and/or sympathetic nerve endings is specifically responsible in this regard. Researchers exploring their consequences put forth that excessive discharge of epinephrine and norepinephrine is believed to induce many of the pathogenic states associated with the perception of stress. These states include hemodynamic effects, such as increased blood pressure and heart rate [30]; suppression of cellular immune function [31]; production of neurochemical imbalances that contribute to the development of psychiatric disorders [32]; and ventricular arrhythmias provoking variations in normal heart rhythms that may also lead to sudden death in some cases [33].

Hypothalamic–Pituitary–Adrenocortical Axis

One of the pioneers of the physiological research paradigm in stress research, Selye [3, 34] elaborated upon the hormonal responses of the hypothalamic–-pituitary–adrenocortical axis (HPA) in his milestone description of stress as a nonspecific (general) physiological reaction that occurs in response to excessive stimulation. According to Selye, all kinds of stressors including pathogens, physical stressors, and psychosocial stressors elicit common pattern of physiological response which proceeds in characteristic three stages of *alarm, resistance*, and *exhaustion*. He referred to this response pattern as the general adaptation syndrome (GAS). During the first stage of the GAS, the *alarm stage, the* organism's physiological changes reflect the initial reactions necessary to meet the demands made by the stressor agent. This approach asserts that when an individual is faced with excessive stimulation or any stressor, at the initial stage the anterior pituitary gland secretes adrenocorticotrophic hormone (ACTH), which then activates the adrenal cortex to secrete cortisol. During this stage, the hormone output from the adrenal cortex increases rapidly. After the initial stage of alarm reactions, the second stage of *resistance* comes into play with an intention to get consequent improvement or disappearance of symptoms by means of resisting or having a complete adaptation to the stressor. For the obvious reason of resistance or adaptation, the output of corticosteroids (cortisol) remains high but stable during the resistance stage. Finally, if the stressors acting on the individual are so severe and prolonged that individual's resources and somatic defenses get depleted, the third stage of *exhaustion* occurs. Since secretion of corticosteroid hormones has been constantly high during the resistance stage, over prolonged exposure, the anterior pituitary and the adrenal cortex lose their capacity to secrete hormones, and the organism can no longer adapt to the stressor. Hence, stress symptoms reappear, and if the stress response continues at same degree without getting subsided, vulnerable organs as determined by individual's genetic and surrounding environmental factors are likely to break down. This breakdown results in illness and ultimately death. Selye's paradigm of GAS claims that any stressors or noxious agent, physical or psychosocial in nature, would mobilize a similar GAS response [35].

Selye's model got a huge momentum and promoted many physiological researches on similar lines; however, in contrast, critiques of Selye's model disagree with the presence of GAS and suggest that each stressor elicits its own distinct physiological reactions [36, 37]. Though these researchers agree that there is a nonspecific physiological response to stressors, they emphasize on the appraisal of any stimulation as stressful and claim that the physiological response is a concomitant of the *emotional reaction* that occurs when situations are appraised as stressful. When conditions are designed to reduce the psychological threat that might be engendered by laboratory procedures, there is no nonspecific reaction to a physical stressor [37]. It has been found that by minimizing competitive concerns and avoiding severe exertion, the danger of young men getting threatened by treadmill exercise was reduced, and as the associated emotional reactions were removed, the GAS pattern was not found [37, 38]. In his later works, Selye [35, 39] acknowledged that there are both specific as well as general (nonspecific) factors in physiological response to a stressor. He also suggested that the GAS does not occur or is at least not destructive in response to all kinds of stressors; however, he maintained that the nonspecific response is not mediated psychologically always [8].

Taking the bio-physiological aspects of stress further, since the late 1970s, interest in the biological bases of psychiatric disorders has stimulated an alternative focus on the HPA and the possible role of HPA dysregulation in development of depression and other psychiatric disorders has been explored. It has been seen that relatively pronounced HPA activation is common in depression, with episodes of cortisol secretion being more frequent and of longer duration [40]. However, not much of research evidence exists to clarify whether the hyperactivation of HPA is a cause or effect of depressive disorders. HPA regulation may play a role in other psychiatric disorders as well.

Other Associated Physiological Changes

Although majority of researches discuss hormones of the SAM and HPA as the biochemical substances involved in stress responses, a range of other hormones, neurotansmitters, and brain substances have also been found to react in response to different kinds of stressors, and they also play significant role in determining the influence of stress on different aspects of health. These include stressor-associated elevations in growth hormone and prolactin secreted by the pituitary gland and in the natural opiates beta endorphin and enkephalin released in the brain [8, 41].

Developmental Perspective

Some of the researchers interested in life-span development focus on the role of life events with regard to development. Researchers with the developmental perspective believe that "life events are as integral to life-span development theory as are atoms and other lesser particles to physical theory" [42, p. 368]. By such a group of researchers, life events are considered as agents of disequilibrium that generate the possibility of positive development. This view is guided by an implicit model of growth which is based upon the cognitive theory of schema building. Stress is considered as a necessary factor in learning by cognitive theories. In order to adjust the thinking of the child for including new information, the world must become unbalanced and a state of disequilibrium must arise. Cognitive-developmental theorists assume that all type of human behavior is the product of "thinking" and that this thinking is a process of adapting and reconstruing cognitive structures. The scientists supporting this perspective believe that when children get to know that their cognitive schemas do not cover the situation, a state of disequilibrium arises that, in order to achieve equilibrium, is followed by a readjustment to the thinking patterns of the children [43].

Pointing on some of the specific focus areas of the researchers interested in the approach to study stress from a life-span development perspective, Robson [43] has cited that the group of stress researchers who are fascinated by life-span development [42, 44, 45] have recognized several characteristics of life events that appear to be quite important in the study of adolescent stress but appear to have been ignored in the research. Few such characteristics include (a) the age-relatedness of many biological and social events, for example, physical growth, changes in the endocrine system, development in the brain and central nervous system, changes in social roles, and family and school transitions; (b) emphasis upon the social distribution of events, for example, economic depression; and (c) a highlighting of historical or cohort effects upon events, for example, the difference in an adolescent's experience of life events during the second world war and another's experience during the "swinging sixties" [46]. The importance of life events in the investigation of the stress process in children from this focus seems very relevant and connects with importance of developmental issues such as children's response to parental separation, coping in achievement contexts, repression or sensitization, and developing resilience [43].

It has been seen that major theories of growth and development have two main focuses. One approach concentrates on child as an individual and the other one views development as a product of social interaction. As pointed out earlier, stress is considered as a necessary factor for learning by the cognitive theories. Supporters of psychodynamic approach believe that babies are born with internal instinctive drives that need to be discharged and stress is the outcome of the interplay the tension of an unmet drive and energy spent in the release of a drive. If it is not possible for the child to achieve an optimal balance between need gratification and delay, they tend to become disorganized under stress [43]. Compared to the psychodynamic perspective, behavioral theories assume stress coming from the external environment. The supporters of this perspective maintain that stress is learned, reinforced, and maintained by environmental consequences. Providing an explanation for a child's response to stress, the social learning theory [47] asserts that adolescents' problems are a result of our culture and social learning and the primary determinant of successful coping is the ability to feel in the control of the situation, thereby indicating a developmental and learning-based cognitive explanation rather than a simple stimulus–response model.

Environmental Stress Perspective

Stressful life events have derived the interest of researchers in studying the role of stressors in human disease. The initial seminal work of Adolf Meyer in 1930s culminated interest in understanding the role of life events in stress and other resultant illnesses. Meyer advocated the use of a life chart by physicians to record the life events of the patients as part of preliminary medical examination in order to check the etiologic importance of the elicited life events for a variety of physical illnesses [48]. Meyer's ideas were highly influential and led to a substantial body of research which established that stressful life events were associated with a variety of physical illnesses [49]. Focusing on other environmental (external) factors, a longitudinal study in telephone operators brought out that rather than normal periods, illness was much more likely to occur during periods of inordinate demands, frustrations, and losses [50]. The interest in the research on stressful life events advanced with the development of "Schedule of Recent Experiences" (SRE; [51]) as an effort to systematize Meyer's life chart. The researchers conducted by using the SRE documented associations between stressful life events and illnesses including skin diseases and heart diseases [20]. Studies in this direction advanced with modification of SRE as Social Readjustment Rating Scale (SRRS; [52]) that incorporated standardized weighted events based on degree of difficulty required to adjust to the event. The SRRS provides a summary measure of environmental stressors as a sum of life change units. The SRRS furthered the research and found associations between stressful life events and illnesses like sudden cardiac death [53]. These researches documented that the effects of stress operate largely through the creation of excessive adaptive demands, and researchers became more concerned with the magnitude of life change rather than whether the change was positive or negative [38]. The environmental perspective-based researches of stress continued with taking subjective experiences into account, development of investigator ratings based on life event interview, and attempts were made to estimate the impact of an event in a specific context for the average person, avoiding individual subjective reactions [54]. Scales were also developed to assess stressful events in specific populations like children, adolescents, and the elderly. The basic research has further continued substantively using newer stressful life event measures to document the effects of stressful events on a variety of physical and mental health outcomes [38].

Psychological Perspectives on Stress

On the basis of researches being conducted in their domain, the psychological perspectives on stress emphasize on specific personality traits, psychological states, and behavioral responses. Few such psychological focal notions that promote different types of researches may include that individuals with some particular personality traits may be more prone to stress and/or its effects, certain specific psychological and social states may predispose an individual towards stress, and various types of covert and overt behavioral responses can occur as reactions to the stressors and the resultant stress [5]. Psychological research paradigms on stress are primarily focused on – studying the types of external stimulation and circumstances that provoke psychological stress, understanding the types of mechanisms that poses these circumstances as stressors and connects them to stress, and exploring the contextual, buffering/exacerbating, and mediating and/or moderating factors that influence the stress processes [55]. Different conceptual models attempt to document the processes of "buffering" the relationship between stressors and the experience of stress. In such a psychological perspective-based stress-suppression model, when the individual is under excessive demands, then that stress exposure mobilizes a "resource," which then alleviates stress by affecting the appraisal of the stressor, by modifying the responses to the stressor, by checking further stress proliferation, and/or by controlling the relation between stress and ill-health [13]. These factors can be viewed as moderators of the stress process, and the approach focuses more on studying the possible moderators and the moderation process. Another such research model portrays resources as causally antecedent to stress in a way that resources decrease the exposure to stress, rather than its impact on health [18]. Such resource factors are viewed as mediators of the stress process. Mediation process and the mediator variables are of special attention in such psychological researches. In another research model, stress and resources are considered to remain completely independent of each other and having separate and opposite effects. According to this model, resources counterbalance the stressor rather than buffering it since support operates independently even in the absence of stress [11]. Researchers studying the resources acting as stress buffers/exacerbators have primarily centered on personality traits, coping strategies and styles, and external resources available with the individual [13].

The personality factor paradigm has focused on exploration of two types of personality traits – the ones that act as buffer against stress and the others that increase the vulnerability of the individual. Hardiness, optimism, self-control, self-efficacy, self-esteem, ego strength, sense of coherence, extraversion, humor, and conscientiousness are among those that buffer against stress and act as the protective factors. On the other hand, negative affectivity (including anger, anxiety, hopelessness, helplessness, depression), lack of emotional stability, pessimistic explanatory style [13], type A behavior pattern, and external locus of control [11, 55, 56] are aspects of one's personality that make him or her vulnerable to stress experience.

Further the psychological stress paradigm places emphasis on the *organism's perception and evaluation* of the potential harm posed by objective environmental experiences. Individuals label themselves as stressed and experience a concomitant negative emotional response when they perceive that environmental demands are exceeding their abilities to cope. Psychological models of stress argue that events influence only those persons who appraise or perceive them as stressful. It is also assumed to be quite significant by the psychological perspective that stress appraisals are determined not solely by the stimulus condition or the response variables, rather they are determined by persons' interpretations of their relationships to their environments. That is, one's perception of experiencing stress is a product of the interpretation of the meaning of an event and the evaluation of the adequacy of own coping resources [8]. One of the most influential model of the appraisal process claims that [20, 57] an appraisal of a stimulus as threatening or benign, termed *primary appraisal,* occurs between stimulus presentation and stress reaction. Lazarus [36, 58] has later asserted that a situation will also result in a stress reaction if it is evaluated as a harm/loss, threat, or challenge.

This perspective assumes two classes of antecedent conditions for primary appraisal. These conditions include the perceived features or characteristics of the stimulus situation (e.g., magnitude, intensity, imminence, or durations) and the psychological structure of the individual (e.g., belief system, thought pattern, and personality dispositions). When a stimulus is appraised stressful and as requiring coping response, individuals evaluate their coping resources to eliminate or control stress. This process of evaluation of coping resources and thoughts of controlling stress is termed as "secondary appraisal" [20]. Researchers of this perspective claim that there may be two types of models for coping. Coping responses or strategies may involve actions designed to directly alter the stressors (e.g., problem confrontation) or thoughts or actions whose goals are to relieve the emotional responses (i.e., body or psychological disturbances). The former is considered as "problem-focused" coping and the latter is referred to as "emotion-focused" coping [20]. This approach explains that if one perceives that effective coping responses are available, then there is no stress response, and if, on the other hand, one is uncertain about the coping with a situation appraised as stressful, threatening, or demanding, then stress is experienced. The psychological perspective of the cognitive appraisals of the situation and coping resources also attempts to explore and understand the kind of responses elicited by the appraisal of threat.

In addition to the mentioned two types of coping strategies, other coping strategies that have been the subject matter of the psychological stress realm include self-control, positive reappraisal, accepting responsibility, and seeking social support [13]. Other psychological paradigms devote their efforts in understanding individual specific habitual preferences for approaching problems termed as "coping styles" [12]. The general two types of coping styles can be confrontative and avoidance styles. Different researchers at times posit to have different opinions about specific types of coping styles, but coping styles, in general, appear to mediate the relationship between personality traits and the stress effects [13]. Social support has also been studied as a coping supplement that acts as buffers of the harmful effects of stressors [55]. Furthermore, the research from the psychological perspective has led to a number of insights into the understanding of stress including the reciprocal relationship between life events and the stress process [13]. Researchers are also devoting to understand various psychological factors that can enable people to respond appropriately to stressors and successful coping strategies that can mitigate negative psychological responses and avoid health-damaging effects. The psychological perspective considers an individual as an active agent involved in the stress process and appreciating the individual as activist provides challenging opportunity to explore more fully the interplay between personal agency and structural constraints [12].

Sociological Perspective

Looking at stress research from sociological or anthropological perspectives, it is evident that stress research in these disciplines has tended to focus on analyzing differences in group vulnerability, and the issues of gender, class, racial, and cultural differences are found at the center of those researches. Researchers with an anthropological bent of mind tend to have an integrative view, and they emphasize on embedded system of stress and disease in the complex interplay of social organization, cultural context, and historical change [4, 55]. Citing sociology researchers [12, 16], Paradies [5] has put forth that the stress construct provides a potentially valuable bridge linking large-scale organization with individual experience and action and it has the ability to absorb the far-reaching notion of inseparability between the circumstances of social life and individual functioning [5].

The attempt is made by the sociological perspective to provide an overall understanding of why social inequalities among people, in terms of the social distribution of stress wherein chronic stressors are generally unevenly distributed in society, reflect health inequalities among people [55]. Well-being is deeply affected by socially patterned differences in life circumstances. Hence, since social roles and socioeconomic position signify differential exposure and vulnerability to life stressors, they have significant consequences on one's health [59].

From a sociological perspective, stress process can be understood in the sense that any individuals' respective position in the social structure reflects inequality in resources, status, and power; due to this inequality, different people are differentially exposed to stressors which in turn can damage their physical and psychological health, and further, this health damage is generally moderated by one's psychological resources and coping strategies which are again socially patterned in such ways that members of disadvantaged groups, at least partially, remain more vulnerable to the harmful effects of stress [12, 16, 17]. The existing research draws attention to the significance of social roles, socioeconomic positions, and social patterning of health according to major institutionalized roles and unequal distribution of resources, but very limited attention has been drawn towards the cause of existence of such a pattern [59]. Some researchers believe that sociological approaches to stress seek stressors in the organization of lives and in the structure of experience rather than among unrelated "risk factors" [16]. Exclusion from full participation in the social system and the kind of participation that does not provide expected returns are two major pathways that link social structure with stress [11]. Another perspective in this domain proposes that to the extent that socioeconomic class, race and ethnicity, gender, and age are systems that embody the unequal distribution of resources and opportunities; a low status within them may itself be a source of stress [16].

Considering the structural perspective on social causation, stress is understood as a consequence of location in the social system as well as a determinant of the psychological "distress." There is a focal relationship between social position and psychological distress – stressful life experience is but one pathway linking structure to emotional well-being. "Location in the social system influences the risk of encountering stressors, which in turn influences the chances of becoming emotionally distressed" [11, p. 19, as cited by 5]. Another research approach emphasizes the link between a person's social and subjective existence and challenges various dichotomies such as mind/body, culture/nature, and society/biology [55]. Supporters of this approach consider it useful in understanding that why some social groups are more prone to stress-induced physical illness rather than to psychological distress [11]. Some sociological researchers on stress have challenged the view that people are passive respondents to external circumstances [55] by highlighting that people may learn new strategies from stressful experiences which can be mobilized to meet the demands of the situations [4].

Occupational Stress Perspective

Stress is an all prevailing phenomenon. The occupational perspective on stress devotes researches in finding the occupational or organizational situations or structures acting as stressors. Although, again like the psychological perspective, the role of certain individual specific factors cannot be denied, the nature of stressors and the nature of their resultant influences are somewhat different in case of organizations. The occupational stress research paradigm primarily attempts to explore the elements of job that cause terms of job stress among employees, and efforts are directed to optimize the organizational productivity by means of reducing employee stress. The other occupational aspects that draw attention of researchers include types of role, role overload, role conflict, work-overload, sense of belongingness and security, employee empowerment, interpersonal conflicts, work–family interference and balance, and managerial shortcomings. Researchers with occupational perspective demonstrate apt concern to the nature of stressors, most of which are controllable by organization level interventions rather than individual coping efforts. At the target of the occupational research outcomes are the mutually related aspects of stress-free working environment and satisfied employees, which in turn increase organizational productivity. Some of the researchers are inclined to create socially harmonious organizations by studying the factors that inculcate cohesiveness and generate drive to the common organizational goals, whereas others draw their researches on the lines of understanding and promoting the factors that are individual employee specific and which generate sense of competition for higher relative achievements. In general, though some of the variables studied by occupational researchers are particular to organizational settings, the occupational perspective utilizes a mixed of the psychological and sociological approaches.

Conclusive Words

The concept of stress has long been the focus of researches in various biological, physiological, and behavioral science disciplines. Enormous numbers of efforts have been devoted by the scholars to explore and comprehend various dynamics of stress. Even after around a century long devoted innumerable research efforts, there exist numerous crucial questions that still need answer. New queries are emerging day by day and so are the research paradigms. Today the dedicated scientific and research community is trying to not leave any stone unturned. Advanced research methodologies are being applied. Though the stress research has been utilizing robust research designs and appropriate analytical methods, there has been a relative isolation as far as application of more than one methodological approach is concerned. Though it is well evident that the existing researches have tried to add to the epistemological understanding of stress by means of using laboratory-based experimental designs, multifactor comparative factorial designs, and correlational designs, different aspects of the stress domain have lesser been subjected to more than one design. Similarly, there have been substantial number of cross-cultural studies; however, the list of the aspects that have been explored across different cultures remains relatively small. Though there have been many longitudinal studies and plenty of cross-sectional studies, the combination of the two is yet to be seen. Majority of the existing researches have used either of the two types of data, that is, either quantitative or qualitative, and there is a paucity of researches that support their evidence by both kinds of data – qualitative as well as quantitative. Though some of the earlier researches have attempted to explore the mediators and moderators of stress and its reactions, future researches should go beyond to explore the mediated-moderation and moderated mediation. There exist different interdisciplinary explanations of various stress conceptualizations, which, though may be adequate to the respective disciplinary realm, keeps a lay reader somewhat confused in his comprehension. As it is evident that even after year-long research efforts some aspects of the stress domain do not reflect consensus among different researchers, we call for more researches with combined methodological approaches tested with multivariate analytic methods. Future researches should further simplify the understanding of the concept of stress from a multidisciplinary perspective so that all the underlying dynamics to face and eliminate can become easily comprehendible.

References

1. Pollock C (1988) On the nature of social stress: production of a modern mythology. Soc Sci Med 26:381–392
2. Lazarus RS (1999) Stress and emotion: a new synthesis. Springer, New York
3. Selye H (1974) Stress without distress. Hodder & Stoughton, London
4. Mulhall A (1996) Cultural discourse and the myth of stress in nursing and medicine. Int J Nurs Stud 33(5):455–468
5. Paradies Y (2004) A review of the relationship between psychosocial stress and chronic disease for indigenous and African American peoples. Cooperative Centre for Aboriginal Health, Darwin

6. Young A (1980) The discourse on stress and the reproduction of conventional knowledge. Soc Sci Med Part B: Med Anthropol 14(3):133–146
7. Doublet S (2000) The stress myth. Science & Humanities Press, Chesterfield
8. Cohen S, Kessler RC, Gordon LU (1995) Strategies for measuring stress in studies of psychiatric and physical disorder. In: Cohen S, Kessler RC, Gordon LU (eds) Measuring stress: a guide for health and social scientists. Oxford University Press, New York, pp 3–26
9. Quick JC, Nelson DL, Quick JD, Orman DK (2001) An isomorphic theory of stress: the dynamics of person-environment fit. Stress Health 17:147–157
10. Wheaton B (1999) Social stress. In: Aneshensel CS, Phelan JC (eds) Handbook of the sociology of mental health. Plenum Press, New York, pp 277–300
11. Aneshensel CS (1992) Social stress: theory and research. Ann Rev Sociol 18:15–38
12. Thoits PA (1995) Stress, coping, and social support processes: Where are we? What next? J Health Soc Behav 35(Extra issue):53–79
13. Taylor SE, Aspinwall LG (1996) Mediating and moderating processes in psychological stress: appraisal, coping, resistance, and vulnerability. In: Kaplan HB (ed) Psychological stress. Academic, San Diego, pp 71–110
14. Wheaton B (1994) Sampling the stress universe. In: Avison WR, Gotlib IH (eds) Stress and mental health: contemporary issues and prospects for the future. Plenum Press, New York, pp 77–114
15. Kanner AD, Coyne SC, Schafer C, Lazarus R (1981) Comparison of two modes of stress measurement: daily hassles and uplifts versus major life events. J Behav Med 4:1–39
16. Pearlin LI (1989) The sociological study of stress. J Health Soc Behav 30:241–256
17. Turner RJ, Blair W, Lloyd D (1995) The epidemiology of social stress. Am Sociol Rev 60:104–125
18. Pearlin LI (1999) The stress concept revisited: reflections on concepts and their Interrelationships. In: Aneshensel CS, Phelan JC (eds) Handbook of the sociology of mental health. Plenum Press, New York, pp 395–415
19. Wheaton B (1996) The domains and boundaries of stress concepts. In: Kaplan HB (ed) Psychosocial stress: perspectives on structure, theory, life-course, and methods. Academic, San Diego, pp 29–70
20. Lazarus RS, Folkman S (1984) Stress, appraisal and coping. Springer, New York
21. Janusek LW, Cooper DT, Mathews HL (2012) Epigenetics and stress: a life course perspective. In: Rice VH (ed) Handbook of stress coping and health: implications for nursing research, theory, and practice, 2nd edn. Sage Publications, Inc., Thousand Oaks, pp 71–96
22. Feinberg AP (2008) Epigenetics as the epicenter of modern medicine. JAMA 299(11):1345–1350
23. Mathews HL, Janusek LW (2011) Epigenetics and psychoneuroimmunology: mechanics and models. Brain Behav Immun 25(1):25–39
24. Zhang TY, Meaney MJ (2010) Epigenetics and the environmental regulation of the genome and its function. Annu Rev Psychol 61(439–466):C431–C433
25. Nanduri J, Makarenko V, Reddy VD, Yuan G, Pawar A, Wang N, Khan SA, Zhang X, Kinsman B, Peng YJ, Kumar GK, Fox AP, Godley LA, Semenza GL, Prabhakar NR (2012) Epigenetic regulation of hypoxic sensing disrupts cardiorespiratory homeostasis. Proc Natl Acad Sci U S A 109(7):2515–2520
26. Cannon WB (1932) The wisdom of the body. W. W. Norton, New York
27. Levi L (1972) Stress and distress in response to psychosocial stimuli. Pergamon Press, New York
28. Dunbar HF (1954) Emotions and bodily changes: a survey of literature on psychosomatic interrelationships, 1910–1953, 4th edn. Columbia University Press, New York
29. Raab W (1971) Cardiotoxic biochemical effects of emotional-environmental stressors – fundamentals of psychocardiology. In: Levi L (ed) Society, stress and disease, vol 1. Oxford University Press, London, pp 331–337
30. McCubbin JA, Richardson J, Obrist PA, Kizer JS, Langer AW (1980) Catecholaminergic and hemodynamic responses to behavioural stress in young adult males. Paper presented at the twentieth annual meeting of the Society for Psychophysiological Research. Vancouver, British Columbia
31. Rabin BS, Cohen S, Ganguli R, Lyle DT, Cunnick JE (1989) Bidirectional interaction between the central nervous system and immune system. CRC Crit Rev Immunol 9:279–312
32. Anisman H, Zacharko RM (1992) Depression as a consequence of inadequate neurochemical adaptation in response to stressors. Br J Psychiatr 160:36–43
33. Herd JA (1978) Physiological correlates of coronary prone behaviour. In: Dembroski TM, Weiss SM, Shields JL, Haynes SG, Feinleib M (eds) Coronary-prone behaviour. Springer, New York, pp 129–136
34. Selye H (1956) The stress of life. McGraw-Hill, New York
35. Selye H (1974) Stress without distress. J. B. Lippincott, Philadelphia
36. Lazarus RS (1977) Cognitive and coping processes in emotion. In: Monat A, Lazarus RS (eds) Stress and coping. Columbia University Press, New York, pp 145–158
37. Mason JW (1975) A historical view of stress field, part −2. J Human Stress 1:22–36
38. Cohen S, Kessler RC, Gordon LU (1995) Measuring stress: a guide for health and social scientists. Oxford University Press, New York
39. Selye H (1980) Stress, aging, and retirement. J Mind Behav 1:93–110
40. Stokes PE (1987) The neuroendocrine measurement of depression. In: Marsella AJ, Hirschfeld RMA, Katz MM (eds) The measurement of depression. Guilford Press, New York, pp 153–195
41. Baum A, Grunberg NE, Singer JE (1982) The use of psychological and neuroendocrinological

measurements in the study of stress. Health Psychol 1:217–236
42. Brim O, Ryff C (1980) On the properties of life events. In: Baltes P, Brim O (eds) Life-span development and behaviour, vol 3. Academic, New York, pp 367–388
43. Robson MA (1997) An exploration of stress and its perception in childhood. Durham theses, Durham University. Available at Durham e-theses online: http://etheses.dur.ac.uk/5061/
44. Danish SJ, Smyer MA, Nowak CA (1980) Developmental intervention: enhancing life-event processes. In: Baltes P, Brim O (eds) Life-span development and behaviour, vol 3. Academic, New York, pp 339–366
45. Hultsch D, Plemons J (1979) Life events and life span development. In: Baltes P, Brim O (eds) Life-span development and behaviour, vol 2. Academic, New York, pp 1–36
46. Compas B (1987) Stress and life events during childhood and adolescence. Clin Psychol Rev 7:275–302
47. Bandura A, Walters R (1963) Social learning and personality development. Holt, Rinehart & Winston, New York
48. Meyer A (1951) The life chart and the obligation of specifying positive data in psychopathological diagnosis. In: Winters EE (ed) The collected papers of Adolf Meyer, vol 3, Medical teachings. The Johns Hopkins University Press, Baltimore, pp 52–56
49. Wolff HG, Wolf SG Jr, Hare CC (1950) Life stress and bodily disease. Williams & Wilkins, Baltimore
50. Hinkle LE Jr, Wolff HG (1974) Ecologic investigations of the relationship between illness, life experiences and the social environment. Ann Intern Med 49:1373–1388
51. Hawkins NG, Davies R, Holmes TH (1957) Evidence of psychosocial factors in the development of pulmonary tuberculosis. Am Rev Tuberc 75:768–780
52. Holmes TH, Masuda M (1974) Life changes and Illness susceptibility. In: Dohrenwend BS, Dohrenwend BP (eds) Stressful life events: their nature and effects. Wiley, New York, pp 45–72
53. Rahe RH, Lind E (1971) Psychosocial factors and sudden cardiac death: a pilot study. J Psychosom Res 15:19–24
54. Brown GW, Harris TO (1989) Life events and illness. Guilford Press, New York
55. Bartley M, Blane D, Smith GD (1998) The sociology of health inequalities. Blackwell Publishers, Oxford
56. Wheeler RJ, Frank MA (1988) Identification of stress buffers. Behav Med 14(2):78–89
57. Lazarus RS (1966) Psychological stress and the coping process. McGraw-Hill, New York
58. Lazarus RS (1981) The stress and coping paradigm. In: Eisdorfer C, Cohen D, Kleinman A, Maxim P (eds) Models for clinical psychopathology. Spectrum, New York, pp 177–214
59. McDonough P, Walters V, Strohschein L (2002) Chronic stress and the social patterning of women's health in Canada. Soc Sci Med 54(5):767–782

Therapeutic Potential of Intermittent Hypoxia: Lessons from Respiratory Motor Plasticity

A. Navarrete-Opazo, E.A. Dale, and Gordon S. Mitchell

Abstract

Intermittent hypoxia (IH) is a subject of considerable interest since it has both beneficial and adverse effects. Unfortunately, a lack of consistency in the use of the term "intermittent hypoxia" has led to considerable confusion in the field. In reviewing available literature, the physiological and pathological impact of IH appears to be highly associated with the effective IH "dose." IH consisting of modest hypoxic episodes ($\geq$9 % inspired O_2) and lesser numbers of hypoxia/reoxygenation events per day ($\leq$15 cycles/day) is generally associated with beneficial effects in multiple body systems. In contrast, severe hypoxic episodes (<9 % inspired oxygen) and more frequent hypoxic episodes per day (40–2,400 cycles/day) shift the balance towards morbidity. In accordance, the impact of IH on the neural system controlling breathing is critically dependent on variables including the pattern of hypoxia (intermittent versus sustained), the severity of hypoxia within episodes, and the overall duration of IH exposure (minutes to years). A low IH "dose" (few episodes, moderate hypoxia) elicits serotonin-dependent spinal, respiratory motor plasticity that may be harnessed as a therapeutic approach to improve respiratory function in clinical conditions that impair breathing, such as cervical spinal injury. With a similar protocol but more severe hypoxic episodes, a distinct adenosine-dependent mechanism of spinal respiratory motor plasticity is observed. The cumulative effectives of repeated, low-dose IH (metaplasticity) suggest that repetitive, acute IH may represent a simple, safe, and effective treatment to promote meaningful therapeutic benefit in a range of clinical conditions that compromise respiratory (and nonrespiratory) somatic motor function.

A. Navarrete-Opazo • G.S. Mitchell (✉)
Department of Comparative Biosciences,
University of Wisconsin-Madison,
2015 Linden Drive, Madison 53706, WI, USA
e-mail: mitchell@svm.vetmed.wisc.edu

E.A. Dale
Department of Physiological Sciences, UCLA,
Los Angeles, CA, USA

S.B. Singh et al. (eds.), *Translational Research in Environmental and Occupational Stress*,
DOI 10.1007/978-81-322-1928-6_4, © Springer India 2014

Introduction

Intermittent hypoxia (IH) consists of repeated hypoxic episodes, interspersed with intervals of normoxia. In recent years, considerable attention has been paid to IH, in part because of its role in the pathology experienced by individuals with sleep-disordered breathing (e.g., hypertension, learning deficits, metabolic syndrome, etc.) [1–5]. On the other hand, considerable literature also suggests that IH has beneficial effects. For example, "intermittent hypoxia" increases aerobic athletic performance [6] and promotes recovery of lost respiratory and nonrespiratory somatic motor function with traumatic [7–9] or neurodegenerative spinal injury [10]. Here we consider distinctions between the IH protocols giving rise to pathology versus beneficial effects and then consider the mechanisms and implications of respiratory and nonrespiratory motor plasticity elicited by modest protocols of IH. In specific, the goals of this brief review are to (1) differentiate between the experimental IH protocols giving rise to beneficial versus pathological effects in multiple body systems, (2) review the impact of modest IH protocols on respiratory and nonrespiratory motor function, (3) review recent advances in our understanding of mechanisms giving rise to that IH-induced motor plasticity, and (4) consider the clinical implications of IH-induced motor plasticity.

A Diversity of Experimental IH Protocols

The specific protocol or "dose" of IH differs appreciably among research fields (and even groups), leading to discrepancies in terminology and making it difficult to make generalizations concerning the beneficial versus pathological effects of IH. For example, the inspired oxygen (O_2) during hypoxic episodes ranges from 2 to 16 %, causing large differences in the severity of hypoxemia during each episode. The duration of hypoxic episodes can be as brief as 15–30 s or as much as 12 h. The number of hypoxia/reoxygenation cycles varies from 3 to more than 2,400 (in aggressive protocols intended to simulate obstructive sleep apnea). The overall duration of the entire IH protocol can be less than 1 h to between 2 and 90 days.

Moderate hypoxia ($\geq$9 % inspired O_2) and lower numbers of hypoxia/reoxygenation cycles per day ($\leq$15 cycles/day) are generally associated with reported beneficial and/or therapeutic IH effects [6–8, 11–24]. In contrast, severe (<9 % inspired O_2) and/or more frequent (40–2,400 cycles) presentations of hypoxia/reoxygenation are generally used in studies reporting deleterious/pathological effects [1, 2, 4, 5, 25–32]. In other words, the biological impact of IH is a matter of "dose" [33].

Although cellular/molecular mechanisms distinguishing IH protocols that differ in "dose" are undoubtedly complex, and not completely understood, some have hypothesized that recurrent hypoxia/reoxygenation is similar to repetitive ischemia-reperfusion injury caused by increased reactive oxygen species (ROS) formation during each reoxygenation event [34]. Accordingly, IH protocols characterized by more frequent hypoxic episodes are expected to elicit detrimental effects from greater oxidative stress. Conversely, mild or moderate IH protocols are less likely to cause ROS-related injury, enabling the expression of beneficial effects that may be harnessed for therapeutic advantage. Indeed, accumulating evidence demonstrates that a moderate IH protocols represent a simple, safe, and effective means of promoting meaningful therapeutic benefits in diverse clinical disorders.

Multiple Body Systems Exhibit Benefits and Pathology After IH

Moderate IH has been reported to induce therapeutic benefits in clinical disorders associated with immune, metabolic, learning, cardiovascular, and neural (including respiratory and nonrespiratory somatic motor systems) pathology. Conversely, severe IH protocols are more likely to cause/exacerbate morbidity/pathology in these same systems.

Cardiovascular Function

Intense protocols of chronic intermittent hypoxia (CIH) are strongly associated with hypertension and heart disease [1, 2, 25–27, 35]. On the other hand, moderate IH has beneficial cardiovascular effects in humans and animal models [14, 36]. For example, abundant literature describes therapeutic benefits of moderate IH in the treatment of systemic hypertension; IH consisting of 10, 5-min episodes per day (10–14 % O_2; 5-min normoxic intervals) reduced both systolic and diastolic blood pressure, heart rate, and peripheral resistance in 56 patients with stages I–II hypertension [14]. Mechanisms postulated to explain these antihypertensive effects of moderate IH include reduced sympathetic nervous activity [37, 38], prevention of calcium overload of vascular smooth muscle [39], improved water and salt metabolism [40], increased antioxidant enzyme activity [41], and increased synthesis of the angiogenic growth factors vascular endothelial growth factor (VEGF) and fibroblast growth factor (FGF) [42].

Immune Function

Although CIH (severe IH) is associated with systemic and central neural inflammation [31, 34], there is some evidence to support beneficial effects and possible clinical utility of moderate IH protocols in immunologically compromised patients. For example, moderate IH enhances the innate immune system, but also has anti-inflammatory effects [19, 43, 44]. Exposure of cultured human hematopoietic stem progenitor cells to between 1 and 3 % O_2 activates hematopoiesis [43, 44]. In healthy subjects, exposure to 4, 5-min episodes of 10 % O_2 interspersed with 5-min normoxic intervals for 14 days decreases circulating hematopoietic stem progenitor cells, activates complement, increases circulating platelets, augments phagocytic and bactericidal activities of neutrophils, and suppresses proinflammatory mediators such as TNF-α and IL-4 by >90 % [19]. These responses, which persist at least 7 days post-IH, may augment the body's immune defenses while suppressing inflammation.

Metabolic Function

Moderate IH protocols have beneficial effects on metabolic functions, including effects on body weight, cholesterol, blood sugar levels, and insulin sensitivity. For example, IH (10–12 % inspired O_2, 3 times per week, 3–6 weeks) with or without physical activity (20-min strength/resistance exercise and 30-min high-intensity aerobic exercise) is reported to be an efficient way to lose weight and increase aerobic capacity [6]. Such moderate IH may reduce body weight by increasing leptin expression and release [45]. Although leptin is involved in body weight regulation, it also plays key roles in inflammation, immunity, tissue repair, and angiogenesis. Thus, IH-induced leptin upregulation may improve peripheral tissue repair [46]. Moreover, moderate IH (12 h 14 % O_2, 7 days/week, 4 weeks), with or without training, improves glucose tolerance and increases glucose transporter (GLUT4) expression in rats [11]. Increased GLUT4 levels and activity may increase glucose removal from the blood [11, 47–49].

Central Nervous System Function

Intense CIH causes cognitive impairment in rodents, and this effect is greatest when CIH is experienced early or late in life [4, 28, 50–52]. CIH is associated with cognitive impairment in humans [5]. The pathological effects of CIH on learning and memory are associated with neuronal apoptosis and cytoarchitectural disorganization in the hippocampal CA1 subfield as well as the frontoparietal cortex [4]. In contrast to CIH, moderate IH protocols do not elicit detectable central nervous system pathology. For example, in rats exposed to 10, 5-min episodes of 10.5 % inspired O_2 (5-min normoxic intervals) for seven consecutive days [7, 53] or 3 days per week for 10 weeks [7, 53, 54], there are no signs of hippocampal apoptosis, reactive gliosis, or systemic hypertension.

In contrast to CIH, moderate IH may increase brain development in early life, leading to increased learning capacity [12, 16, 55, 56]. For example, neonatal mice exposed to IH (10.8 % inspired O_2, 4 h/day) from birth to 4 weeks of age exhibit enhanced Morris water maze and 8-arm radial maze performance [12, 16]. In addition to the beneficial/deleterious actions of IH on brain/cognitive function, IH also triggers respiratory and nonrespiratory motor plasticity [33]. This topic will be the focus of the remainder of this brief review.

Moderate Acute Intermittent Hypoxia (AIH) Induces Respiratory Motor Plasticity

In contrast to earlier views, the respiratory motor control system exhibits considerable plasticity [57]. One of the most extensively studied forms of respiratory plasticity is phrenic long-term facilitation (pLTF) induced by acute intermittent hypoxia (AIH) [58]. In anesthetized, paralyzed, and ventilated rats, AIH (3, 5-min episodes of 10 % O_2 separated by 5 min of 50 % O_2) elicits a progressive and sustained increase in phrenic motor output lasting more than 90 min post-hypoxia (i.e., pLTF; [59]). pLTF is pattern sensitive since continuous hypoxia of an equal cumulative duration does not elicit the response [60]. A slightly longer AIH protocol (8–10, 5-min episodes of 10.5 % O_2, 5-min intervals) elicits long-lasting increases in inspiratory EMG activity of the diaphragm, genioglossus [61–63], and second external intercostal muscle [64].

Similar IH protocols also elicit ventilatory long-term facilitation (vLTF) in multiple species [59, 65]. Ventilatory LTF has been reported in dogs [66], goats, [67], ducks [68], rabbits [69], rats [70], and mice [71]. In humans, ventilatory LTF is evident only if carbon dioxide is sustained slightly above baseline levels during wakefulness [72, 73], but can be observed during normocapnic conditions when asleep [74].

AIH-induced LTF is a central neural mechanism in normal animals. Indeed, phrenic LTF was first reported in anesthetized cats while recording phrenic nerve activity after episodic stimulation of the cut, central end of the carotid sinus nerve [75, 76]; this finding alone demonstrates that pLTF can result exclusively from central neural mechanisms. Although activation of the peripheral arterial chemoreceptors plays a clear role in at least some forms of LTF in normal animals [59], the plasticity itself arises in within individual respiratory motor nuclei [77–79]. Although AIH-induced pLTF is attenuated in rats with surgical [80] or chemical carotid chemodenervation [81], it is not eliminated, suggesting that at least some pLTF can occur without functional carotid body chemoreceptors. Plasticity within the carotid body chemoreceptors per se does not normally contribute to LTF since carotid chemosensory LTF is not observed in rodents unless they have been pretreated with chronic intermittent hypoxia [82, 83].

In longer time domains, CIH or repetitive AIH (rAIH) reveals an enhanced capacity to express AIH-induced pLTF (i.e., metaplasticity; [57, 84]). In rats, enhanced AIH-induced pLTF is observed after pretreatment with (1) cervical dorsal rhizotomy [85], (2) CIH [86], and (3) repetitive AIH (rAIH) [53, 87, 88]. For example, CIH pretreatment (5-min episodes of 11–12 % O2; 5-min normoxic intervals; 12 h/day, 7 days) nearly doubles pLTF in anesthetized rats [86] and ventilatory LTF in unanesthetized rats [89]. rAIH elicits similar pLTF metaplasticity (but without detrimental effects associated with CIH) [53, 87, 88]. Similarly, rats exposed to rAIH (ten 5-min episodes/day, 3 days/week, 4 weeks) enhance subsequent AIH-induced pLTF [87, 88] without signs of hippocampal pathology [54]. Repetitive AIH appears to be a safe and effective means of eliciting respiratory motor plasticity; thus, rAIH may represent a novel therapeutic approach in the treatment of clinical disorders associated with respiratory insufficiency [7, 33, 90].

Cellular Mechanisms of AIH-Induced Respiratory Motor Plasticity

In recent years, we have made considerable progress towards an understanding of cellular mechanisms giving rise to respiratory motor plasticity. Such an understanding is essential to minimize pathology and maximize benefits as we

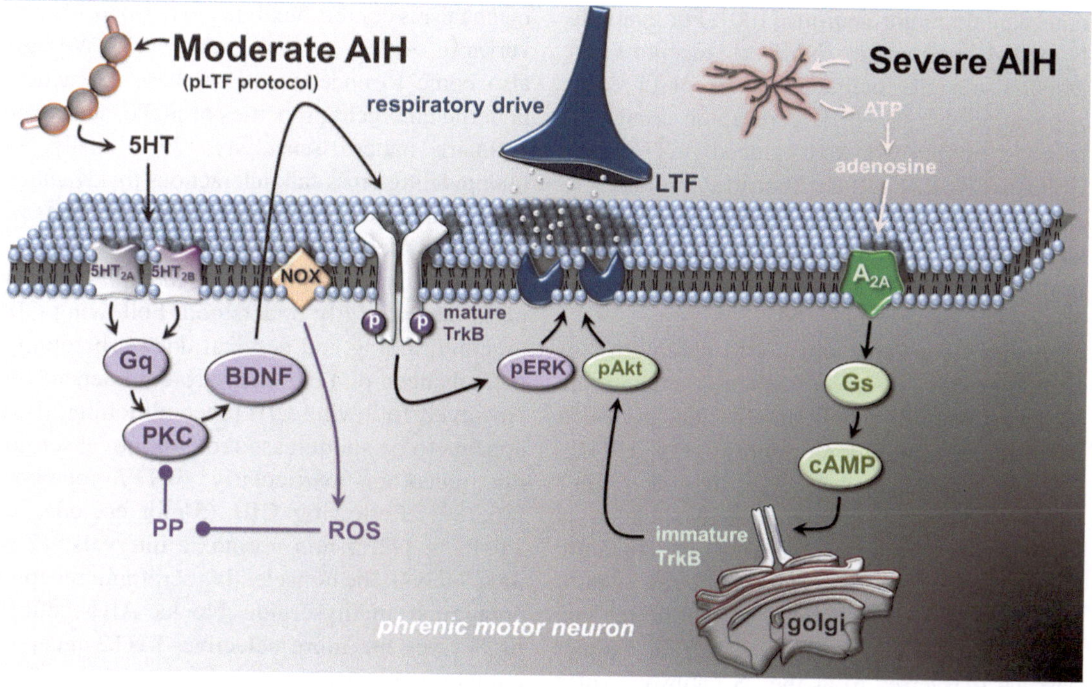

Fig. 1 Diverse cellular mechanisms of phrenic motor facilitation (Modified from Dale et al. [33]). In our working model, moderate acute intermittent hypoxia (AIH) stimulates serotonin (5-HT) release from raphespinal projections near phrenic motor neurons. Activation of serotonin type 2 (5-HT2A and/or 2B) receptors (both Gq protein coupled) increases PKC activity, thereby stimulating new BDNF synthesis within phrenic motor neurons. BDNF is released extracellularly, binding and activating its high-affinity receptor, mature TrkB. Activated TrkB phosphorylates and activates ERK/MAP kinases (pERK), subsequently leading to functional enhancement of glutamatergic synaptic inputs from bulbospinal respiratory motor neurons through an uncertain mechanism (possibly involving glutamate receptor trafficking). This intracellular cascade is referred to as the Q pathway to phrenic motor facilitation and is responsible for phrenic LTF after moderate AIH protocols. In contrast, severe hypoxia during AIH episodes triggers greater ATP release from nearby glia, leading to extracellular adenosine accumulation. By activating Gs protein-coupled A_{2A} receptors on phrenic motor neurons, increased cyclic AMP formation triggers new synthesis of an immature TrkB isoform; this immature isoform autophosphorylates (without BDNF) and signals from the (intracellular) Golgi apparatus via protein kinase B/Akt activation (pAkt); Akt activity subsequently leads to functional enhancement of glutamatergic synaptic inputs from bulbospinal respiratory neurons by an uncertain mechanism. This intracellular cascade is referred to as the S pathway to phrenic motor facilitation and is responsible for phrenic LTF with severe AIH protocols. These diverse pathways confer flexibility in the phrenic response to IH and may be harnessed for therapeutic advantage

test/develop IH as a viable therapeutic tool. What we have come to realize is that the picture is complex and that multiple, distinct cellular mechanisms give rise to IH-induced respiratory motor plasticity depending on the severity, pattern, and duration of hypoxic exposure.

Multiple cellular mechanisms elicit phenotypically similar respiratory motor plasticity (Fig. 1) [33, 78, 79]. Brief AIH protocols with mild to moderate hypoxic episodes (arterial $PO_2 > 35$ mmHg) trigger episodic release of serotonin in or near the phrenic motor nucleus [91, 92], thereby activating 5-HT2 receptors [77, 93–95]. Episodic 5-HT2 receptor activation subsequently activates phospholipase C/protein kinase C theta [96], triggers new synthesis of brain-derived neurotrophic factor (BDNF) [97], and activates the high-affinity BDNF receptor TrkB [97, 98] and then signals via ERK MAP kinase activation [53, 99]. Although downstream events are less clear, activated ERK may strengthen excitatory glutamatergic synapses onto phrenic motor neurons [100]. For example, N-methyl-D-aspartate (NMDA) receptors are essential for the maintenance of pLTF since application of an NMDA receptor antagonist reverses established AIH-induced pLTF [101]. Collectively, this cellular cascade is referred to as the "Q pathway" since the initiating receptor is a Gq protein-coupled metabotropic receptor, and multiple receptors signaling via Gq proteins elicit a similar response [78].

A distinct cellular cascade to phrenic motor facilitation is observed when metabotropic receptors coupled to Gs proteins are activated on or near phrenic motor neurons (Fig. 1) [78]. For example, adenosine 2A [102] or 5-HT7 receptor activation in the cervical spinal cord [103] triggers new synthesis of an immature TrkB isoform that auto-dimerizes, auto-phosphorylates, and signals from within the cell via protein kinase B (i.e., Akt). This cellular cascade is referred to as the "S pathway" since multiple Gs protein-coupled metabotropic receptors elicit the response [78].

Although AIH with mild to moderate hypoxic episodes (arterial $PO_2 > 35$–40 mmHg) elicits pLTF by the spinal, serotonin-dependent Q pathway, phenotypically similar pLTF is observed when severe hypoxic episodes (arterial $PO_2 < 30$ mmHg) are used. However, in this case, the facilitation arises from a different adenosine-dependent (serotonin-independent) mechanism [104]. Thus, pLTF results from completely distinct mechanisms, depending on the severity of hypoxia within each episode of the AIH protocol (Q pathway with moderate but S pathway with severe hypoxia).

The Q and S pathways to phrenic motor facilitation interact via cross-talk inhibition. For example, with moderate AIH, either cervical adenosine 2A [105] or 5-HT7 receptor inhibition [106] enhances pLTF. Thus, subthreshold S pathway activation inhibits the predominant Q pathway-dependent pLTF, and this cross-talk inhibition results from a PKA-dependent mechanism [106]. By understanding interactions between cellular cascades and phrenic motor facilitation, we have come to realize that the diversity of potential mechanisms confers flexibility to respond when IH varies in severity and/or duration [33]. We have also come to understand that these interactions underlie emergent properties of pLTF, such as its hallmark pattern sensitivity [79]. Finally, by manipulating cross-talk interactions to advantage, we may be able to enhance the impact of IH as we harness its therapeutic potential.

Mechanisms giving rise to metaplasticity in pLTF are not fully understood. Following CIH preconditioning and cervical dorsal rhizotomy, the enhanced pLTF is still serotonin dependent. However, following CIH preconditioning, there appears to be an increased role for novel serotonin receptors, particularly 5-HT7 receptors [86, 89]. Following CIH (5-min episodes of 11–12 % O2; 5-min normoxic intervals; 12 h/day, 7 days), the nonselective serotonin receptor antagonist methysergide blocks AIH-induced pLTF, yet the more selective 5-HT2 receptor antagonist ketanserin is no longer able to block the response completely [86]. The "missing" serotonin receptor appears to be of the 5-HT7 receptor subtype since antagonists that block this receptor eliminate residual pLTF observed after 5-HT2 receptor inhibition [86].

Following repetitive AIH (10, 5-min episodes/day, 3 days/week, 4 weeks), 5-HT2A receptor expression, BDNF, TrkB, and phospho-ERK within the phrenic motor nucleus are all increased [54]. In this case, the role of different serotonin receptors in pLTF metaplasticity has not been investigated. However, in addition to these Q-pathway molecules, rAIH also increases HIF-1α-regulated proteins capable of eliciting phrenic motor facilitation within the phrenic motor nucleus, such as VEGF [107] and erythropoietin [108]. One study attempting to define the role of these proteins in enhanced pLTF following rAIH was inconclusive [109]. Considerable work must be done before we will understand the significance and mechanistic basis of pLTF metaplasticity following IH preconditioning. However, the existence of metaplasticity in pLTF suggests the possibility that IH effects may accumulate, amplifying the efficacy of IH with repetitive exposures.

Therapeutic Potential of Moderate IH After Spinal Cord Injuries

Although further studies are needed to assess the therapeutic potential of moderate IH in individuals with cardiovascular, immune, metabolic, and/or neural (cognitive or motor) disorders, moderate IH appears to be safe, simple to administer, and effective, offering considerable promise for clinical application. One promising area for therapeutic application is the treatment of motor deficits caused by traumatic spinal injury.

More than one half of all spinal injuries (56 %) occur at the cervical level [110], interrupting descending bulbospinal pathways to respiratory motor neurons and causing respiratory and nonrespiratory somatic muscle paresis/paralysis below the injury. Although our understanding of CNS regeneration continues to advance, we are far from harnessing this knowledge to treat persons with chronic spinal injuries. In the short term, a more realistic approach is to develop treatments that restore function based on the intrinsic capacity for spinal plasticity in spared neural pathways (most spinal injuries are incomplete). Moderate IH has considerable potential to promote spinal motor plasticity after spinal injury, thereby increasing respiratory (and nonrespiratory) motor functions [17, 33, 78, 90].

Cervical (C2) hemisection (C2HS) from the midline to lateral edge of the cervical spinal cord has been used extensively as a model to study respiratory plasticity after cervical injury [111–115]. C2 hemisection interrupts descending bulbospinal pathways from the medulla to ipsilateral phrenic motor neurons, paralyzing the ipsilateral hemidiaphragm. Slow, limited spontaneous recovery of phrenic/diaphragm activity occurs below the hemisection, a well-studied phenomenon known as the crossed phrenic phenomenon [112]. However, with respect to overall breathing capacity, compensation may also occur in contralateral (i.e., intact) phrenic and intercostal (or other accessory) motor pools [116–119]. Our approach is to harness moderate IH to induce additional plasticity in intact neural pathways to respiratory motor pools.

Plasticity in spared neural pathways can occur through changes in (functional) synaptic connections, such as the activation of previously silent crossed-spinal synaptic pathways to phrenic motor neurons [112, 120]. Other adaptations also increase breathing capacity in the face of spinal injury [57, 111, 119]. All of these adaptations are susceptible to IH-induced plasticity, amplifying the recovery of breathing capacity [7]. For example, with chronic C2 hemisection, AIH increases ipsilateral phrenic motor output (i.e., pLTF; [121]) and ipsilateral diaphragm activity (Navarrete and Mitchell, unpublished observations). AIH also increases contralateral hemidiaphragm activity, further contributing towards restoring breathing capacity (Navarrete and Mitchell, unpublished observations).

Repetitive IH protocols, such as daily AIH (dAIH), further enhance functional recovery after spinal injuries, both in terms of the magnitude and duration of effects. Specifically, rats with C2 hemisections increase respiratory capacity after dAIH [7, 17], restoring much of the lost capacity to increase tidal volume [7, 122]. Initial trials suggest that AIH at least partially restores lost breathing capacity in humans with chronic cervical spinal injuries [123], suggesting that this is a promising avenue for further research.

In addition to its effects on breathing capacity, repetitive AIH also improves nonrespiratory motor function after cervical spinal injury. For example, dAIH elicits long-lasting improvement in a skilled walking task in rats with limited cervical hemisections; specifically, rats improve performance on a horizontal ladder walking task when daily AIH is paired with ladder walking training for 7 days [7]. Similarly, AIH restores ankle strength in humans with chronic, incomplete spinal injuries (American Spinal Cord Injury Association Impairment Scale C or D); a single AIH presentation (15, 1-min episodes of 9 % O2; 1-min normoxic intervals) increases voluntary plantar flexion torque generation and gastrocnemius EMG activity by more than 80 % for more than 90 min post-AIH [8]. A recent randomized, double-blind, placebo-controlled, crossover study tested the effect of dAIH (15, 90-s episodes of 9 % inspired O2, 60-s intervals,

5 consecutive days) combined with overground walking in persons with chronic, incomplete spinal injuries (>1 year post injury; AIS D) [9]. In this study, dAIH alone improved walking speed, but dAIH paired with walking training increased both speed and endurance (37 %) for at least several days post-training [9].

Although the mechanisms underlying AIH-induced plasticity after spinal injuries have not been completely elucidated, tantalizing (but incomplete) evidence suggests that AIH may induce functional recovery via different mechanisms at different times post-injury (i.e., Q versus S pathway). Similarly, we have only a rudimentary understanding of cellular mechanisms giving rise to IH-induced recovery of nonrespiratory somatic motor function. Our working hypothesis is that it reflects the same serotonin- and adenosine-dependent mechanisms characteristic of IH-induced respiratory motor plasticity (Fig. 1) [33].

Currently, there are no approved therapies for respiratory or nonrespiratory somatic motor function with chronic spinal injuries. Repetitive AIH is a promising new strategy to enhance function in these underserved individuals. Although CIH elicits respiratory motor plasticity after cervical injury [124], it also elicits significant morbidity (e.g., hypertension, learning deficits; see above). Repetitive "low-dose" AIH may significantly improve function without the morbidities attendant to CIH. The potential application of low-dose IH to promote health and/or restore function in multiple clinical disorders is an intriguing area for future research.

Acknowledgements A. Navarrete Opazo supported by a Fulbright Fellowship

References

1. Brooks D, Horner RL, Kozar LF, Render-Teixeira CL, Phillipson EA (1997) Obstructive sleep apnea as a cause of systemic hypertension. Evidence from a canine model. J Clin Invest 99:106–109
2. Lesske J, Fletcher EC, Bao G, Unger T (1997) Hypertension caused by chronic intermittent hypoxia–influence of chemoreceptors and sympathetic nervous system. J Hypertens 15:1593–1603
3. Fletcher EC (2000) Effect of episodic hypoxia on sympathetic activity and blood pressure. Respir Physiol 119:189–197
4. Gozal D, Daniel JM, Dohanich GP (2001) Behavioral and anatomical correlates of chronic episodic hypoxia during sleep in the rat. J Neurosci 21:2442–2450
5. Champod AS, Eskes GA, Foster GE, Hanly PJ, Pialoux V, Beaudin AE, Poulin MJ (2013) Effects of acute intermittent hypoxia on working memory in young healthy adults. Am J Respir Crit Care Med 187:1148–1150
6. Urdampilleta A, Gonzalez-Muniesa P, Portillo MP, Martinez JA (2012) Usefulness of combining intermittent hypoxia and physical exercise in the treatment of obesity. J Physiol Biochem 68:289–304
7. Lovett-Barr MR, Satriotomo I, Muir GD, Wilkerson JE, Hoffman MS, Vinit S, Mitchell GS (2012) Repetitive intermittent hypoxia induces respiratory and somatic motor recovery after chronic cervical spinal injury. J Neurosci 32:3591–3600
8. Trumbower RD, Jayaraman A, Mitchell GS, Rymer WZ (2012) Exposure to acute intermittent hypoxia augments somatic motor function in humans with incomplete spinal cord injury. Neurorehabil Neural Repair 26:163–172
9. Hayes HB, Jayaraman A, Herrmann M, Mitchell GS, Rymer WZ, Trumbower RD (2013) Daily intermittent hypoxia enhances walking after chronic spinal cord injury: a randomized trial. Neurology 82:104
10. Nichols NL, Gowing G, Satriotomo I, Nashold LJ, Dale EA, Suzuki M, Avalos P, Mulcrone PL, McHugh J, Svendsen CN, Mitchell GS (2013) Intermittent hypoxia and stem cell implants preserve breathing capacity in a rodent model of amyotrophic lateral sclerosis. Am J Respir Crit Care Med 187:535–542
11. Chiu LL, Chou SW, Cho YM, Ho HY, Ivy JL, Hunt D, Wang PS, Kuo CH (2004) Effect of prolonged intermittent hypoxia and exercise training on glucose tolerance and muscle GLUT4 protein expression in rats. J Biomed Sci 11:838–846
12. Zhang JX, Chen XQ, Du JZ, Chen QM, Zhu CY (2005) Neonatal exposure to intermittent hypoxia enhances mice performance in water maze and 8-arm radial maze tasks. J Neurobiol 65:72–84
13. Tam CS, Wong M, Tam K, Aouad L, Waters KA (2007) The effect of acute intermittent hypercapnic hypoxia treatment on IL-6, TNF-alpha, and CRP levels in piglets. Sleep 30:723–727
14. Serebrovskaya TV, Manukhina EB, Smith ML, Downey HF, Mallet RT (2008) Intermittent hypoxia: cause of or therapy for systemic hypertension? Exp Biol Med (Maywood) 233:627–650
15. Haider T, Casucci G, Linser T, Faulhaber M, Gatterer H, Ott G, Linser A, Ehrenbourg I, Tkatchouk E, Burtscher M, Bernardi L (2009)

Interval hypoxic training improves autonomic cardiovascular and respiratory control in patients with mild chronic obstructive pulmonary disease. J Hypertens 27:1648–1654
16. Lu XJ, Chen XQ, Weng J, Zhang HY, Pak DT, Luo JH, Du JZ (2009) Hippocampal spine-associated Rap-specific GTPase-activating protein induces enhancement of learning and memory in postnatally hypoxia-exposed mice. Neuroscience 162:404–414
17. Vinit S, Lovett-Barr MR, Mitchell GS (2009) Intermittent hypoxia induces functional recovery following cervical spinal injury. Respir Physiol Neurobiol 169:210–217
18. Basovich SN (2010) The role of hypoxia in mental development and in the treatment of mental disorders: a review. Biosci Trends 4:288–296
19. Serebrovskaya TV, Nikolsky IS, Nikolska VV, Mallet RT, Ishchuk VA (2011) Intermittent hypoxia mobilizes hematopoietic progenitors and augments cellular and humoral elements of innate immunity in adult men. High Alt Med Biol 12:243–252
20. Tsai YW, Yang YR, Wang PS, Wang RY (2011) Intermittent hypoxia after transient focal ischemia induces hippocampal neurogenesis and c-Fos expression and reverses spatial memory deficits in rats. PLoS One 6:e24001
21. Wang ZH, Chen YX, Zhang CM, Wu L, Yu Z, Cai XL, Guan Y, Zhou ZN, Yang HT (2011) Intermittent hypobaric hypoxia improves postischemic recovery of myocardial contractile function via redox signaling during early reperfusion. Am J Physiol Heart Circ Physiol 301:H1695–H1705
22. Querido JS, Sheel AW, Cheema R, Van Eeden S, Mulgrew AT, Ayas NT (2012) Effects of 10 days of modest intermittent hypoxia on circulating measures of inflammation in healthy humans. Sleep Breath 16:657–662
23. Yokhana SS, Gerst DG 3rd, Lee DS, Badr MS, Qureshi T, Mateika JH (2012) Impact of repeated daily exposure to intermittent hypoxia and mild sustained hypercapnia on apnea severity. J Appl Physiol (Bethesda, MD: 1985) 112:367–377
24. Tsai YW, Yang YR, Sun SH, Liang KC, Wang RY (2013) Post ischemia intermittent hypoxia induces hippocampal neurogenesis and synaptic alterations and alleviates long-term memory impairment. J Cereb Blood Flow Metab 33:764–773
25. Fletcher EC, Lesske J, Culman J, Miller CC, Unger T (1992) Sympathetic denervation blocks blood pressure elevation in episodic hypoxia. Hypertension 20:612–619
26. McGuire M, Bradford A (2001) Chronic intermittent hypercapnic hypoxia increases pulmonary arterial pressure and haematocrit in rats. Eur Respir J 18:279–285
27. Tahawi Z, Orolinova N, Joshua IG, Bader M, Fletcher EC (2001) Altered vascular reactivity in arterioles of chronic intermittent hypoxic rats. J Appl Physiol 90:2007–2013, discussion 2000
28. Douglas RM, Miyasaka N, Takahashi K, Latuszek-Barrantes A, Haddad GG, Hetherington HP (2007) Chronic intermittent but not constant hypoxia decreases NAA/Cr ratios in neonatal mouse hippocampus and thalamus. Am J Physiol Regul Integr Comp Physiol 292:R1254–R1259
29. Savransky V, Nanayakkara A, Li J, Bevans S, Smith PL, Rodriguez A, Polotsky VY (2007) Chronic intermittent hypoxia induces atherosclerosis. Am J Respir Crit Care Med 175:1290–1297
30. Zoccal DB, Bonagamba LG, Oliveira FR, Antunes-Rodrigues J, Machado BH (2007) Increased sympathetic activity in rats submitted to chronic intermittent hypoxia. Exp Physiol 92:79–85
31. Arnardottir ES, Mackiewicz M, Gislason T, Teff KL, Pack AI (2009) Molecular signatures of obstructive sleep apnea in adults: a review and perspective. Sleep 32:447–470
32. Rasche K, Keller T, Tautz B, Hader C, Hergenc G, Antosiewicz J, Di Giulio C, Pokorski M (2010) Obstructive sleep apnea and type 2 diabetes. Eur J Med Res 15(Suppl 2):152–156
33. Dale EA, Ben Mabrouk F, Mitchell GS (2014) Unexpected benefits of intermittent hypoxia: enhanced respiratory and nonrespiratory motor function. Physiology (Bethesda) 29(1):39–48
34. Lavie L (2005) Sleep-disordered breathing and cerebrovascular disease: a mechanistic approach. Neurol Clin 23:1059–1075
35. Nattie EE, Bartlett D Jr, Johnson K (1978) Pulmonary hypertension and right ventricular hypertrophy caused by intermittent hypoxia and hypercapnia in the rat. Am Rev Respir Dis 118:653–658
36. Shatilo VB, Korkushko OV, Ischuk VA, Downey HF, Serebrovskaya TV (2008) Effects of intermittent hypoxia training on exercise performance, hemodynamics, and ventilation in healthy senior men. High Alt Med Biol 9:43–52
37. Melin A, Fauchier L, Dubuis E, Obert P, Bonnet P (2003) Heart rate variability in rats acclimatized to high altitude. High Alt Med Biol 4:375–387
38. Pshennikova M, Malyshev I, Manukhina E, Meerson F (2005) Distinction between stress resistance and protective effects of adaptation in rats of different genetic strains: role of regulatory systems. In: Hargens A (ed) Adaptation biology and medicine, vol 4. Narosa, New Delhi, pp 29–40
39. Cohen RA, Weisbrod RM, Gericke M, Yaghoubi M, Bierl C, Bolotina VM (1999) Mechanism of nitric oxide-induced vasodilatation: refilling of intracellular stores by sarcoplasmic reticulum Ca2+ ATPase and inhibition of store-operated Ca2+ influx. Circ Res 84:210–219
40. Behm R, Honig A, Griethe M, Schmidt M, Schneider P (1984) Sustained suppression of voluntary sodium intake of spontaneously hypertensive rats (SHR) in hypobaric hypoxia. Biomed Biochim Acta 43:975–985
41. Asha Devi S, Subramanyam MV, Vani R, Jeevaratnam K (2005) Adaptations of the antioxidant system in erythrocytes of trained adult rats: impact of intermittent hypobaric-hypoxia at two altitudes.

Comp Biochem Physiol C Toxiol Pharmacol 140:59–67
42. Wang JS, Chen LY, Fu LL, Chen ML, Wong MK (2007) Effects of moderate and severe intermittent hypoxia on vascular endothelial function and haemodynamic control in sedentary men. Eur J Appl Physiol 100:127–135
43. Cipolleschi MG, Dello Sbarba P, Olivotto M (1993) The role of hypoxia in the maintenance of hematopoietic stem cells. Blood 82:2031–2037
44. Danet GH, Pan Y, Luongo JL, Bonnet DA, Simon MC (2003) Expansion of human SCID-repopulating cells under hypoxic conditions. J Clin Invest 112:126–135
45. Ling Q, Sailan W, Ran J, Zhi S, Cen L, Yang X, Xiaoqun Q (2008) The effect of intermittent hypoxia on bodyweight, serum glucose and cholesterol in obesity mice. Pak J Biol Sci 11:869–875
46. Fantuzzi G, Faggioni R (2000) Leptin in the regulation of immunity, inflammation, and hematopoiesis. J Leukoc Biol 68:437–446
47. Cartee GD, Douen AG, Ramlal T, Klip A, Holloszy JO (1991) Stimulation of glucose transport in skeletal muscle by hypoxia. J Appl Physiol (Bethesda, MD: 1985) 70:1593–1600
48. Azevedo JL Jr, Carey JO, Pories WJ, Morris PG, Dohm GL (1995) Hypoxia stimulates glucose transport in insulin-resistant human skeletal muscle. Diabetes 44:695–698
49. Mackenzie R, Maxwell N, Castle P, Brickley G, Watt P (2011) Acute hypoxia and exercise improve insulin sensitivity (S(I) (2*)) in individuals with type 2 diabetes. Diabetes Metab Res Rev 27:94–101
50. Tagaito Y, Polotsky VY, Campen MJ, Wilson JA, Balbir A, Smith PL, Schwartz AR, O'Donnell CP (2001) A model of sleep-disordered breathing in the C57BL/6 J mouse. J Appl Physiol (Bethesda, MD: 1985) 91:2758–2766
51. Row BW, Kheirandish L, Cheng Y, Rowell PP, Gozal D (2007) Impaired spatial working memory and altered choline acetyltransferase (CHAT) immunoreactivity and nicotinic receptor binding in rats exposed to intermittent hypoxia during sleep. Behav Brain Res 177:308–314
52. Perry JC, D'Almeida V, Lima MM, Godoi FR, Vital MA, Oliveira MG, Tufik S (2008) Intermittent hypoxia and sleep restriction: motor, cognitive and neurochemical alterations in rats. Behav Brain Res 189:373–380
53. Wilkerson JE, Mitchell GS (2009) Daily intermittent hypoxia augments spinal BDNF levels, ERK phosphorylation and respiratory long-term facilitation. Exp Neurol 217:116–123
54. Satriotomo I, Dale EA, Dahlberg JM, Mitchell GS (2012) Repetitive acute intermittent hypoxia increases expression of proteins associated with plasticity in the phrenic motor nucleus. Exp Neurol 237:103–115
55. Shao G, Zhang R, Wang ZL, Gao CY, Huo X, Lu GW (2006) Hypoxic preconditioning improves spatial cognitive ability in mice. Neurosignals 15:314–321
56. Martin N, Pourie G, Bossenmeyer-Pourie C, Jazi R, Gueant JL, Vert P, Daval JL (2010) Conditioning-like brief neonatal hypoxia improves cognitive function and brain tissue properties with marked gender dimorphism in adult rats. Semin Perinatol 34:193–200
57. Mitchell GS, Johnson SM (2003) Neuroplasticity in respiratory motor control. J Appl Physiol 94:358–374
58. Bach KB, Mitchell GS (1996) Hypoxia-induced long-term facilitation of respiratory activity is serotonin dependent. Respir Physiol 104:251–260
59. Mitchell GS, Baker TL, Nanda SA, Fuller DD, Zabka AG, Hodgeman BA, Bavis RW, Mack KJ, Olson EB Jr (2001) Invited review: Intermittent hypoxia and respiratory plasticity. J Appl Physiol (Bethesda, MD: 1985) 90:2466–2475
60. Baker TL, Mitchell GS (2000) Episodic but not continuous hypoxia elicits long-term facilitation of phrenic motor output in rats. J Physiol 529 (Pt 1):215–219
61. Mateika JH, Fregosi RF (1997) Long-term facilitation of upper airway muscle activities in vagotomized and vagally intact cats. J Appl Physiol (Bethesda, MD: 1985) 82:419–425
62. McKay LC, Janczewski WA, Feldman JL (2004) Episodic hypoxia evokes long-term facilitation of genioglossus muscle activity in neonatal rats. J Physiol 557:13–18
63. Ryan S, Nolan P (2009) Episodic hypoxia induces long-term facilitation of upper airway muscle activity in spontaneously breathing anaesthetized rats. J Physiol 587:3329–3342
64. Navarrete-Opazo A, Mitchell G (2013) Inspiratory intercostal long term facilitation in unanesthetized rats. FASEB J 27:934–936
65. Mateika JH, Sandhu KS (2011) Experimental protocols and preparations to study respiratory long term facilitation. Respir Physiol Neurobiol 176:1–11
66. Cao KY, Zwillich CW, Berthon-Jones M, Sullivan CE (1992) Increased normoxic ventilation induced by repetitive hypoxia in conscious dogs. J Appl Physiol (Bethesda, MD: 1985) 73:2083–2088
67. Turner DL, Mitchell GS (1997) Long-term facilitation of ventilation following repeated hypoxic episodes in awake goats. J Physiol 499 (Pt 2):543–550
68. Mitchell GS, Powell FL, Hopkins SR, Milsom WK (2001) Time domains of the hypoxic ventilatory response in awake ducks: episodic and continuous hypoxia. Respir Physiol 124:117–128
69. Sokolowska B, Pokorski M (2006) Ventilatory augmentation by acute intermittent hypoxia in the rabbit. J Physiol Pharmacol 57(Suppl 4):341–347
70. Olson EB Jr, Bohne CJ, Dwinell MR, Podolsky A, Vidruk EH, Fuller DD, Powell FL, Mitchel GS (2001) Ventilatory long-term facilitation in unanesthetized rats. J Appl Physiol (Bethesda, MD: 1985) 91:709–716

71. Terada J, Nakamura A, Zhang W, Yanagisawa M, Kuriyama T, Fukuda Y, Kuwaki T (2008) Ventilatory long-term facilitation in mice can be observed during both sleep and wake periods and depends on orexin. J Appl Physiol (Bethesda, MD: 1985) 104:499–507
72. Harris DP, Balasubramaniam A, Badr MS, Mateika JH (2006) Long-term facilitation of ventilation and genioglossus muscle activity is evident in the presence of elevated levels of carbon dioxide in awake humans. Am J Physiol Regul Integr Comp Physiol 291:R1111–R1119
73. Wadhwa H, Gradinaru C, Gates GJ, Badr MS, Mateika JH (2008) Impact of intermittent hypoxia on long-term facilitation of minute ventilation and heart rate variability in men and women: do sex differences exist? J Appl Physiol (Bethesda, MD: 1985) 104:1625–1633
74. Pierchala LA, Mohammed AS, Grullon K, Mateika JH, Badr MS (2008) Ventilatory long-term facilitation in non-snoring subjects during NREM sleep. Respir Physiol Neurobiol 160:259–266
75. Millhorn DE, Eldridge FL, Waldrop TG (1980) Prolonged stimulation of respiration by a new central neural mechanism. Respir Physiol 41:87–103
76. Millhorn DE, Eldridge FL, Waldrop TG (1980) Prolonged stimulation of respiration by endogenous central serotonin. Respir Physiol 42:171–188
77. Baker-Herman TL, Mitchell GS (2002) Phrenic long-term facilitation requires spinal serotonin receptor activation and protein synthesis. J Neurosci 22:6239–6246
78. Dale-Nagle EA, Hoffman MS, MacFarlane PM, Mitchell GS (2010) Multiple pathways to long-lasting phrenic motor facilitation. Adv Exp Med Biol 669:225–230
79. Devinney MJ, Huxtable AG, Nichols NL, Mitchell GS (2013) Hypoxia-induced phrenic long-term facilitation: emergent properties. Ann N Y Acad Sci 1279:143–153
80. Bavis RW, Mitchell GS (2003) Intermittent hypoxia induces phrenic long-term facilitation in carotid-denervated rats. J Appl Physiol (Bethesda, MD: 1985) 94:399–409
81. Sibigtroth CM, Mitchell GS (2011) Carotid chemoafferent activity is not necessary for all phrenic long-term facilitation following acute intermittent hypoxia. Respir Physiol Neurobiol 176:73–79
82. Peng YJ, Overholt JL, Kline D, Kumar GK, Prabhakar NR (2003) Induction of sensory long-term facilitation in the carotid body by intermittent hypoxia: implications for recurrent apneas. Proc Natl Acad Sci U S A 100:10073–10078
83. Peng YJ, Yuan G, Jacono FJ, Kumar GK, Prabhakar NR (2006) 5-HT evokes sensory long-term facilitation of rodent carotid body via activation of NADPH oxidase. J Physiol 576:289–295
84. Abraham WC, Bear MF (1996) Metaplasticity: the plasticity of synaptic plasticity. Trends Neurosci 19:126–130
85. Kinkead R, Zhan WZ, Prakash YS, Bach KB, Sieck GC, Mitchell GS (1998) Cervical dorsal rhizotomy enhances serotonergic innervation of phrenic motoneurons and serotonin-dependent long-term facilitation of respiratory motor output in rats. J Neurosci 18:8436–8443
86. Ling L, Fuller DD, Bach KB, Kinkead R, Olson EB Jr, Mitchell GS (2001) Chronic intermittent hypoxia elicits serotonin-dependent plasticity in the central neural control of breathing. J Neurosci 21:5381–5388
87. MacFarlane P, Vinit S, Roopra A, Mitchell G (2010) Enhanced phrenic long-term facilitation following repetitive acute intermittent hypoxia: role of glycolytic flux. FASEB J 24:15
88. Vinit S, MacFarlane PM, Satriotomo I, Mitchell GS (2010) Enhanced phrenic long-term facilitation (pLTF) following repetitive acute intermittent hypoxia. FASEB J 24:14
89. McGuire M, Zhang Y, White DP, Ling L (2003) Chronic intermittent hypoxia enhances ventilatory long-term facilitation in awake rats. J Appl Physiol (Bethesda, MD: 1985) 95:1499–1508
90. Mitchell G (2007) Respiratory plasticity following intermittent hypoxia: a guide for novel therapeutic approaches to ventilatory control disorders. In: Gaultier C (ed) Genetic basis for respiratory control disorders. Springer, New York, pp 291–311
91. Kinkead R, Bach KB, Johnson SM, Hodgeman BA, Mitchell GS (2001) Plasticity in respiratory motor control: intermittent hypoxia and hypercapnia activate opposing serotonergic and noradrenergic modulatory systems. Comp Biochem Physiol A Mol Integr Physiol 130:207–218
92. MacFarlane PM, Mitchell GS (2009) Episodic spinal serotonin receptor activation elicits long-lasting phrenic motor facilitation by an NADPH oxidase-dependent mechanism. J Physiol 587:5469–5481
93. Kinkead R, Mitchell GS (1999) Time-dependent hypoxic ventilatory responses in rats: effects of ketanserin and 5-carboxamidotryptamine. Am J Physiol 277:R658–R666
94. Fuller DD, Zabka AG, Baker TL, Mitchell GS (2001) Phrenic long-term facilitation requires 5-HT receptor activation during but not following episodic hypoxia. J Appl Physiol (Bethesda, MD: 1985) 90:2001–2006, discussion 2000
95. MacFarlane PM, Vinit S, Mitchell GS (2011) Serotonin 2A and 2B receptor-induced phrenic motor facilitation: differential requirement for spinal NADPH oxidase activity. Neuroscience 178:45–55
96. Devinney M, Fields D, Mitchell G (2013) Spinal PKC-θ activity is required for phrenic long-term facilitation after acute intermittent hypoxia. FASEB J 27:930–934
97. Baker-Herman TL, Fuller DD, Bavis RW, Zabka AG, Golder FJ, Doperalski NJ, Johnson RA, Watters JJ, Mitchell GS (2004) BDNF is necessary and sufficient for spinal respiratory plasticity following intermittent hypoxia. Nat Neurosci 7:48–55

98. Dale E, Fields D, Mitchell G (2013) Intrapleural siRNAs targeting TrkB attenuate acute intermittent hypoxia (AIH)-induced phrenic long-term facilitation. FASEB J 27:930–935
99. Hoffman MS, Nichols NL, Macfarlane PM, Mitchell GS (2012) Phrenic long-term facilitation after acute intermittent hypoxia requires spinal ERK activation but not TrkB synthesis. J Appl Physiol 113:1184–1193
100. Fuller DD, Bach KB, Baker TL, Kinkead R, Mitchell GS (2000) Long term facilitation of phrenic motor output. Respir Physiol 121:135–146
101. McGuire M, Zhang Y, White DP, Ling L (2005) Phrenic long-term facilitation requires NMDA receptors in the phrenic motonucleus in rats. J Physiol 567:599–611
102. Golder FJ, Ranganathan L, Satriotomo I, Hoffman M, Lovett-Barr MR, Watters JJ, Baker-Herman TL, Mitchell GS (2008) Spinal adenosine A2a receptor activation elicits long-lasting phrenic motor facilitation. J Neurosci 28:2033–2042
103. Hoffman MS, Mitchell GS (2011) Spinal 5-HT7 receptor activation induces long-lasting phrenic motor facilitation. J Physiol 589:1397–1407
104. Nichols NL, Dale EA, Mitchell GS (2012) Severe acute intermittent hypoxia elicits phrenic long-term facilitation by a novel adenosine-dependent mechanism. J Appl Physiol (Bethesda, MD: 1985) 112:1678–1688
105. Hoffman MS, Golder FJ, Mahamed S, Mitchell GS (2010) Spinal adenosine A2(A) receptor inhibition enhances phrenic long term facilitation following acute intermittent hypoxia. J Physiol 588:255–266
106. Hoffman MS, Mitchell GS (2013) Spinal 5-HT7 receptors and protein kinase A constrain intermittent hypoxia-induced phrenic long-term facilitation. Neuroscience 250:632–643
107. Dale-Nagle EA, Satriotomo I, Mitchell GS (2011) Spinal vascular endothelial growth factor induces phrenic motor facilitation via extracellular signal-regulated kinase and Akt signaling. J Neurosci 31:7682–7690
108. Dale EA, Satriotomo I, Mitchell GS (2012) Cervical spinal erythropoietin induces phrenic motor facilitation via extracellular signal-regulated protein kinase and Akt signaling. J Neurosci 32:5973–5983
109. Dale E, Satriotomo I, Mitchell G (2012) Spinal VEGF induced phrenic motor facilitation is unaffected by pretreatment with repetitive acute intermittent hypoxia. FASEB J 26:704–710
110. NSCISC (2005) Annual report for the model spinal cord injury care systems. National Spinal Cord Injury Statistical Center, Birmingham
111. Nantwi KD, El-Bohy AA, Schrimsher GW, Reier PJ, Goshgarian HG (1999) Spontaneous functional recovery in a paralyzed hemidiaphragm following upper cervical spinal cord injury in adult rats. Neurorehabil Neural Repair 13:225–234
112. Goshgarian HG (2003) The crossed phrenic phenomenon: a model for plasticity in the respiratory pathways following spinal cord injury. J Appl Physiol 94:795–810
113. Fuller DD, Baker-Herman TL, Golder FJ, Doperalski NJ, Watters JJ, Mitchell GS (2005) Cervical spinal cord injury upregulates ventral spinal 5-HT2A receptors. J Neurotrauma 22:203–213
114. Fuller DD, Golder FJ, Olson EB Jr, Mitchell GS (2006) Recovery of phrenic activity and ventilation after cervical spinal hemisection in rats. J Appl Physiol 100:800–806
115. Lane MA, White TE, Coutts MA, Jones AL, Sandhu MS, Bloom DC, Bolser DC, Yates BJ, Fuller DD, Reier PJ (2008) Cervical prephrenic interneurons in the normal and lesioned spinal cord of the adult rat. J Comp Neurol 511:692–709
116. Teitelbaum J, Borel CO, Magder S, Traystman RJ, Hussain SN (1993) Effect of selective diaphragmatic paralysis on the inspiratory motor drive. J Appl Physiol 74:2261–2268
117. Katagiri M, Young RN, Platt RS, Kieser TM, Easton PA (1994) Respiratory muscle compensation for unilateral or bilateral hemidiaphragm paralysis in awake canines. J Appl Physiol 77:1972–1982
118. Brichant JF, De Troyer A (1997) On the intercostal muscle compensation for diaphragmatic paralysis in the dog. J Physiol 500(Pt 1):245–253
119. Johnson RA, Mitchell GS (2013) Common mechanisms of compensatory respiratory plasticity in spinal neurological disorders. Respir Physiol Neurobiol 189:419–428
120. Golder FJ, Fuller DD, Davenport PW, Johnson RD, Reier PJ, Bolser DC (2003) Respiratory motor recovery after unilateral spinal cord injury: eliminating crossed phrenic activity decreases tidal volume and increases contralateral respiratory motor output. J Neurosci 23:2494–2501
121. Golder FJ, Mitchell GS (2005) Spinal synaptic enhancement with acute intermittent hypoxia improves respiratory function after chronic cervical spinal cord injury. J Neurosci 25:2925–2932
122. Lovett-Barr MR, Sibigtroth CM, Mitchell GS (2007) Daily acute intermittent hypoxia improves respiratory function in rats with chronic cervical hemisection. FASEB J 21:A1292–A1293
123. Tester NJ, Fuller DD, Fromm JS, Spiess MR, Behrman AL, Mateika JH (2014) Long-term facilitation of ventilation in humans with chronic spinal cord injury. Am J Respir Crit Care Med 189(1):57–65
124. Fuller DD, Johnson SM, Olson EB Jr, Mitchell GS (2003) Synaptic pathways to phrenic motoneurons are enhanced by chronic intermittent hypoxia after cervical spinal cord injury. J Neurosci 23:2993–3000

High-Altitude Research and Its Practical Clinical Application

Gustavo Zubieta-Castillo Sr. and Gustavo Zubieta-Calleja Jr.

Abstract

Upon arriving above 2,500 m, the organism compensates the diminished inspired oxygen partial pressure by increasing ventilation and cardiac output. The pneumodynamic pump moves more rarefied air into the alveoli through an increase of the respiratory frequency and the tidal volume. Likewise, the hemodynamic pump increases both the cardiac frequency and the stroke volume, as if exercise were performed at sea level. The two pumps, one for air and the other for blood, carry out the essential role of supplying sufficient oxygen to the tissues and increasing the energy consumption until the red blood cells take over. The acid–base status, adequately interpreted at high altitude through the titratable hydrogen ion difference, along with the adaptation formula and multiple cellular changes, gives rise to adaptation. The tolerance to hypoxia formula reflects and explains the paradoxical concept that resistance to hypoxia grows as one goes higher. The possibility that man can adapt to live in the hypoxic environment of the summit of Mt. Everest is exposed. Furthermore, the knowledge of life at high altitude is proposed as an alternative to the environment of space travel. Herein, we describe our 43 years of experience and discoveries with fundamental concepts that change the way disease is treated in the hypoxic environment of high altitude.

Introduction

The Earth is surrounded by an atmosphere that has an altitude of approximately 20 km. The air composition is the same at any altitude: 20.95 % oxygen, 78.09 % nitrogen, 0.93 % argon, and 0.039 % carbon dioxide. Small amounts of other gases are also present which are released upon industrial combustion and other natural events such as volcanic gas eruptions. It is actually the partial pressure of gases that decrease upon going higher. Nitrogen is an inert gas and its importance in respiration is transcendental, although currently lacking awareness. If the atmosphere was hypothetically only composed of oxygen (i.e., 100 %), the incidence of alveolar collapse, medically known as atelectasis, would

G. Zubieta-Castillo Sr. • G. Zubieta-Calleja Jr. (✉)
High Altitude Pulmonary and Pathology Institute (IPPA), La Paz, Bolivia
e-mail: gzubietajr@altitudeclinic.com

S.B. Singh et al. (eds.), *Translational Research in Environmental and Occupational Stress*, DOI 10.1007/978-81-322-1928-6_5, © Springer India 2014

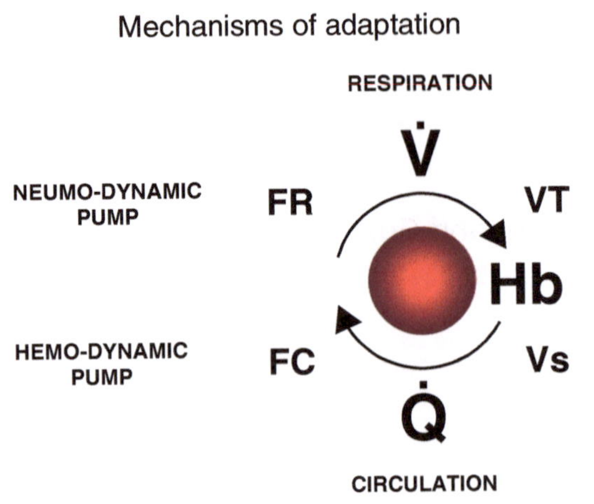

Fig. 1 The mechanisms of adaptation showing in the first stage: the increase of V = ventilation per minute composed of both the FR = respiratory frequency and Vt = tidal volume of the pneumodynamic pump and the Q = cardiac output, composed of both the FC = cardiac frequency and Vs = stroke volume of the hemodynamic pump. In the second stage: the increase in Hb = hemoglobin in the middle plays the fundamental role in the adaptation to high altitude

greatly increase. Nitrogen and the lung surfactant aid in preventing this.

Adaptation to high altitude has three stages: (1) the *acute* altitude adaptation stage lasting 1 or 2 days, (2) the gradual *subacute* stage, and (3) the final *chronic* altitude adaptation stage [17]. In the acute and subacute stages, the immediate response to a diminished inspired oxygen tension (PIO_2) is an increase in the pneumodynamic and hemodynamic pumps (Fig. 1).

Both require a high energy demand that gradually decreases over time. Subsequently if the subject remains at the same altitude, the final optimal adaptation is achieved over a fixed amount of time. This time is directly proportional to the altitude. The adaptation formula described below is derived from these observations. In the second and third stages, molecular changes with a gradual increase in the number of red blood cells and hemoglobin occur. Concomitantly, the steroids and other biochemical metabolic changes play a secondary role. Chronic adaptation is achieved by reducing the expenditure of energy [18]. Hemoglobin increases but to a mean level suitable for the majority of the high-altitude residents, in the absence of illness. However, in some subjects with pulmonary shunts or uneven ventilation (sequel of pulmonary disease), the hemoglobin levels are higher. This has been previously referred to as "excessive polycythemia," a term we are not in agreement with. As a consequence of these pulmonary alterations, blood flows through the organism with a lower oxygen tension. Upon reaching the kidney, erythropoietin production is stimulated which in turn induces an increase of red blood cells in the bone marrow.

Medicine is changing with time. Historically, we observed a much higher incidence of increased hemoglobin, which resulted from untreated and frequent bacterial pneumonias. The pulmonary alveoli were damaged, but the vessels and pulmonary capillary circulation remained unaffected. This resulted in shunts and subsequent low oxygen partial pressures in the blood. This has changed fundamentally due to the use of antibiotics. Currently, pneumococcus is treatable; however, new emerging diseases caused by virus that may not be as dramatic play a role not yet fully identified. Consequently, there are now fewer cases of what we call polyerythrocythemia formerly called polycythemia, increased polycythemia, erythrocythemia, or excessive erythrocytosis. The most appropriate term that we now use, based on the Greek terminology, is polyerythrocythemia: poly (many), erythrocyte (red blood cells), and hemia (in the blood).

Cardiorespiratory diseases at high altitude along with others have changed pathologically. For example, syphilis that affected a large number of people no longer exists. It turned out that *Treponema pallidum* was very sensitive to penicillin and was literally wiped out.

Much has been said about chronic mountain sickness. In this regard, we have had many discussions with world-renowned experts that still accept the term. We have stated that there is no such a thing as "chronic mountain sickness" at high altitude [5]. Diseases are the same, as at sea level, but they have been confused as chronic mountain sickness at altitude. These are mostly lung diseases of different etiopathogenesis, fundamentally with shunt production. The increase of hemoglobin above the normal at high altitude is due to respiratory insufficiency. Polyerythrocythemia has also been observed in cardiac arrhythmias, and after a pacemaker implantation, the number of red blood cells returned to normal values for that specific altitude.

Naturally, there may be very rare hematologic diseases, some yet undiscovered, but these would be exceptions. The carotid body disease can present with polyerythrocythemia. There is evidence of an increased incidence of carotid glomus at high altitude. Also, mistakenly it was thought that there was a relationship between age and polyerythrocythemia, but it has widely been disregarded. Some kidney diseases with renal artery narrowing also produce increased hematocrit. Some authors have sought other alterations in the gonads with hypoventilation, which led them to attempt treatments with progesterone. This only resulted in feminization of men without an effective and practical decrease in the number of red blood cells.

Others attempted to physically remove the RBCs. They even used cytotoxic drugs such as phenylhydrazine [19, 25]. Many died (and unfortunately continue to do so) due to this iatrogenic disease instead of the polyerythrocythemia. In Bolivia, one of the authors had to fight against the use of this internationally banned drug. He was initially misunderstood creating a social whirl but eventually won, resulting in the recognition and comprehension that the use of this drug is obviously prohibited by the World Health Organization.

Phlebotomy is still used as a treatment in these patients in order to physically reduce the number of RBCs. Our institute, IPPA, has never supported phlebotomy and because the fundamental concept is not to alter the perfect (energy-efficient) physiological balance of chronic adaptation [11]. When a phlebotomy is performed, with a consequent reduction of the optimal number of RBCs, it induces a condition similar to acute adaptation in the high-altitude patient, as if the high-altitude resident had gone to a higher altitude. The resulting increase in heart rate and ventilation is misinterpreted as an improvement. Patients with polyerythrocythemia have been found to hypoventilate. This was erroneously misinterpreted as the cause instead of an energy-efficient mechanism of oxygen transport [18]. After phlebotomy, there is an extra energy expenditure that is gradually reduced by again increasing the RBCs.

The high viscosity of blood is often feared as a possible cause of cerebrovascular accidents. Yet, an increase of RBCs does not necessarily lead to the formation of thrombus. This is somewhat conflicting with our studies. There are reports of hypercoagulability in other latitudes and longitudes, but we have not seen this phenomenon in Bolivia. It may partly be due to dehydration in newcomers to high altitude or the presence of some sort of undetected infection or even genetic tendencies.

The result of many years of study on polyerythrocythemia has led to a better understanding and hence improved therapy significantly reducing the concept of this previously considered evil disease. This was achieved in the laboratory, jumping to a practical and pragmatic use for health improvement of the inhabitants of high altitude. The essential concepts of our interpretation of hypoxia follow in the following paragraphs.

Chronic Mountain Sickness (CMS)

We consider that the increase of the hematocrit (above that of the normal residents) found in some high-altitude residents is not a disease in itself but rather a sign, i.e., a hematologic response to disease in the hypoxic environment of hypobaria at high altitude [21, 23]. What was previously known as chronic mountain sickness (CMS), a condition in which the hematocrit is increased above the normal level in residents at high altitude, is now referred to as polyerythrocythemia (PECH) from the Greek poly = increase, erythro = red, cyt = cells, and hemia = in the blood. Consequently, where the abbreviation PECH is used throughout, it refers to what was previously named CMS [18].

Chronic mountain sickness can occur in some long-term residents of altitudes above 2,500 m – the Andes and elsewhere. The number of red cells in the blood (Polyerythrocythemia) and the hematocrit develops to exceptionally high values. The high hematocrit and viscosity of the blood do not necessarily lead toward the formation of emboli, which occurs when there is concurrent phlebitis or other vascular disease. In patients with lung disease, pulmonary vasoconstriction maintained over the years can lead to pulmonary hypertension and enlargement of the right ventricle, resulting eventually in right-sided congestive heart failure – Latin: cor pulmonale. A large fraction of the cardiac output is shunted through vessels with hypoventilated alveoli, so the patient is cyanotic with congested ear lobes and some with finger clubbing. CMS is an inadequate denomination for pulmonary disease at high altitude, associated frequently to pulmonary shunts with increased erythropoietin and polyerythrocythemia.

In the article "Consensus Statement on Chronic and Subacute High Altitude Diseases" in 2005, we as coauthors held a minority point of view [5]. Our work of over 50 years in the city of La Paz located between 4,100 and 3,100 m gave us a lens through which we observed accurately and inquisitively, high-altitude pathology, on a daily basis.

In the aforementioned article, two essential points were a source of discussion:

1. The use of a questionnaire to evaluate the symptoms of CMS.
2. The use of the term "loss of adaptation" as opposed to "adaptation to disease in the hypoxic environment." The concept "loss of adaptation" was incorrect since it was a mere attempt to avoid the explanation of the etiopathophysiology of the disease.

We opine that CMS is rather an adaptive reaction to an underlying malfunction of some organs and no specific symptoms could be quantified. To substantiate our line of reasoning, we reviewed 240 CMS cases seen at the High Altitude Pathology Institute in La Paz. Patients who had a high hematocrit (>58 %) underwent pulmonary function studies in search for the cause of hypoxia: hypoventilation, diffusion alteration, shunts, and uneven ventilation–-perfusion. The tests included arterial blood gas tests, chest x-rays, spirometry, hyperoxic tests, flow–volume curves, ventilation studies at rest and during exercise, ECG, exercise testing, and Doppler color echocardiography to assess heart structure and function. When correlated with clinical history, these results revealed that CMS is practically always secondary to some type of anomaly in cardiorespiratory or renal function and very rarely other hematologic disease. Therefore, a questionnaire that tried to catalog symptoms common to many types of diseases that lead to hypoxia is flawed because it leads to incomplete diagnosis and inappropriate treatment. CMS, once again, was shown to be an adaptation of the blood transport system to a deficient organs' function due to diverse disease processes, the adaptation aimed at sustaining normoxia at the cellular level in the hypoxic environment at high altitude.

Quoting Prof. Zubieta-Castillo (Sr): "The organic systems of human beings and all other species tend to adapt to any environmental change and circumstance, and never tend towards regression which would inevitably lead to death."

The Triple Hypoxia Syndrome

Many patients with PECH at high altitude in the city of La Paz (3,600 m), with hematocrit greater than 55 % but not greater that 70 %, apparently can function normally. They work, play soccer, develop intellectual activities, and frequently perform better than sedentary normal people. They request medical attention, only when they present symptoms similar to those of acute mountain sickness (AMS), such as headache, dyspnea, nausea, lassitude, and indigestion. Without going higher, they have been said to experience "soro jchi (soroche) AMS in bed." Their arterial blood gases may show extreme hypoxia with an oxygen arterial tension (PaO_2) near 20 mmHg, with or without hypercapnia and a normal or acidotic pH. We have previously named this complication of PECH (formerly referred to as chronic mountain sickness (CMS)) triple hypoxia syndrome (THS). It is due to the addition of three hypoxias: (1) normal high-altitude adaptation to hypoxia, (2) PECH hypoxia (CMS), and (3) an acute additional hypoxia that can be reversed by oxygen. This is similar to "surviving" in the summit of Mt. Everest at a much lower altitude. It may be caused by viral infections (gripe) or some other acute respiratory disease, with malaise that lasts several days without treatment and typically is reversed by 24 h of oxygen to PaO_2 baseline values of their chronic condition with polyerythrocythemia. The diagnosis is important, since THS is an acute transitory condition that when not recognized and treated with oxygen can possibly lead to cardiac, pulmonary, or cerebral complications and even death [22, 26].

The Increase in Hematocrit During the High-Altitude Adaptation Process

High-altitude adaptation is altitude and time dependent following the simplified equation:

$$\text{Adaptation} = \frac{\text{Time}}{\text{Altitude}}$$

where each concept stands for:

High-altitude adaptation = Time at altitude (days) / Altitude in kilometers (km) [17].

For a fixed altitude, the only variable that changes is time. Immediately upon ascent, the organism senses the lowering of the oxygen partial pressure due to a diminished barometric pressure. The initial acute phase usually lasts between 1 and 2 days, varying according to the health conditions of the subject and if he or she has previously been exposed to high altitude and learned how to "handle" it in complex macrosystemic and microsystemic mechanisms. The former implies cardiorespiratory compensation and the latter adaptation processes at the cellular and molecular level. The macrosystemic mechanisms attempt to raise the PaO_2 to "sea level values" through increased ventilation and higher cardiac frequency, never being able to achieve it, thereby reducing the $PaCO_2$ resulting in respiratory alkalosis, a negative action at the cellular level. The microsystemic mechanisms return the pH to normal, through kidney function, reducing the negative symptoms of acute mountain sickness through adequate cellular function. However, a permanent and stable adaptation is only achieved at around 4 weeks when the hematocrit reaches the optimal level for the altitude (3600 m). Consequently, high-altitude adaptation is defined as having three stages:

(A) Acute, first 72 h, where acute mountain sickness (or high-altitude pulmonary edema, HAPE, and high-altitude cerebral edema, HACE) can occur.

(B) Subacute, from 72 h until the slope of increase of the hematocrit rise with time is zero; here high-altitude subacute heart disease can occur, if excessive exercise is performed.

(C) Chronic, where the hematocrit level is constant and the healthy high-altitude residents achieve their optimal hematocrit.

We have measured hematocrit changes in one high-altitude resident traveling several times between La Paz (3,510 m) and Copenhagen (35 m above sea level) for 3 years. We have also studied the fall in hematocrit values in three lowlanders traveling once from La Paz to Copenhagen [17].

A complete and optimal hematocrit adaptation is only achieved at around 40 days for a subject going from sea level to 3,510 m in La Paz as shown in Fig. 2. The time in days required to achieve full adaptation to any altitude, ascending from sea level, can be estimated by multiplying the adaptation factor of 11.4 by the altitude in km. Subjects with different cardiopulmonary deficiencies increase the hematocrit to levels above those of normal individuals.

Conversely, descending from high altitude in La Paz to sea level in Copenhagen, the hematocrit response is a linear fall over 18–23 days as shown in Fig. 3.

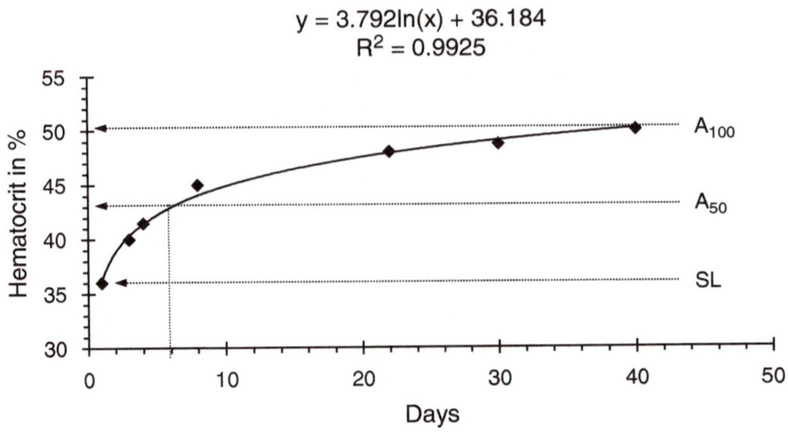

Fig. 2 Hematocrit changes after altitude travel from sea level (35 m) to high altitude (3,510 m) in a male high-altitude native (first author) that spent 3 months at sea level. A_{100} = 100 % adaptation and A_{50} = 50 %

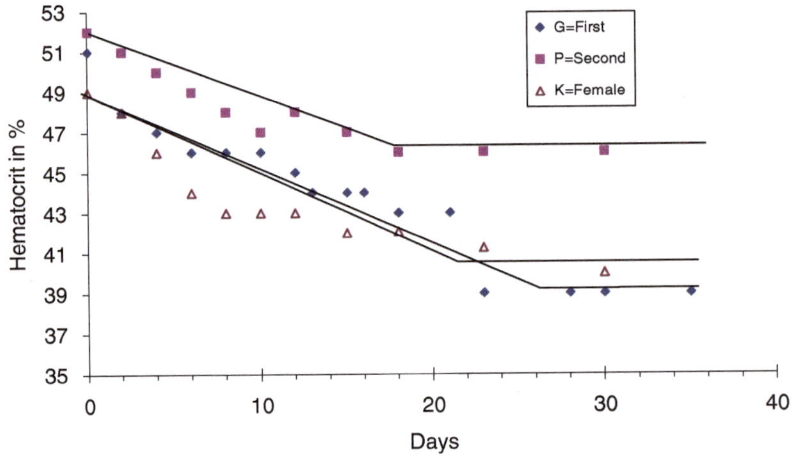

Fig. 3 Hematocrit changes on descent from high altitude (3,510 m) to sea level (35 m) in three subjects

Hypoventilation in Chronic Mountain Sickness

Patients diagnosed with chronic mountain sickness or rather PECH have repeatedly been found to hypoventilate. Low saturation in PECH is attributed to hypoventilation. Although this observation seems logical, a further understanding of the exact mechanism of hypoxia is mandatory. An exercise study using the Bruce Protocol in PECH subjects ($n = 13$) compared with normals N ($n = 17$), measuring ventilation (VE), pulse (P), and saturation by pulse oximetry (SaO_2) was performed. Ventilation at rest while standing, prior to exercise in a treadmill, was indeed lower in PECH subjects (8.37 l/min compared with 9.54 l/min in N). However, during exercise at stages one (3′) through four (12′), ventilation and cardiac frequency both remained higher than in N. In spite of this, SaO_2 gradually decreased. Although PECH subjects increased ventilation and heart rate more than N, saturation was not sustained, suggesting respiratory insufficiency. The degree of venoarterial shunting of blood is obviously higher in the PECH patients both at rest and during exercise as judged from the SaO_2 values. The higher shunt fraction is probably due to a larger degree of trapped air in the lungs with uneven ventilation of the PECH patients.

One can infer that hypoventilation at rest is an energy-saving mechanism of the pneumodynamic and hemodynamic pumps [18]. Increased ventilation would cause an unnecessary high SaO_2 at rest (low metabolism). This is particularly true during sleep.

A Biological and Mechanical Energy-Efficient Space Travel Alternative: Hypobaric Cabin Pressure

Astronauts in the microgravity environment of space suffer with conditions such as anemia [10]. Thus far unexplained, neocytolysis has been described as the possible underlying mechanism [2, 9]. The adaptation to space capsules involves less use of muscles and changes in ventilation–perfusion at lung level so that the organism finds it convenient to reduce the hematocrit following the least energy expenditure concept [18]. The knowledge and understanding of polycythemia and anemia resulting from altitude shifts [17] allows for a logical proposal to bloodletting of high-altitude residents on travel to sea level for periods longer than 20 days (an outstanding humanitarian blood resource) and similarly in astronauts when going into space. The logic is that they would economize energy, avoiding the destructive phase of adaptation. However, in astronauts, upon return to sea level, reinfusion of the phlebotomized blood could return the hematocrit to normal levels. Erythropoietin administration is also an option but not favored due to associated complications.

Original space flights were in a pure oxygen environment and one-third the sea level pressure until serious fire accidents were encountered. Currently, the cabin pressure is normal sea level pressure at 760 Torr with 20 % oxygen and 80 % nitrogen (NASA) [6].

An alternative to the complication of anemia would be to reduce the ambient oxygen tension within space vehicles, down to 2/3 the sea level pressure (similar to the altitude of the city of La Paz with over 1.5 million inhabitants), in order to maintain a hypoxic stimulus and sustain the number of red blood cells for re-entry to Earth. Furthermore, the weightlessness space conditions require less oxygen consumption as there is less muscular use and hence tolerance to hypoxia can be increased. Likewise, the *Extravehicular Mobility Unit* could benefit from a lower oxygen tension, less pressure difference between outside and within the space capsule, a speedier preparation, and additionally more autonomy. Long space flights in the near future would require less wasted resources in excess oxygen production. The return to the normal (relative hyperoxic) environment of sea level would ease adaptation, as there would presumably be no reduction of the hematocrit during space flight [15].

Man may undertake space travel in the future where oxygen tensions will be a crucial life-saving variable, and this proposal provides a physiological and physical energy-efficient alternative.

Adaptation to Life at the Extreme Altitude of Mt. Everest

Short permanence of humans, under acute conditions, with partial and incomplete adaptation with eminent risk of death demonstrates that even at the highest places of the planet Earth, enough mechanisms of tissue oxygenation are still present. The only objective of the ascent is the conquest of the summit, even for just a few minutes, demonstrating the extraordinary physical capacity of human beings. This is done even under adverse climatic conditions like extreme cold and storms.

As knowledge and technology improve, man is able to accomplish the most difficult challenges previously considered impossible. The conquest of the summit of Mt. Everest (8,848 m), the highest point of the planet, for example, was originally thought to be impossible. Careful preparation of the expeditions, experience, persistence, and extraordinary stamina allowed Sir Edmund Hillary and Tenzing to stand on top of the mountain. Up to the year 2006, hundreds have reached such extreme altitude and, after Messner and Habeler, some of them without supplementary oxygen. But the question remains: What is the highest possible point of permanent residence?

Based on the observation of man's highest habitats, most scientists believe 5,000 m above sea level is the limit for adaptation to hypoxia. Likewise, it is well known that acute exposure of humans to this altitude and above, without any kind of adaptation for tissue oxygenation, can result in loss of conscience and even death.

In the study entitled "A consideration of the Possibility of Ascending Mt. Everest," Kellas proposes that the "limits of permanent acclimatization to high altitudes," based on his own experience, was 6,096 m. Curiously, several decades later, most physiologists believed it was much lower. In the north of Chile, the Aucanquilcha mine is at 5,950 m. The Indian Armed Forces have set up military posts in the border with Pakistan at around 6,000 m, residing there for long periods up to 6 months or more. Inder Anand reported that several soldiers suffered subacute heart disease. This can now be attributed to a fast ascent and inadequate time of adaptation (not disregarding possible previous cardiac pathologies not adequately diagnosed prior to the ascent, in some)

The hypothesis "man can adapt to the altitude of the summit of Mt. Everest" was conceived based on experience gained over 36 years (at the time of publication). Ever since its creation, in 1970, the High Altitude Pathology Institute, Clinica IPPA (3,510 m), has studied and treated patients (permanent residents and newcomers) in the cities of La Paz and El Alto (3,100–4,100 m) and the surrounding areas in Bolivia (4). It was estimated that in Bolivia, more than 5,000,000 people (2/3 of the total population) live above 2,000 m. Additionally, the proposed hypothesis was also based on a study of the history of high life and exponentially growing scientific literature on high altitude around the world.

With adequate and gradual adaptation, life is possible even at the hypoxic levels of the summit of Mt. Everest [20, 24, 27]. We describe some further examples that provide proof of such a statement:

1. Life at high altitude in the city of La Paz between 3,100 and 4,100 m where people sustain variable degrees of tissue hypoxia having a very low arterial oxygen tension ($PaO_2 = 37$ mmHg) due to respiratory and/or cardiac disease.
2. It is possible to perform maximal work at extreme altitudes, as evidenced by a soccer match played at 6,542 m on the summit of Mount Sajama.
3. Severe high-altitude pulmonary edema (HAPE) occurred in a rugby player within 72 h at 3,600 m (PaO_2 of 27 mmHg and a SaO_2 of 45 %) upon arrival from Portugal.
4. In the triple hypoxia syndrome (THS) where polyerythrocythemic (CMS) patients with gradual adaptation to hypoxia, with a high hematocrit, can occasionally tolerate a PaO_2 of 30 mmHg for a week or longer, a severe hypoxic condition similar to that on the summit of Mt. Everest [22].
5. Human fetus under normal conditions develop at oxygen tension values equal to the altitude of Mt. Everest until delivery ($PaO_2 = 28$ mmHg). They are naturally capable of living in hypoxic environmental conditions present on our planet.

Consequently, normal subjects with full capacity for adaptation will show that life is possible at any existing altitude on planet Earth, provided that the following conditions are met: adequate environmental temperatures, heated lodging, adequate food, and slow and progressive adaptation to increasing altitudes. This seems possible in only one generation in a young perfectly healthy individual, as the human organism is provided with the adequate compensation mechanisms. Once adapted, the capacity for reproduction on site seems feasible.

Pregnancy and delivery, at the highest altitude on planet Earth, occurred in the city of La Paz. A young girl had an unwanted pregnancy at the city of El Alto (4,100 m), and consequently, she moved to an isolated place in Chacaltaya at 5,300 m for the rest of her pregnancy. She worked at a refuge there until the time of delivery that was carried out uneventfully (precariously at the resourceless refuge) giving birth to a healthy female.

Extreme Hypoxia in Newcomers to High Altitude: How Can It Be Tolerated?

Patients suffering from the acute effects of hypobaric hypoxia can have extremely low arterial oxygen tensions (PaO_2), which are quite well tolerated. They can come to consultation with a PaO_2 between 30 and 40 mmHg. Their recovery after a few days is uneventful following the efficient treatment of the underlying cause.

At sea level, these very low oxygen tensions are not tolerable. A patient presenting a PaO_2 of 60 mmHg can be sent to an intensive care unit, as his or her life could be at risk. However at high altitude in the city of La Paz of 3,600 m, the normal acid–base values are as follows: $PaO_2 = 60 \pm 2$ mmHg, $PaCO_2 = 30 \pm 2$ mmHg and $pH = 7.40 \pm 0.02$, and $SpO_2 = 91\ \% \pm 1\ \%$. The latter oscillates with irregular breathing and deep breaths can achieve even 98 % as has been previously described [12].

One out of four subjects, arriving to La Paz, has some form of acute mountain sickness. This implies headaches, shortness of breath, loss of appetite, malaise, nausea that can evolve to vomiting, and more severe neurological alterations in HACE or pulmonary alterations in HAPE.

Some can present extreme hypoxia. The pathologies that we have recently seen associated with this extreme hypoxia are pneumonia upon ascent, pulmonary thromboembolism, high-altitude pulmonary or cerebral edema, and several others.

For example, a 25-year-old Frenchman climbed Huayna Potosi at 6,000 m 2 days after arriving to La Paz 3,600 m from Paris. On the way down, he felt shortness of breath and was unable to sleep over the night. He came to consultation walking. A blood gas analyses reported a $PaO_2 = 35$ mmHg, $PaCO_2 = 29$ mmHg, and a $pH = 7.53$. This was diagnosed as severe hypoxia and respiratory alkalosis in high-altitude pulmonary edema. Would he have been alive if at sea level? What is the explanation for this extreme hypoxia tolerance?

The alkaline pH during acute high-altitude exposure shifts the oxygen dissociation curve to the left, allowing more capture and transport of oxygen. The other variable that can allow for the tolerance to extremely low PaO_2 values is the normal low (relative to sea level) $PaCO_2$. The acid–base balance in the human body is calculated by the Van Slyke equation based on sea level measurements. The maintenance of blood pH within a fairly strict range at/around pH 7.4, with due consideration of the effect of hyperventilation, is essential for cellular function at any altitude. This is because various chemical processes occurring in the body, e.g., those involving proteins and enzymes, are pH dependent. As is well known, oxygen and carbon dioxide partial pressures get lower as the altitude gets higher. At permanent low levels of the arterial carbon dioxide partial pressure ($PaCO_2$), the acid–base balance begins to change. Mountaineering physiologists, unfortunately, assumed that the sea level equation for A–B balance would be equally applicable, without any critical thought, and never thought of appropriate equations valid for high-altitude calculations [14]. The pH effects are inextricably linked critically with hemoglobin and oxygen status that can be crucial at high altitudes. Therefore, for a more precise recalculation of the "titratable hydrogen ion difference," that should use a Hb and HCO_3^- values for a particular altitude, we have derived our modified Van Slyke equation [8]. An adequate acid–base balance is probably the fundamental metabolic adaptation that allows for mountaineers to tolerate extreme hypoxia and even reach the summit of Mt. Everest.

The Tolerance to Hypoxia Formula

People living at high altitude in the city of La Paz (3,600 m), with diverse types of lung disease, exhibit extremely low arterial oxygen tensions (PaO_2). Their arterial oxygen partial pressure of oxygen (PaO_2) can range between 30 and 40 mmHg (normal values—$PaO_2 = 60 \pm 2$ mmHg, arterial carbon dioxide tension ($PaCO_2$) $= 30 \pm 2$ mmHg, and $pH = 7.40 \pm 0.02$;

oxyhemoglobin saturation by pulse oximetry (SpO_2) is 91 % ± 1 %). The SpO_2 oscillates with irregular breathing and taking a deep breath can result in even 98 % (like at sea level) provided there is a normal pulmonary function, as previously described [12]. This results from a decrease in the ratio between pulmonary dead space and alveolar ventilation. When the medical reports from people in La Paz were shown to physicians at sea level, they often asked, "Were these people conscious?" This clearly shows that people at sea level cannot tolerate such low arterial PO_2. A patient presenting a PaO_2 below 60 mmHg at the sea level is usually sent to an intensive care unit, as his life could be in peril.

Importance of Arterial PCO_2: The First Factor

The other variable that can allow for the tolerance to extremely low PaO_2 values is the low $PaCO_2$ values at high altitude relative to sea level. As is well known, oxygen and carbon dioxide partial pressures descend as the altitude increases. The distribution of $PaCO_2$ in arterial blood gases at the city of La Paz (3,510 m) shows clearly that the average is around 30 mmHg. Furthermore, it is evident that the highest $PaCO_2$ reached is 72 mmHg in an isolated critically terminally ill patient. The great majority hardly reach a $PaCO_2$ above 53 mmHg. At high altitude, high $PaCO_2$ levels as those seen at sea level are not compatible with life.

Hypocapnia and the ensuing alkaline pH during high-altitude exposure shift the oxygen dissociation curve to the left, in an attempt to increase the efficiency of the capture and transport of oxygen. In reality, the curve's P-50 is dynamic moving to the left at the pulmonary capillaries and to the right at the tissue level. The acid–base balance in the human body is calculated by the Van Slyke equation based on sea level measurements. The maintenance of blood pH within a fairly strict range at around pH 7.4, with due consideration of the effect of hyperventilation, is essential for cellular function at any altitude. This is because various chemical processes occurring in the body, e.g., those involving proteins and enzymes, are pH dependent. With chronic low levels of the arterial carbon dioxide partial pressure ($PaCO_2$), the acid–base balance begins to change. Mountaineering physiologists employed the sea level equation for estimations of acid–base balance in high-altitude subjects without critical appraisal of its validity. The pH effects are inextricably linked critically with hemoglobin and oxygen status that can be crucial at high altitudes.

Therefore, for a more precise recalculation of the "titratable hydrogen ion difference" (THID) that should use Hb and HCO_3^- values for a particular altitude, we have derived our modified Van Slyke equation: THID in eECF $= (1 - [Hb]/43) \times (\Delta[HCO_3^-] + ßB) \times (pH\text{-}7.4)$ [8]. An adequate acid–base balance is probably the fundamental metabolic adaptation that allows for mountaineers to tolerate extreme hypoxia and even reach the summit of Mt. Everest [14].

Hemoglobin: The Other Factor

Hemoglobin (Hb) increases with altitude as shown in Fig. 4. This data is from permanent residents at different altitude cities or towns. From the trend curve, one can attempt to estimate the optimal hemoglobin value upon the summit of Mt. Everest, if complete adaptation could be achieved.

It becomes evident from the observation of a longer breath holding time at high altitude in polyerythrocythemia (PECH) patients with low PaO_2, compared with normal-high altitude subjects, that they are more tolerant to hypoxia [13]. The greater oxygen content of blood in PECH, as a result of the increase in Hb, allows for a better tolerance to hypoxia. This is also demonstrated by plotting the oxygen consumption of yeast cells against time after full saturation with 100 % oxygen [7]. The resulting oxygen dissociation curve of PECH patients has a much broader oxygen content area when exposed to 100 % oxygen [16].

The above observations on $PaCO_2$ and Hb prompted us to propose the hypothesis [16] that

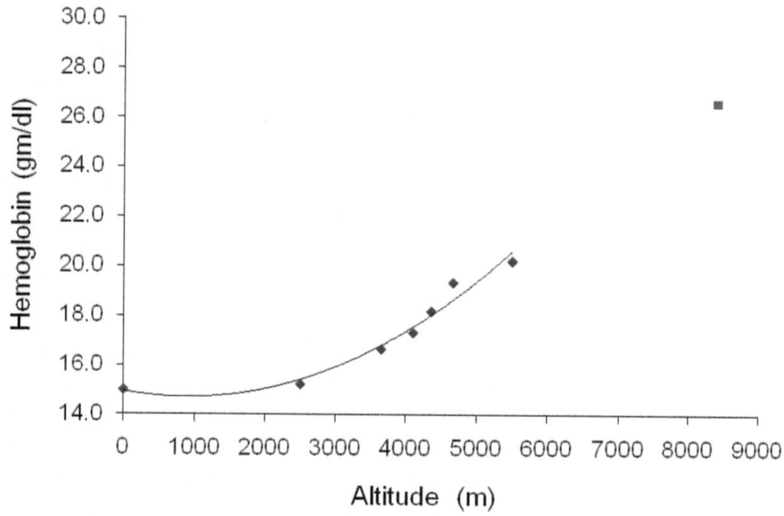

Fig. 4 Hemoglobin values found at different altitudes in resident populations. The red dot is the calculated Hb from the trend equation: Hb = 2E-07x^2 - 0.0003x + 14.45 with an $R = 0.9718$. The Hb values for residents at 2,500, 3,600, and 4,100 m for the Bolivia cities of Cochabamba, La Paz, and El Alto, respectively, are the normal values in our labs. The Hb values of residents at 4,355, 4,660, and 5,500 m were obtained from bibliographic data [4]

Table 1 Tolerance to hypoxia calculated at different altitudes from the hypoxia tolerance formula. The values of Hb and $PaCO_2$ are obtained from [3, 17]

	Altitude in m	Hb/PaCO2	HT
Sea level	0	13/40 * 3.01	1
La Paz, Bolivia	3,510	16.6/30 * 3.01	1.7
Mt. Everest	8,842	21/13 * 3.01	4.86

the tolerance to hypoxia (TH) formula can be defined as

$$TH = \frac{Hb}{PaCO_2} \times 3.01$$

The constant factor is obtained by using the Hb and $PaCO_2$ normal sea level values and equating to 1. This way, the tolerance to hypoxia for comparison purposes becomes 1 at sea level. From then on, the values at different altitudes are calculated as shown in Table 1. In this table, the Mt. Everest value is obtained from the paper by Grocott et al. [3] taken on the Cauldwell Expedition to Mt. Everest. Blood gases were measured at 8,400 m, and the Hb values were obtained from four subjects averaging the measurements at 5,300 m before and after the climb to the summit. This value should have been measured at higher altitudes and without averaging before and after climb as it reduces the true value. Only the value obtained after the ascent could be a more appropriate measure. However since the subjects were climbing and changing altitudes and the body has a fixed production of red blood cells, there was not enough time for full hematologic adaptation [17]. An optimal hemoglobin value for the summit of Mt. Everest is not known. However, based on Fig. 4, it is estimated that the optimal hemoglobin for oxygen transport at the summit of Mt. Everest would be around 26 g%. This roughly corresponds to a hematocrit of 78 %. In our medical practice, we have seen patients with PECH even above 80 % [12, 18, 19]. Hence, although it seems that these are surprising high values, they are within biological limits for humans. It is noteworthy that the maximum possible increase of Hb is double the sea level value, hence the tolerance to hypoxia formula is also is ruled by this.

Our formula links Hb with $PaCO_2$, the hematologic and respiratory (including acid–base parameter) responses to hypoxia, respectively. These biological responses to hypoxia in humans will also apply to animals. This is clearly demonstrated when a normal average sea level Hb in males of 13.3 g% rises to an average of 16.6 g% in the city of La Paz, Bolivia, at 3,510 m of altitude, for example. The formula also includes the $PaCO_2$ where it is showing the hyperventilatory response to hypoxia. For example, a sea level normal $PaCO_2$ of 40 mmHg upon arrival to high altitude immediately is decreased to a $PaCO_2$ of 30 mmHg, again, in the city of La Paz (3,510 m). Although this is a tolerance to hypoxia formula, the PIO_2 is not included in the formula as it is directly related to the barometric pressure, as originally described by Paul Bert in 1878 [1]. The barometric pressure, a physical atmospheric parameter, is indirectly included with the $PaCO_2$ which is equal to the $FaCO_2 * PB - PH_20$.

Other factors of adaptation to hypoxia, such as HIF, VEGF, increase in the density of the mitochondria, increase in capillary density, increase in pulmonary artery pressure, and increase in heart rate and ventilatory rate, are not included in this formula, as they are linked to the two biological factors included. Adaptation to hypoxia is complex, but this formula uses only two essential variables sufficient for the adequate interpretation of the concept in our criteria.

Conclusion

Adaptation to life at high altitude is possible, even at extreme environments such as the summit of Mt. Everest. The practice of medicine and observation over 44 years has taught us much about hypoxia. During a congress in Lhasa, Tibet, several physicians visited an intensive care unit where there was a subject suffering from pulmonary edema. He was in bed and receiving oxygen by mask. We wanted to see what his SpO_2 would be when breathing ambient air. After a few minutes, the SPO_2 started decreasing, and much to our surprise, one of the physicians who never had any experience in dealing with patients living in high altitude panicked and immediately ordered that the oxygen be restored. He was terribly alarmed by the ensuing hypoxia. This is understandable at sea level. A normal PaO_2 of the city of La Paz, 60 mmHg if seen at sea level, would produce tremendous alarm and possibly immediate hospitalization and intensive care. Paradoxically, the higher one goes, the more tolerant the subject is to hypoxia. Please don't misunderstand us, as we are talking of chronic adapted individuals and not acute exposure that can certainly produce symptomatology and even progress to high-altitude pulmonary edema or high-altitude cerebral edema. Life under chronic hypoxia is normal as at sea level. Diseases at high altitude, particularly in the cardiorespiratory areas, behave differently because the organism adapts to chronic hypoxia, even deviating from the normal and optimal status of good health.

References

1. Bert P (1978) Barometric pressure. Undersea Medical Society, Bethesda
2. De Santo NG, Cirillo M, Kirsch KA, Correale G, Drummer C, Frassl W et al (2005) Anemia and erythropoietin in space flights. Semin Nephrol 25:379–387
3. Grocott MP, Martin DS, Levett DZ, McMorrow R, Windsor J, Montgomery HE (2009) Arterial blood gases and oxygen content in climbers on Mount Everest. N Engl J Med 360:140–149
4. Leon-Velarde F, Gamboa A, Chuquiza JA, Esteba WA, Rivera-Chira M, Monge CC (2000) Hematological parameters in high altitude residents living at 4,355, 4,660, and 5,500 meters above sea level. High Alt Med Biol 1:97–104
5. Leon-Velarde F, Maggiorini M, Reeves JT, Aldashev A, Asmus I, Bernardi L et al (2005) Consensus statement on chronic and subacute high altitude diseases. High Alt Med Biol 6:147–157
6. NASA Crew Compartment Cabin Pressurization (2007) spaceflight.nasa.gov/shuttle/reference/shutref/orbiter/eclss/cabinpress.html. Accessed 15 Jun 2014
7. Neville J (1974) Hemoglobin oxygen affinity measurement using biotonometry. J Appl Physiol 37:961–971
8. Paulev PE, Zubieta-Calleja GR (2005) Essentials in the diagnosis of acid-base disorders and their high altitude application. J Physiol Pharmacol 56(Suppl 4):155–170

9. Rice L, Alfrey CP (2005) The negative regulation of red cell mass by neocytolysis: physiologic and pathophysiologic manifestations. Cell Physiol Biochem 15:245–250
10. Tavassoli M (1982) Anemia of spaceflight. Blood 60:1059–1067
11. Zubieta-Calleja G (2004) Bloodletting: a medical resource since the stone age. J Qinghai Med Coll 25:273–275
12. Zubieta-Calleja G, Zubieta-Castillo G (1998) Changes in oximetry during breath holding in normal residents of high altitude. In: Hideki Ohno, Toshio Kobayashi, Shigeru Masuyama and Michiro Nakashima (eds) Progress in mountain medicine and high altitude physiology. Press Committee of the 3rd World Congress on Mountain Medicine and High Altitude Physiology, Matsumoto, pp 343–348
13. Zubieta-Calleja G, Zubieta-Castillo G, Zubieta-Calleja L, Zubieta N (2002) Measurement of circulatory time using pulse oximetry during breath holding in chronic hypoxia. HAMB 3:115
14. Zubieta-Calleja GJ, Paulev P-E, Mehrishi JN, Zubieta-Castillo G Sr (2012) Extremely high altitude hypoxic conditions during Mount Everest expeditions, residence at South Pole stations, in Tibet and among the Andes: Van Slyke equation modification is crucially important for acid–base measurements. JBPC 12:103–112
15. Zubieta-Calleja GR (2007) Human adaptation to high altitude and to sea level: acid-base equilibrium, ventilation, and circulation in chronic hypoxia. Thesis. University of Copenhagen, Copenhagen
16. Zubieta-Calleja GR, Ardaya G, Zubieta N, Paulev PE, Zubieta-Castillo G (2013) Tolerance to hypoxia. J Fisiol 59(4): 65–72
17. Zubieta-Calleja GR, Paulev P-E, Zubieta-Calleja L, Zubieta-Castillo G (2007) Altitude adaptation through hematocrit changes. J Physiol Pharmacol 58:811–818
18. Zubieta-Calleja GR, Paulev PE, Zubieta-Calleja L, Zubieta-Calleja N, Zubieta-Castillo G (2006) Hypoventilation in chronic mountain sickness: a mechanism to preserve energy. J Physiol Pharmacol 57(Suppl 4):425–430
19. Zubieta-Calleja GR, Zubieta-Castillo G, Zubieta-Calleja L (1995) Inadequate treatment of excessive erythrocytosis. Acta Andina 4:123–126
20. Zubieta-Castillo G Sr, Zubieta-Calleja GR Jr, Zubieta-Calleja L, Zubieta N (2004) How chronic mountain sickness (cms) contributed to the theory of life at the hypoxic levels of the summit of mount Everest. HAMB 5:291–292
21. Zubieta-Castillo G Sr, Zubieta-Calleja GR Jr, Zubieta-Calleja L (2006) Chronic mountain sickness: the reaction of physical disorders to chronic hypoxia. J Physiol Pharmacol 57(Suppl 4):431–442
22. Zubieta-Castillo G, Zubieta-Calleja G (1996) Triple hypoxia syndrome. Acta Andina 5:15–18
23. Zubieta-Castillo G, Zubieta-Calleja G, Arano E, Zubieta-Calleja L (1998) Respiratory disease, chronic mountain sickness and gender differences at high altitude. In: Ohno H, Kobayashi T, Masuyama S, Nakashima M (eds) Progress in mountain medicine and high altitude physiology. Press Committee of the 3rd World Congress on Mountain Medicine and High Altitude Physiology, Matsumoto, pp 132–137
24. Zubieta-Castillo G, Zubieta-Calleja GR (2007) Facts that prove that adaptation to life at extreme altitude (8848 m) is possible. In: Lukyanova L, Takeda N, Singal PK (eds) Adaptation biology and medicine: health potentials, vol 5. Narossa Publishing House, New Dehli
25. Zubieta-Castillo G, Zubieta-Calleja GR (1988) Iatrogenic disease in a polycythemic patient at altitude. Am Rev Respir Dis 137:509
26. Zubieta-Castillo G, Zubieta-Calleja GR (1988) The triple hypoxia syndrome at altitude (abstract). Am Rev Respir Dis 137:509
27. Zubieta-Castillo G, Zubieta-Calleja GR, Zubieta-Calleja L, Zubieta C, Nancy (2003) Adaptation to life at the altitude of the summit of Everest. Fiziol Zh 49:110–117

Nanomaterials in Healthcare

Thomas Tumillo, Avijit Roy, Sahana Pentyala, Pooja Mysore, and Srinivas N. Pentyala

Abstract

The strength of a nation invariably depends upon the strength of every individual. Healthy citizens make a healthy and strong society. As a society, we started looking at expensive macroscopic high-tech solutions to solve many of our problems. With growing interests in the field of nanotechnology, looking microscopically at the problem to come up with solutions has generated a paradigm shift. Nanomaterials are currently being used in every branch of science, and particularly their use in healthcare opened up a new horizon to provide quality care to patients. The tools and therapies for diagnosis and prognosis of diseases and symptoms revolve around three important aspects – diagnostics, devices and drugs. Nanomaterial applications are now found in all these three aspects and are being effectively used in health care. New discoveries and inventions in the field of nanohealth care point to a bright future for the wellbeing of mankind. An introduction to nanomaterials in healthcare has been presented in this review.

Introduction

Nanotechnology-based applications are now being widely used in the field of healthcare. Drugs, devices, and diagnostics are the three key components that help a healthcare provider deliver quality care to the patient. Nanomaterials are being used in all these three aspects of providing healthcare to prevent morbidity and mortality. Many new potential nanomaterials are being discovered and developed, and novel applications are being designed for the use of these nanomaterials in modern healthcare. In this review, we present a synopsis of different nanomaterial-based technologies in healthcare with emphasis on therapeutics, medical devices, and diagnostics.

What Are Nanomaterials?

Nanomaterial is a term used to define all nanosized materials which include human-engineered nanomaterials and incidental

T. Tumillo • A. Roy • S. Pentyala • P. Mysore
S.N. Pentyala (✉)
Department of Anesthesiology, Stony Brook Medical Center, Stony Brook, New York, USA
e-mail: Srinivas.Pentyala@stonybrookmedicine.edu

S.B. Singh et al. (eds.), *Translational Research in Environmental and Occupational Stress*,
DOI 10.1007/978-81-322-1928-6_6, © Springer India 2014

nanomaterials, the ones found in nature. A material with at least one external and internal structure in the range of 0–200 nm (nm) is considered a nanomaterial. They may be in the form of tubes, rods, particles, or fibers and contain devices or systems made by manipulating individual atoms or molecules. In healthcare, nanomaterials have a lot of potential because they are particles which can directly interact with or influence living cells. Because the physical and chemical properties differ at the nanoscale from bulk material, we observe increased chemical activity and the ability to cross tissue barriers leading to new drug targeting and delivery systems [1]. The physical and chemical properties of nanomaterials differ from bulk materials, and hence nanomaterials call for special risk assessment before being used in healthcare settings. Reversible recoveries of the damages caused by mechanical and irradiation exposure have been demonstrated [2]. Currently in vitro and in vivo assays are being performed to assess the short-term effects of nanoparticles in the body, but no substantial evidence or tests exist yet to determine the long-term effects of nanomaterials. Nanotechnology is making it possible to stimulate the body's own mechanisms like never before to detect and repair diseased or damaged tissues, which may replace the need for transplants or bioengineered organs. Potentially nanoparticles can be used to search and destroy a single malignant cell, taking us closer to ultimate disease prevention. Nanoparticles are effectively being used in disease diagnosis and prognosis in many different formats. Nanomedicine is now a key branch of nanotechnology that includes the development of nanostructures and nanoanalytical systems for various medical applications [3].

Nanomaterials and Drugs

Nanotechnology has facilitated the surge of new techniques that are more effective in treating various diseases. Nanomedicine is likely to enable delivery of pharmaceuticals through different routes of administration with safety and efficacy compared to the conventional methods used today. In recent years, there has been a lot of emphasis on the utilization of nanotechnology in oncology, which raises new possibilities in cancer diagnosis and treatment. These nanomedicine formulations improve the treatment of systemically applied chemotherapeutic drugs. To improve the biodistribution of cancer drugs, nanoparticles have been designed with optimal size and surface characteristics to increase their circulation time in the bloodstream [4]. Doped nanomaterial might be a promising therapy owing to its high specific surface area, low resistances, high catalytic activity, and attractive electrochemical and optical properties [5]. The use of such materials has helped deliver drugs across the blood-brain barrier, alleviate allergy symptoms, specifically target cancer or HIV cells, and more [6]. Nanomaterials come in a variety of materials, shapes, and sizes, which causes various effects on different systems in the body. These effects are currently being investigated to help improve the efficiency [7]. Common effects of nanomaterial use are inflammatory responses and an increased production of reactive oxygen species (ROS) [6].

Nanomedicine is already being practiced. Nanometer-sized carrier materials are used to increase the drug tissue bioavailability [8]. Doxil®, a PEGylated liposomal doxorubicin formulation, was the first nanosized therapeutic introduced in 1995 in the market and was used as an effective treatment for metastatic breast cancer and recurrent ovarian cancer [7, 9]. There are many other nanomaterial-based therapies that are in various stages of clinical trials. For example, a targeted therapeutic nanoparticle, named BIND-014, accumulates in tumors while avoiding uptake by the healthy cells [7, 10]. It is foreseen that nanomedicine will facilitate the development of personalized medicine and will have a major impact on healthcare [11]. Also, the availability of ligand-functionalized therapeutic nanomaterial will have an impact on the individualized treatment of diseases [7]. The impact of these nanoformulations will depend upon the degree of toxicity reduction. The passive targeting approach takes advantage of the inherent bio-physicochemical properties of the nanoparticles [7]. Nanoparticles are being investigated as a means to provide personalized treatment for a variety of diseases, most commonly cancer. If tumor accumulation is found to correlate with the patient's susceptibility to treatment, a passive targeting approach could be used to identify individuals with lesions possessing leaky vasculature who would benefit the most from nanosized formulation [7]. However, there are growing concerns about potential health hazards that may arise due to nanoparticles' unique physical and chemical properties. Nanomaterials have dimensions similar to those of cell organelles and, thus, can potentially interfere with vital cell functions [7]. Because of their metallic nature and long half-life, it is possible that some nanoparticles would not be cleared from the body for several years, which could result in toxicity. For example, titanium dioxide can cause toxicity because of its crystalline structure [7]. At present, nanomaterials have significant advantages due to their unique nanoscale properties, but there are still significant challenges in the improvement and development of nanoformulations with composites and other materials. Efforts are being made toward improving the balance between the efficacy and the toxicity of therapeutic interventions through different routes of administration and by designing safer and efficacious nanomedicine [8].

Drug Delivery

Metallic particles on the nanoscale such as ferrous oxide, gold, or silver are being used for hyperthermia, cancer treatment, magnetic bioseparations, membrane transport studies, and cell-specific antigen studies. Modified hollow gold nanoparticles such as tubes and boxes have promising applications in the fields of drug delivery and therapy in oncology due to their unique optical and photothermal properties and their ability to modify the surface and conjugate drugs [5]. These metallic particles are especially useful in drug delivery due to their DNA-binding properties. The bound DNA acts as a bridge between the gold nanotube and drugs such as chemotherapy drugs. When the gold nanotubes are injected into the cancerous area and light is applied, the light is absorbed by the gold nanotube generating heat which releases the chemotherapeutic drug on site, making it much more effective. Gold shell nanoparticles are found to improve the solubility of drugs and permit further conjugation of those drugs when in the body [12].

Nanoparticle delivery systems act specifically on cancerous cells to combat cancer. As drug carriers, they are able to efficiently protect the therapeutic agent and allow for sustained drug release. In addition, their surface can be easily manipulated with the addition of special ligands, which are responsible for enhancing tumor-specific nanoparticle permeability [13]. Chemotherapy drugs such as camptothecin along with the Herceptin antibody can be attached to nanoparticles to target breast cancer cells. Camptothecin use has been limited by poor water solubility and the instability of the lactone moiety despite its strong antitumor activity. Formulating the drug into nanocarrier systems can directly combat these problems associated with drug delivery [14]. Nanoparticles have also helped protect surrounding areas from radiation damage. It was found that siRNA-mediated gene knockdown of proapoptotic mediators can reduce cellular apoptosis in salivary glands in vitro. Nanoparticles that are pH responsive electrostatically interact with the siRNAs and under osmolytic conditions release siRNA into the salivary glands, thereby protecting the target genes from nuclease attack [15].

Carbon nanomaterials such as fullerenes and carbon nanotubes are now being made into structures called "buckyballs." Buckyballs are football-shaped structure made of 60 carbon atoms. They are used in drug delivery systems for optimal transport and release of the drugs to the right target. Nanoparticles coat the vaccine and drugs in many cases and protect them, eliciting a stronger immune response. The carbon nanotubes are now being used to coat prosthetics and surgical implants. These coatings improve the function of implants such as vascular stents and help with the delivery. The effects of the electrosensitive transdermal drug delivery system were enhanced by the carbon nanotubes [16]. The carbon nanotubes are also used for gene therapy as a strand of DNA can be bonded to a carbon nanotube. This opens up many genetic applications with carbon nanotubes. Through techniques such as electroporation, nanoholes can be made in the plasma membrane of cells which carbon nanotubes can fit and deliver drugs [17]. These carbon nanomaterials can also help with drugs being targeted through the lymphatic system. It plays a role in transporting extracellular fluid to maintain homeostasis and is able to avoid first-pass metabolism, thus acting as a bypass route for compounds with lower bioavailability [18].

Nanocapsules are also being developed to combat insulin-dependent diseases such as diabetes. A new method to release insulin uses a spongelike matrix that contains insulin and nanocapsules containing an enzyme. When the glucose level rises in the body, the nanocapsule releases hydrogen ions which bind to the fibers of the sponge matrix [19]. The hydrogen ions end turning the sponge into a cation, all positively charged. They repel each other creating openings in the matrix allowing insulin to be released through.

Dendrimers are polymerized macromolecules which are highly branched structures with interior nanocavities or channels with different properties on the interior and exterior. These channels can be used in the body as a carrier for a variety of drugs with the capacity to improve solubility and delivery of poor biomolecules which interact with many other compounds. Used in these dendrimers are new nanoparticles called ceramic nanoparticles. These ceramic particles are inorganic systems used as drug vehicles and are often used in cosmetic applications [20].

Quantum dots are nanocrystals made up of semiconductive materials and are now being used to tag multiple biomolecules to monitor cellular changes and events associated with disease. Because of the conductive properties, quantum dots allow electrical waves and patterns to be generated and used in the body as an indicator of activity. These quantum dots are also being developed for optics technology and disease screening tests [21]. However, due to toxicity in the human body, quantum dots are now being made with silicon instead of cadmium. Silver nanoparticles have been found to be very strong antimicrobial agents and are now being incorporated into a wide range of medical devices, bone cement, surgical instruments, and masks.

Researchers are currently looking into the use of nanoparticles as a means of delivering therapeutic drugs to the brain. Many therapeutic drugs cannot cross the blood-brain barrier. The development of new strategies using nanoparticles could be of great benefit to those with central nervous system diseases [22].

Therapy Techniques

Food is also a form of therapy as specific drugs can be incorporated in food and targeted to larger populations. Nanoliposome technology is the latest technique in nutraceutical nanotechnology. Nanoliposomes are microscopic vesicles made of phospholipid bilayers entrapping one or more aqueous compartments. They can provide controlled release of biochemical agents including ingredients in food and nutraceuticals at the right place and time. They increase efficacy and cellular uptake of the encapsulated material. Nanoliposome technology presents opportunities in areas such as encapsulation and controlled release of food materials, as well as enhanced bioavailability, stability, and shelf-life of sensitive ingredients [23].

Researchers are intrigued by the use of nanoparticles composed of polyethylene glycol hydrophilic carbon clusters (PEG-HCC) which have been shown to absorb free radicals at a much higher rate than the proteins and antioxidants the body uses for this very purpose. After a brain injury, a plethora of free radicals are released, and the use of the PEG-HCCs may reduce the harm caused to the body after a traumatic brain injury. Photo-based nanoparticles have come into prominence because of their ability to act as nanocarriers to deliver fluorescent dyes or photosensitizers for photoimaging and therapeutic applications. They can be used for MRI tests, photothermal therapy, and chemotherapy. Materials used range from gold, silver, and silica nanoparticles to polymer-based PEGs. Researchers have developed a polyethylene glycol shell-sheddable magnetic nanomicelle as the carrier of doxorubicin (Dox). Due to the detachment of PEG shell in the presence of dithiothreitol (DTT), the magnetic nanomicelles showed accelerated in vitro release of the Dox and enhanced cellular uptake ability. Compared with free Dox, the Dox-loaded magnetic nanomicelles showed decreased cytotoxicity against Bel-7402 cell line [24]. The nanocarrier's ability to target cells using specific ligands and improving drug delivery makes it a promising application for the future.

Nanomaterials and Devices

Nanoparticles are being used to treat cases where infected wounds do not respond to standard antimicrobial treatment and care. The number of these incidences has been rapidly increasing. In order to address this issue, a novel antimicrobial magnetic thermotherapy platform has been developed where a high-amplitude, high-frequency, alternating magnetic field is used to rapidly heat magnetic nanoparticles that are bound to *Staphylococcus aureus*. An antibody-targeted magnetic nanoparticle bound to *S. aureus* was found to be effective at thermally inactivating *S. aureus* and achieving accelerated wound healing without causing tissue injury. This procedure has been proven successful in in vitro culture models of *S. aureus* biofilm as well as in mouse models of cutaneous *S. aureus* infection [25]. Nanofibers are created by electrospinning and have unique characters in terms of morphology, scaffold composition, functional groups, and hydrophilicity. The combination of a functional nanofibrous scaffold seeded with stem cells and composed of natural polymers and cross-linked with a natural cross-linking agent, phytic acid, may prove to be a novel therapeutic device for the treatment of myocardial infarction. This functional scaffold assists in the regeneration of the ischemic myocardium [26]. The active targeting approach requires modification of the drug carriers with ligands that specifically bind to the tissue of interest. The cell, through receptor-mediated interactions, can internalize the drug. This nanoparticle platform holds particular promise for treatments of targeted blood vessel walls such as catheter- or stent-induced cardiovascular injuries [7]. Biosensors are increasingly becoming an essential part of modern healthcare as personalized medicine becomes the forefront of the industry. Carbon nanotubes (CNTs) are promising building blocks for biosensors due to their unique electronic and optical properties. Carbon nanotubes are rolled-up cylinders of carbon monolayers (graphene) that can be altered in such a way that biologically relevant molecules can be detected with high sensitive and selectivity [27]. CNTs have been designed specifically for each of the major analytes of blood, glucose, cholesterol, triglyceride, and Hb1AC [28]. CNTs enhance the signals derived from the interaction of the enzymes with different analytes in the blood [28].

Nanomaterials are also being used now for therapy techniques on patients to ease pain and improve quality of life in healthcare. Scientists have developed nanosponges which are polymer nanoparticles coated with a red blood cell membrane. The red blood cell membranes lets the nanosponges travel undetected and freely in the bloodstream and attract the desired toxins from the body [29]. The same team has also been developing a method to be used for noninvasive surgery. A lens coated with carbon nanotubes is used which converts light from a laser to focused sound waves. This method will be able to blast tumors or diseases tissue without damaging the healthy tissue around it.

Nanomaterials and Diagnostics

Early detection of cancer cells is a developing field in nanotechnology. Researchers are also developing a diagnostic sensor which can detect 3–5 cancer cells in a one milliliter blood sample. The sensor uses graphene oxide which is an allotrope of carbon nanoparticles. An antibody with fluorescent markers is attached to the graphene oxide which can then attach to cancer cells [30]. So when attached, the fluorescence indicates the presence of cancer cells. Another nanoparticle is being developed to release biomarkers after being attached to cancer cells. These biomarkers contain peptides, so observing an unusually high concentration of peptides indicated early detection of the cancer disease. An early detection method of brain cancer is using magnetic nanoparticles and NMR technology. The magnetic nanoparticles can attach to microvesicles in the bloodstream which develop in brain cancer cells. They are then detected much easier in NMR imaging, which is used for the early detection of brain cancer.

Nanoparticles are increasingly being used to overcome some of the limitations that plague the current diagnostic tests. Some studies have begun using super-paramagnetic nanoparticles in place of the conventional gold bead standard [31]. Surface-enhanced Raman scattering (SERS) nanoparticles are being used as molecular imaging contrast agents by offering physicians biochemical information that they normally wouldn't have access to with the conventional diagnostics. Endoscopic imaging, for example, only provides structural information about the tissues deep within the body [32]. An accessory, noncontact, fiber-optic-based Raman spectroscopy device was designed to be inserted through a clinical endoscope and was created with the potential to provide real-time, multiplexed functional information during routine endoscopy. Using this endoscope as a key part of a multiplexed detection approach could allow endoscopists to distinguish between normal and precancerous tissues rapidly and to identify flat lesions that are otherwise missed [32].

Biomarkers are becoming increasingly important in the management of diseases; however, our ability to discover new biomarkers remains limited by our dependence on endogenous molecules. "Synthetic biomarkers" composed of mass-encoded peptides conjugated to nanoparticles have been developed to leverage intrinsic features of human disease and physiology for noninvasive urinary monitoring. These agents target the sites of disease, sample dysregulated protease activities, and emit mass-encoded reporters into host urine for multiplexed detection by mass spectrometry. They have been shown to noninvasively monitor liver fibrosis and resolution without the need for invasive core biopsies, and they can also improve early detection of cancer [33].

Another class of nanomaterials currently being used for theranostic application is lanthanide-doped hollow nanomaterials (LDHNs). Theranostics fuses diagnostic and therapeutic functions, empowering early diagnosis, targeted drug delivery, and real-time monitoring of treatment effect into one step [34]. Theranostic nanoparticles can serve as useful tools to explore the fundamental process of drug release after cellular internalization of nanoparticles, which could provide key insights into the rational design of targeted nanocarriers for personalized treatment [7]. The LDHNs present outstanding fluorescent and paramagnetic properties, which allow them to be used as bioimaging agents. In addition, LDHNs have huge interior cavities that are able to store and deliver therapeutic agents [34].

Rapid Diagnostics

Biomarkers are important tools for disease detection and monitoring. They serve as hallmarks for the physiological status during the disease process [35, 36]. A highly effective, clinically useful biomarker for a specific disease should be measurable in a readily accessible body fluid, such as serum, urine, or saliva [37]. The search for biomarkers for early disease detection has included proteins, metabolites, and other biological molecules that are altered and secreted as a consequence of the disease process and are shed into body fluids. After collecting these body fluids, the next step is to isolate and identify the marker that will give an indication of the disease process. Unfortunately, this approach is laborious and time-consuming, as specific candidate biomarker(s) must be identified from among the thousands of intact and altered molecules in the collected body fluids. In many disease manifestations, a marker can occur in trace amounts, yet large volumes of fluids are collected (e.g., blood and urine) [35, 38, 39]. It is very difficult and time-consuming to process these samples to concentrate and identify specific markers for diagnosis or disease status.

Biological fluid collection to identify and analyze different markers of diseases and symptoms is a routine procedure in healthcare settings. The fluids are as varied as urine, blood, mucus, cerebrospinal fluid, mucus, tears, semen, etc. The volumes of the collected biological fluids range from microliters (e.g. tears, CSF) to tens and hundreds of milliliters (blood, urine, etc.). Many of the disease markers and profiles of essential molecules are identified by collecting body fluids [40]. With the discovery that markers for abnormal symptoms and diseases are found to be in trace amounts in large volumes of biological fluids, it is very difficult and time-consuming to process these samples to concentrate and identify specific markers for diagnosis as well as prognosis. Many methods and devices using nanomaterials were developed to rapidly capture, concentrate, and identify biomarkers in body fluids.

Any disease or deleterious symptom in the body may result in changes in the expression of protein biomarkers. Sometimes, biomarker levels can increase or decrease; other times, specific markers are expressed and can be detected in body fluids, particularly in the blood and urine [37]. Identifying these biomarkers can lead to determining whether a person has a disease, disorder, or symptom. Body fluid collection to identify and analyze different biomarkers of diseases and symptoms became a routine procedure. The volumes of the collected biological fluids range from microliters (e.g., tears, CSF) to tens and hundreds of milliliters (blood, urine, etc.). Identification of trace markers in large volumes of body fluids may require long times, skilled professionals, and often sophisticated instruments. A widely used diagnostic method and device is a lateral flow format device using nanobeads to capture biomarkers in body fluids. These lateral flow point-of-care devices using nanobeads are being widely used in different areas of healthcare as bedside devices.

The use of nanomaterials in rapid diagnostics is becoming more prevalent because quicker and more precise diagnosis may save the lives of patients who don't have much time. Researchers at MIT have developed a sensor system using carbon nanotubes embedded in gel. This gel can be injected under the skin to monitor the level or nitric oxides in the blood through interactions with carbon nanotube. The level of nitric oxide is important in diagnosis because it is an indicator for the inflammatory response which can confirm an inflammatory disease. Nanoparticles are also being used in the diagnosis of kidney damage using gold nanorod particles. Levels of amyloids in the blood indicate amyloidosis. This is caused by the buildup of beta-2-microglobulin in the blood, which is released by the kidney after damage. The gold nanorods can attach to these proteins, when the protein accumulates the color of the gold changes, indicating kidney damage. The presence of nanomaterials can offer either an inhibitory effect or promotion of amyloid fibrillation, depending on the structural architectures of carbon nanomaterials and the starting amyloid proteins/peptides considered [41].

Nanomaterials have been found to attach to certain drugs which have proapoptotic factors, thus delivering drugs to malignant cancer cells to induce apoptosis when it could not before. Nanomaterials can also be used as biomarkers to attach to cells in the focused area of radiation and chemotherapy. They have been shown to mask surrounding tissue from the disease defense drugs and protect the tissue from damage due to proximity of the treatment.

In the field of diagnostics, nanomaterials are becoming increasingly important in personalizing patient care. Diagnostics has advanced to have short turnaround time and minimal interference, which allows quick clinical management decisions. In some diseases, early diagnosis and rapid initiation of treatment are crucial to the success of the patient, for example, in cases regarding cardiovascular and infectious diseases [42]. Advances in nanotechnology have led to the development of functionalized nanoparticles (NPs) that are covalently linked to biological molecules such as antibodies, peptides, proteins, and nucleic acids. These functionalized NPs allow for the development of novel diagnostic tools and methods, particularly for pathogens, as rapid and sensitive diagnostics are essential for defining the emergence of infection, for determining the period that preventive measures should be applied, for evaluating drug and vaccine efficacy, and for controlling epidemics [43].

A method has been developed that allows for rapid and simple single-cell lysis and analysis. The technique lyses individual cells on silicon nanowire and nanoribbon biological field effect transistors. This method is to be applied in medical diagnostics, proteome analysis, and developmental biology studies; however, the extent to which these applications will be utilized is still uncertain. Analysis of cell-to-cell variation can further the understanding of intracellular processes and the role of individual cell function within a larger cell population [8].

The ability to effectively detect disease-related DNA biomarkers and drug delivery nanoparticles directly in blood is a major challenge for diagnostic and therapy monitoring [44]. A dielectrophoretic (DEP) method has been developed that makes detection of biomarkers and nanoparticles directly from human and rat blood possible. The DNA biomarkers and nanoparticles were concentrated into a high-voltage field region by positive DEP while the blood cells were concentrated into a low-voltage field by negative DEP. After a fluidic wash, which removes the blood cells, the DNA biomarkers and nanoparticles are detectable through fluorescence. The ability to rapidly isolate and detect DNA biomarkers and nanoparticles from undiluted whole blood will benefit many diagnostic applications by significantly reducing sample preparation time and complexity [44].

One of the critical challenges in the fields of disease diagnostics and environmental monitoring is to concentrate extracellular DNA from a sample mixture rapidly. Researchers were able to concentrate extracellular DNA onto a nanostructured tip using the DEP method in conjunction with capillary action [45]. Captured DNA is investigated by fluorescence microscopy, scanning electron microscopy (SEM), and X-ray analysis. Unlike genomic DNA in normal cells, extracellular DNA dissolved in a biological sample can potentially offer crucial information about pathogens and toxins.

Multiplex Rapid Diagnostic Devices Using Nanomaterials

Quick turnaround time and minimal manual interference are some of the key qualities of effective rapid diagnostic devices because they allow for expedited decisions by physicians. Fluorescent microspheres conjugated to biomarkers (nucleic acids, proteins, lipids, carbohydrates) and analyzed on flow cytometer instruments offer a new approach for multiplexed detection platform in a suspension format. Quantum dots encoded into synthetic microspheres have the potentials to improve current screening bioassays and specifically suspension array technology [46]. A reproducible method was developed for the detection of single-stranded DNA with quantum dot-encoded microspheres. In 1 h, the developers were able to conjugate approximately ten thousand microspheres to short amino-modified DNA sequences [46]. Devices such as these are continuing to develop in the field of fluorescent probe technology and molecular diagnostics.

Quantum dots (QDs) have been applied for simultaneous detection of multiple analytes because of the wide range of properties they can exhibit. A single QD-based nanosensor was developed for the detection of HIV-1 and HIV-2. In this single QD-based nanosensor, the QD functions not only as a fluorescence pair for coincidence detection and as a fluorescence resonance energy transfer (FRET) donor for FRET detection but also as a local nanoconcentrator which significantly amplifies the coincidence-related fluorescence signals and the FRET signals. This single QD-based nanosensor takes advantage of a simple "mix and detection" assay with extremely low sample consumption, high sensitivity, and short analysis time and has the potential to be applied for rapid point-of-care testing, gene expression studies, high-throughput screening, and clinical diagnostics [7].

The ability to differentiate each agent by visible color enables multiplexing and simplifies test interpretation. By using lanthanides, elements with atomic numbers 57–71 in the periodic table, brightly colored probes can be generated. Lanthanides can be employed to create unique emission spectra. Trivalent lanthanide ions (Ln^{III})-based optical probes possess excellent luminescence properties with potential applications in biological science [47–49]. The use of europium (Eu) nanoparticles has already been established as a way to enhance the sensitivity of capture assays for biowarfare agents, improving sensitivity by 100-fold [50].

The resulting specific multiplexed capture assay will require only visible light. Even in low or no light, only a flashlight will be needed to enable identification and distinction among different biomarkers. The multiplex devices will be useful under virtually any condition, from the battlefield or primitive field site to the clinic, emergency room, or hospital, providing a highly innovative low-tech solution to a critical need and the rapid identification of disease, symptoms, or even harmful toxins and biological agents. Also, multiplex devices will be simple enough to be used by even untrained personnel.

Future of Diagnostics

A device like "tricorder" will be the natural extension of point-of-care devices. Star Trek's Dr. Leonard "Bones" McCoy's famous "medical tricorder" scans patients and immediately diagnoses their ailments. Tricorder is a multipurpose device used primarily to scan unfamiliar areas, make detailed examination, and record and review technical data [51]. Medical tricorders should have four characters: noncontact, noncooperative, nonsampling, and noninvasive. In pursuit of life beyond planet Earth, the human race has embarked on a journey to explore outer space and succeeded in sending machines beyond the realm of our solar system [52]. Plans are now underway to send humans to Mars [53]. Parallel development of technologies such as spacecraft and lifesaving and sustenance instruments like the "tricorder" to accompany space travelers will make the dream of traveling into space a reality. Many tricorders based on nanomaterials are in different stages of development. On May 10, 2011, the X PRIZE Foundation and Qualcomm Incorporated announced the Tricorder X Prize, a $10 million incentive to develop a mobile device that can diagnose patients as well as or better than a panel of board-certified physicians [54]. An early entrant to the competition is "Scanadu," which was introduced in 2011. Scanadu is a small handheld sensor put next to a patient's forehead to detect vital signs such as heart rate, breathing rate, blood oxygenation, pulse transmit time, and temperature. The unit works in conjunction with a mobile app [55]. While this tricorder is missing some of the features of its science fiction counterpart—namely, the ability to make internal scans and complex diagnoses—it still can be a handy device for medical checkups on the go. There are already "diagnostic medical apps" via tablet computers and smartphones which can be considered as tricorders [56]. US Department of Homeland Security is testing a laser-based "Standoff Patient Triage Tool" to help medics evaluate patients' vital signs wirelessly from 40 ft (12 m) away. The US Air Force is currently testing a handheld medical tricorder to improve situational awareness of directed energy. This device measures laser exposure, enhances laser detection/response, supports night vision, and provides ease of use to detect threats, thereby enhancing combat medicine [57]. There is hope that medical tricorders, based on the principle of nanomaterial-based rapid diagnostics, are going to become a reality. In the future, one can proudly state, "To Boldly Go Where No Medical Response Has Gone Before."

Questions for the Future

Health concerns are associated with nanoparticles because the interactions of the chemicals on the nanoscale are different than the macro scale. Under normal conditions, nanomaterials may form aggregates which are larger than 100 nm.

These aggregates have the same properties as the individual nanoparticles but sometime release nanoparticles which may be dangerous when introduced in the respiratory system and lung fluid. After nanomaterials are exposed in the body, they could be subjected to absorption, distribution, or metabolism. Many scientists pose questions concerning the bioaccumulation of nanoparticles and elimination mechanisms from the cells. They also believe that while nanomaterials may not be dangerous, a more toxic material may attach itself to the nanoparticle and gain entry into the body.

In the body, nanomaterials circulate through the entire body by moving in and out of blood vessels due to their size. They enter cells and interact with biomolecules both on the cell surface and inside the cells. Nanoparticles can enter the body in many ways which cause health hazards. Inhalation is the most common form of exposure to airborne nanoparticles. They can deposit in the respiratory tract and the lungs or cross the epithelium and enter the blood stream to organs. Inhaled nanoparticles have also been found in the brain through the olfactory nerve route. Ingestion of nanoparticles can occur from hand to mouth transfer from contaminated surfaces or food/water. There is also speculation of dermal penetration of nanomaterials because intact skin is a good barrier, but damaged skin may be an opening for nanomaterials [58].

Conclusion

The main benefits of using nanomaterials in healthcare are the solubility of nanomaterials. Many drugs now have the potential to become soluble in the body on the nanoscale when attached to nanoparticles due to the changing chemical properties on the nanoscale [59]. Biocompatible nanomaterials have been used as biological markers, contrast agents for imaging, healthcare products, pharmaceuticals, and drug delivery systems and in detection, diagnosis, and treatment of various types of diseases [60]. In drugs, the focus is on the use of nanoparticulate compositions comprising agents that are useful for a variety of diseases. In diagnostic kits, nanostructures have been patented that are capable of detecting target analytes. In medical devices, methods have been developed with nanocapsules and incorporated into the functionality of the devices to control the release of various chemical agents. Many innovative technologies and methods are being developed using nanomaterials in healthcare. The field of nanomedicine holds a promising future in taking care of the sick and needy.

References

1. Angeli E, Buzio R, Firpo G, Magrassi R, Mussi V, Repetto L, Valbusa U (2008) Nanotechnology applications in medicine. Tumori 94:206–215
2. Kustandi TS, Loh WW, Shen L, Low HY (2013) Reversible recovery of nanoimprinted polymer structures. Langmuir 29:10498–10504
3. Deng L, Ke X, He Z, Yang D, Gong H, Zhang Y, Jing X, Yao J, Chen J (2012) A MSLN-targeted multifunctional nanoimmunoliposome for MRI and targeting therapy in pancreatic cancer. Int J Nanomedicine 7:5053–5065
4. Cho K, Wang X, Nie S, Chen ZG, Shin DM (2008) Therapeutic nanoparticles for drug delivery in cancer. Clin Cancer Res 14:1310–1316
5. Ahmad MZ, Akhter S, Rahman Z, Akhter S, Anwar M, Mallik N, Ahmad FJ (2013) Nanometric gold in cancer nanotechnology: current status and future prospect. J Pharm Pharmacol 65:634–651
6. Syed S, Zubair A, Frieri M (2013) Immune response to nanomaterials: implications for medicine and literature review. Curr Allergy Asthma Rep 13:50–57
7. Zhang CY, Hu J (2010) Single quantum dot-based nanosensor for multiple DNA detection. Anal Chem 82:1921–1927
8. Jokilaakso N, Salm E, Chen A, Millet L, Guevara CD, Dorvel B, Reddy B Jr, Karlstrom AE, Chen Y, Ji H et al (2013) Ultra-localized single cell electroporation using silicon nanowires. Lab Chip 13:336–339
9. Barenholz Y (2012) Nanomedicine: shake up the drug containers. Nat Nanotechnol 7:483–484
10. Zamboni WC, Torchilin V, Patri AK, Hrkach J, Stern S, Lee R, Nel A, Panaro NJ, Grodzinski P (2012) Best practices in cancer nanotechnology: perspective from NCI nanotechnology alliance. Clin Cancer Res 18:3229–3241
11. Zheng S, Li X, Zhang Y, Xie Q, Wong YS, Zheng W, Chen T (2012) PEG-nanolized ultrasmall selenium nanoparticles overcome drug resistance in hepatocellular carcinoma HepG2 cells through induction of mitochondria dysfunction. Int J Nanomedicine 7:3939–3949

12. Niikura K, Iyo N, Matsuo Y, Mitomo H, Ijiro K (2013) Sub-100 nm gold nanoparticle vesicles as a drug delivery carrier enabling rapid drug release upon light irradiation. ACS Appl Mater Interfaces 5:3900–3907
13. Cheng Y, Morshed R, Cheng SH, Tobias A, Auffinger B, Wainwright DA, Zhang L, Yunis C, Han Y, Chen CT et al (2013) Nanoparticle-programmed self-destructive neural stem cells for glioblastoma targeting and therapy. Small 9:4123–4129
14. Loch-Neckel G, Nemen D, Puhl AC, Fernandes D, Stimamiglio MA, Alvarez Silva M, Hangai M, Santos Silva MC, Lemos-Senna E (2007) Stealth and non-stealth nanocapsules containing camptothecin: in-vitro and in-vivo activity on B16-F10 melanoma. J Pharm Pharmacol 59:1359–1364
15. Arany S, Xu Q, Hernady E, Benoit DS, Dewhurst S, Ovitt CE (2012) Pro-apoptotic gene knockdown mediated by nanocomplexed siRNA reduces radiation damage in primary salivary gland cultures. J Cell Biochem 113:1955–1965
16. Im JS, Bai B, Lee YS (2010) The effect of carbon nanotubes on drug delivery in an electro-sensitive transdermal drug delivery system. Biomaterials 31:1414–1419
17. Chakravarty P, Qian W, El-Sayed MA, Prausnitz MR (2010) Delivery of molecules into cells using carbon nanoparticles activated by femtosecond laser pulses. Nat Nanotechnol 5:607–611
18. Ali Khan A, Mudassir J, Mohtar N, Darwis Y (2013) Advanced drug delivery to the lymphatic system: lipid-based nanoformulations. Int J Nanomedicine 8:2733–2744
19. Gruber HE, Hoelscher GL, Ingram JA, Hanley EN Jr (2012) Matrix metalloproteinase-26, a novel MMP, is constitutively expressed in the human intervertebral disc in vivo and in vitro. Exp Mol Pathol 92:59–63
20. Matsudai M, Hunt G (2005) Nanotechnology and public health. Nihon koshu eisei zasshi 52:923–927
21. Bratschitsch R, Leitenstorfer A (2006) Quantum dots: artificial atoms for quantum optics. Nat Mater 5:855–856
22. Biondi M, Guarnieri D, Yu H, Belli V, Netti PA (2013) Sub-100 nm biodegradable nanoparticles: in vitro release features and toxicity testing in 2D and 3D cell cultures. Nanotechnology 24:045101
23. Mozafari MR, Johnson C, Hatziantoniou S, Demetzos C (2008) Nanoliposomes and their applications in food nanotechnology. J Liposome Res 18:309–327
24. Li D, Yang K, Li JS, Ke XY, Duan Y, Du R, Song P, Yu KF, Ren W, Huang D et al (2012) Antitumor efficacy of a novel CLA-PTX microemulsion against brain tumors: in vitro and in vivo findings. Int J Nanomedicine 7:6105–6114
25. Kim HJ, Kim NC, Wang YD, Scarborough EA, Moore J, Diaz Z, MacLea KS, Freibaum B, Li S, Molliex A et al (2013) Mutations in prion-like domains in hnRNPA2B1 and hnRNPA1 cause multisystem proteinopathy and ALS. Nature 495:467–473
26. Ravichandran R, Venugopal JR, Mueller M, Sundarrajan S, Mukherjee S, Pliska D, Wintermantel E, Ramakrishna S (2013) Buckled structures and 5-azacytidine enhance cardiogenic differentiation of adipose-derived stem cells. Nanomedicine (Lond) 8:1985–1997
27. Kruss S, Hilmer AJ, Zhang J, Reuel NF, Mu B, Strano MS (2013) Carbon nanotubes as optical biomedical sensors. Adv Drug Deliv Rev 65:1933–1950
28. Viswanathan V, Wang FY (2012) Theoretical analysis of the effect of particle size and support on the kinetics of oxygen reduction reaction on platinum nanoparticles. Nanoscale 4:5110–5117
29. Subramanian S, Singireddy A, Krishnamoorthy K, Rajappan M (2012) Nanosponges: a novel class of drug delivery system – review. J Phar Pharm Sci 15:103–111
30. Lu YJ, Yang HW, Hung SC, Huang CY, Li SM, Ma CC, Chen PY, Tsai HC, Wei KC, Chen JP (2012) Improving thermal stability and efficacy of BCNU in treating glioma cells using PAA-functionalized graphene oxide. Int J Nanomedicine 7:1737–1747
31. Li Z, Wu X, Li J, Yao L, Sun L, Shi Y, Zhang W, Lin J, Liang D, Li Y (2012) Antitumor activity of celastrol nanoparticles in a xenograft retinoblastoma tumor model. Int J Nanomedicine 7:2389–2398
32. Zavaleta CL, Garai E, Liu JT, Sensarn S, Mandella MJ, Van de Sompel D, Friedland S, Van Dam J, Contag CH, Gambhir SS (2013) A Raman-based endoscopic strategy for multiplexed molecular imaging. Proc Natl Acad Sci U S A 110: E2288–E2297
33. Tobin LA, Xie Y, Tsokos M, Chung SI, Merz AA, Arnold MA, Li G, Malech HL, Kwong KF (2013) Pegylated siRNA-loaded calcium phosphate nanoparticle-driven amplification of cancer cell internalization in vivo. Biomaterials 34:2980–2990
34. Chiang PC, Deng Y, Ubhayaka S, La H, Cui Y, Chou KJ, Ran Y, Wong H (2012) Novel nanoparticles formulation for cassette dosing via intravenous injection in rats for high throughput pharmacokinetic screening and potential applications. J Nanosci Nanotechnol 12:7993–8000
35. Lathrop JT, Hayes TK, Carrick K, Hammond DJ (2005) Rarity gives a charm: evaluation of trace proteins in plasma and serum. Expert Rev Proteomics 2:393–406
36. Worsley GJ, Attree SL, Noble JE, Horgan AM (2012) Rapid duplex immunoassay for wound biomarkers at the point-of-care. Biosens Bioelectron 34:215–220
37. Merenbloom BK, Oberhardt BJ (1995) Homogeneous immunoassay of whole-blood samples. Clin Chem 41:1385–1390
38. Bokenkamp A, Franke I, Schlieber M, Duker G, Schmitt J, Buderus S, Lentze MJ, Stoffel-Wagner B (2007) Beta-trace protein–a marker of kidney function in children: "original research communication-clinical investigation". Clin Biochem 40:969–975

39. Kristensen K, Wide-Swensson D, Schmidt C, Blirup-Jensen S, Lindstrom V, Strevens H, Grubb A (2007) Cystatin C, beta-2-microglobulin and beta-trace protein in pre-eclampsia. Acta Obstet Gynecol Scand 86:921–926
40. Luchini A, Fredolini C, Espina BH, Meani F, Reeder A, Rucker S, Petricoin EF 3rd, Liotta LA (2010) Nanoparticle technology: addressing the fundamental roadblocks to protein biomarker discovery. Curr Mol Med 10:133–141
41. Luk BT, Fang RH, Zhang L (2012) Lipid- and polymer-based nanostructures for cancer theranostics. Theranostics 2:1117–1126
42. Chan CP, Mak WC, Cheung KY, Sin KK, Yu CM, Rainer TH, Renneberg R (2013) Evidence-based point-of-care diagnostics: current status and emerging technologies. Annu Rev Anal Chem 6:191–211
43. Tripp RA, Alvarez R, Anderson B, Jones L, Weeks C, Chen W (2007) Bioconjugated nanoparticle detection of respiratory syncytial virus infection. Int J Nanomedicine 2:117–124
44. Sonnenberg A, Marciniak JY, Krishnan R, Heller MJ (2012) Dielectrophoretic isolation of DNA and nanoparticles from blood. Electrophoresis 33:2482–2490
45. Yeo WH, Chung JH, Liu Y, Lee KH (2009) Size-specific concentration of DNA to a nanostructured tip using dielectrophoresis and capillary action. J Phys Chem B 113:10849–10858
46. Thiollet S, Higson S, White N, Morgan SL (2012) Investigation and development of quantum dot-encoded microsphere bioconjugates for DNA detection by flow cytometry. J Fluoresc 22:685–697
47. Bunzli J-CG, Piguet C (2005) Taking advantage of luminescent lanthanide ions. Chem Soc Rev 34:1048–1077
48. Richardson FS (1982) Terbium(III) and europium(III) ions as luminescent probes and stains for biomolecular systems. Chem Rev 82:541–552
49. Sabbatini N, Guardigli M, Manet I, Ungaro R, Casnati A, Ziessel R, Ulrich G, Asfari Z, Lehnd J-M (1995) Lanthanide complexes of encapsulating ligands: luminescent devices at the molecular level. Pure Appl Chem 67:135–140
50. Tang S, Moayeri M, Chen Z, Harma H, Zhao J, Hu H, Purcell RB, Leppla S, Hewlett IK (2009) Detection of anthrax toxin by an ultrasensitive immunoassay using Europium nanoparticles. Clin Vaccine Immunol 16:408–413
51. Wahl R (2006) Star trek meets the operating room: nurse scotty, beam me up-A tricorder. Ann Surg Oncol: 2
52. Davies P (2001) The origin of life. I: when and where did it begin? Sci Prog 84:1–16
53. Buckey JC Jr (1999) Preparing for Mars: the physiologic and medical challenges. Eur J Med Res 4:353–356
54. Waters H (2011) New $10 million X prize launched for tricorder-style medical device. Nat Med 17:754
55. McKenna P (2013) The quest to put a doctor in your pocket. New Sci 218:46–49
56. Zanni GR (2013) Medical apps worth having. Consult Pharm 28:322–324
57. Boyd AD, Naiman M, Stevenson GW, Preston R, Valenta AL (2013) Technical and operational users' opinions of a handheld device to detect directed energy. Aviat Space Environ Med 84:528–533
58. Kwon JT, Hwang SK, Jin H, Kim DS, Minai-Tehrani A, Yoon HJ, Choi M, Yoon TJ, Han DY, Kang YW et al (2008) Body distribution of inhaled fluorescent magnetic nanoparticles in the mice. J Occup Health 50:1–6
59. Antunes AM, Alencar MS, da Silva CH, Nunes J, Mendes FM (2012) Trends in nanotechnology patents applied to the health sector. Recent Pat Nanotechnol 6:29–43
60. Demetzos C, Pippa N (2013) Advanced drug delivery nanosystems (aDDnSs): a mini-review. Drug Deliv 21:250–257

Nanotoxicity and Cellular Stress Response: Physical and Chemical Properties and Their Link to Translational Research

Nicole M. Schaeublin, Kristen K. Comfort, John J. Schlager, Shashi Bala Singh, and Saber M. Hussain

Abstract

The incorporation of nanomaterials (NMs) in applications and consumer products has grown exponentially over the past 5 years. The distinctive physical and chemical properties of NMs make them ideal candidates for integration into countless products and for application into novel technical approaches for translational research, including advanced electronics, sensors, in situ cellular and tissue imaging, targeted drug delivery, and energetic/reactive systems. NMs can significantly differ in their chemical/physical properties compared to bulk-produced materials. These material property differences produce difficulties in creating a logical prediction of effect/response from exposure at the organism, organ, cellular, and subcellular level. Due to many NMs being stable, solid-phase materials in aqueous systems, their unique surface physical and chemical properties produce many novel activities that disrupt cellular behavior and function. NM may change the cell membrane's ability to serve as a barrier, block a receptor's capability to become stimulated, or disrupt a critical subcellular physiological pathway necessary for proper cell function. As such, it is important to explore the resultant system toxicity and, in turn, determine the key attributes of a nanomaterial's deleterious effects and those that create true biocompatibility before material introduction to public use. The focus of this chapter is to review and demonstrate the importance of evaluating the biocompatibility of these materials. Data and arguments presented here focus to link material physical and chemical characteristics to specific cellular responses. It will be only through careful material synthesis and building strong fundamental understanding of materials in complex and dynamic bioeffects that NMs will be safely incorporated into broad target applications.

N.M. Schaeublin • J.J. Schlager • S.M. Hussain (✉)
Molecular Bioeffects Branch, Bioeffects Division, 711 Human Performance Wing, Human Effectiveness Directorate, Air Force Research Laboratory, Wright Patterson AFB, Dayton, OH 45433, USA
e-mail: saber.hussain@us.af.mil

K.K. Comfort
Department of Chemical and Materials Engineering, University of Dayton, Dayton, OH, USA

S.B. Singh
Defence Institute of Physiology and Allied Sciences (DIPAS), New Delhi, India
e-mail: drshashisingh@gmail.com

S.B. Singh et al. (eds.), *Translational Research in Environmental and Occupational Stress*, DOI 10.1007/978-81-322-1928-6_7, © Springer India 2014

Introduction

Nanomaterials (NMs) have structural features with at least one dimension of 100 nm or less. These materials possess unique physicochemical and biological properties that distinguish them from their bulk counterparts making them ideal for novel advanced science studies and use in advanced engineered systems focused to enhance human health and performance. Currently, NMs are used to make UV-resistant packaging and are incorporated into paint and textiles to inhibit bacterial and fungal growth [1]. NMs have been added to surfaces to make them scratch resistant and provide optically transparent protection from UV radiation [2], and many applications are now in cosmetics, sunscreen, and other skin care products [1]. Several military applications for NMs exist including addition to body armor to enhance resistance to stab impacts [3], bandages containing nanosilver as antibiotic, and negatively charged NMs particles lined in bandages to trigger and augment rapid blood clotting [4]. In basic protein biochemistry and function, NM particles are used to track complex molecular events such as protein interactions, aggregation, and folding elucidated by their colorimetric sensor properties [5, 6]. Nanoparticles (NPs) are advancing cell biology allowing detailed assessment of cellular structures when covalently conjugated to antibodies for cellular microscopy, and these same NP characteristics are exploited for precise in situ imaging for cancer diagnosis [7]. Although the range of applications of synthetically derived NM is growing, there is a lack of information concerning their impact on human health and effects to the environment following bulk distribution and purposeful or untoward material release. Limited studies exist for toxicity risk assessment and for defining the mechanism of response in mammalian systems, their cells, tissues, and organs. Due to their small dimensional size and unique chemical properties, NMs have the potential to affect mammalian physiology at the molecular level affecting cellular response and in turn disrupting and transforming cellular functions through blocking and initiating compensatory pathways associated with their presence or imparting novel physicochemical activity. In this way, the same properties that make NM beneficial in numerous applications may also be responsible for adverse effects in the cell.

NM physiochemical properties linked to cellular responses can be broken down into four main material groups that include size, shape, nanoparticle chemical composition, and surface chemistry (Fig. 1). Furthermore, within each of these groups,

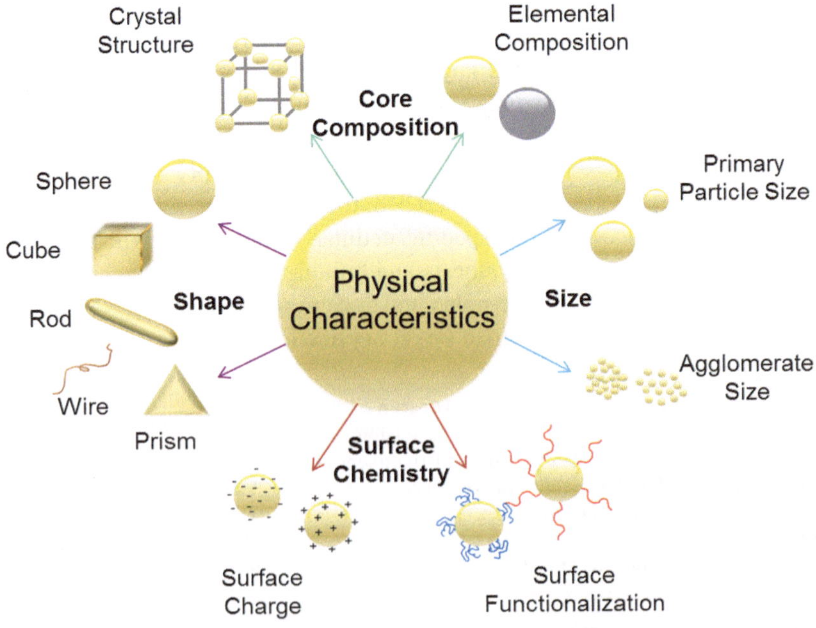

Fig. 1 Four main classification groups of nanomaterial physicochemical properties and their subcategories

multiple classifications occur, introducing more specialized characteristics for examination of NM properties. For example, when addressing material size, both primary particle dimension and material agglomeration propensity play very specific roles in how eukaryotic cells interact with NM. The surface area-to-volume (SA/V) ratio is important for standoff detection based upon NM optical properties. Nanosized particle engineering produces tremendously large SA/V ratios that affect particle diffusion and material chemistry producing increases in bioexposure to solid-phase material and chemistry dosimetry. The shape of the NM also plays a major role in inter-particle forces, behavior in solution, and interactions with a cellular environment. This especially comes into play with noble metallic NM made of gold and silver, which possess unique plasmon-resonance properties, producing distinctive spectral signatures that are applied in construction of sensors, optics, and electronics [8, 9]. NM size, shape, and aspect ratio (AR) designed and synthesized precisely have been tuned to provide a particular optical signature specified by targeted application. Examples include targeting for photothermal ablation of tumors and sensor development that enhances bio-imaging techniques [10].

The third main NM property is the elemental and molecular chemical composition. This includes the core composition as well as the crystal matrix surface structure created during NM solid-phase synthesis. Some NMs are inherently more toxic than others simply based on their chemical core composition. Evidence now shows that this effect is partly due to the dissociation of ions from once thought insoluble materials following particle disbursement in aqueous solutions. Ionization characteristics of these materials change depending on solutes in solution such as complex solutions made for cell culture [11]. As with all material, purity affects NM behavior. Depending on the synthetic method used for preparation, NM may include catalysts and residual impurities, which even at very small molar amounts can alter biocharacteristics that lead to differences in cellular/NM interaction [12]. An example of this is the metal catalysts used to initiate condensation of carbon in nanotube production, where the trace iron catalyst dominates toxicity by oxygen free radical generation.

The last group of physicochemical properties that can play a major role in cellular/NM interactions is surface chemistry including reactivity and catalysis. For example, surfactants are used in synthesis of certain NMs to control dispersion essentially stabilizing the particles to obtain a desired dimensional size. These surface treatments modify the physicochemical characteristics such as solubility and stability of the NMs in both aqueous and organic solution [13, 14]. However, the surfactants can impart their own chemical toxicity to the cells. Many new approaches for creating coatings masking the core of the chemicals only permit NM biostabilities for a short time after biological dosing. For toxicity studies, procedures should address the combined assessment of both surfactant and NM and approaches to remove the surfactant for determining the NM's toxicity for its overall biocompatibility [15, 16]. Additionally, the NM surfaces can be chemically functionalized with countless covalent, chelated, or polymeric chemical and/or biological surface features making their surface extremely functionally diverse. Synthetic engineering can attach site-directed targets for particular tissues, cells, or suborganelles in the body, frequently using specific biopolymer molecules such as RNA, DNA, or proteins [7]. The addition of ligands changes the overall and regional NM surface charge, producing modulation of material and chemical affinities with differing propensities of chemical interactive behavior. In summary, all of these aforementioned physicochemical properties have the potential to dictate the means, mechanism, and strength of the interactions between NMs and biological systems. With small modifications to NM properties, dramatic changes in bioresponses have been discovered. It remains imperative to study NM properties and effects prior to their incorporation into translational research, in order to first understand NM-cellular interactions and toxicological responses.

Primary Particle Size

As one can imagine, size is by far the most significant factor when examining the unique physicochemical properties of NMs and their biological interactions [17]. The size of a material at the nano-level creates an SA/V ratio which greatly alters a material's optical properties and provides a tremendous driving force for particle diffusion [18]. Additionally, NM size can affect the extent and mechanism of NM uptake, which traditionally occurs through a combination of endocytic mechanisms [19, 20]. Silver and gold are two of the most abundantly employed NMs used to study size effects for numerous reasons, including ease of in-lab synthesis, the ability to control precise particle size, and their unique plasmonic properties. These reasons contribute to the fact that both gold and silver are the two NMs most used in translational research and are the focus of this Chapter.

Silver Size Studies

Silver nanoparticles (Ag NPs) are of particular interest as they possess antimicrobial properties and as such are the most abundant NM used in the biomedical fields and consumer products today [21–23]. AgNPs are incorporated into numerous medical devices such as catheters, wound dressings, and antiseptic creams and consumer products including socks, T-shirts, hand sanitizer, cosmetics, and toothpaste [24–26]. AgNPs also possess unique plasmon-resonance optical scattering properties that make them ideal for use in sensors, optics, and electronics [8, 9]. However, despite their widespread use, the chemical and bio-behavior of nanosilver material in a biological system have yet to be fully elucidated. Numerous published studies indicate that primary particle size is a predominant factor in the dependent effects of silver NMs. For example, in a study by Braydich-Stolle et al., small-sized silver spheres (15 nm) were reported to be extremely toxic to the germ line stem cell [27]. Size-dependent toxicity of silver nanospheres was also demonstrated in studies by Hussain et al. and Carlson et al. that found smaller particles were the most toxic and generated higher amounts of reactive oxygen species (ROS), which was most likely the cause of cell death [28, 29]. However, as stated earlier, nanoparticles with the same chemical makeup are not equal in toxicity. One study revealed that 20-nm Ag-lysozyme-linked particles were antimicrobial at a dose that was not toxic to a cultured keratinocyte cell model [23]. Additionally, Speshock et al. found that 10-nm Ag NPs were able to inhibit host cell replication of arena virus at concentrations that were not toxic to the host cells [30]. These results along with other emerging findings in key studies may provide strong cross meta-analyses to determine the many factors affecting Ag NP toxicity.

Gold Size Studies

Like silver, gold is a noble metal that possesses unique optical properties that are tunable, based on size and shape. These distinctive properties have led to their inclusion in extensive biological applications including sensors, drug and gene delivery, photothermal therapy, and contrast agents for imaging [31–35]. Because gold (Au) NPs are relatively biochemically inert compared to silver, Au has been used in many biological and medical applications with the assumptions that no significant negative bioeffects occur when used. However, one must definitively study the potential untoward effects on biological systems such as those coming from repeated, long-term, or long resident time material exposure. Currently, conflicting data are available in the literature regarding the cytotoxicity of Au NPs. A number of studies have demonstrated the biocompatibility of these materials and suggest that exposure to tested levels used is not a concern. For example, Shukla et al. showed that 3.5-nm Au NPs were not toxic to cultured macrophages, and exposure did not initiate stress-induced secretions of the proinflammatory cytokines TNFα or IL-1β [36]. The biocompatibility of Au NPs is explored and annotated in great detail in a recent review by Murphy et al. [37]. In contrast, a more recent publication by Abdelhalim and Jarrar found size-dependent toxicity in rat liver tissue after exposure to different-sized AuNPs. The authors determined that this size-dependent toxicity was caused by an increase in sinusoidal Kupffer cells and an increase in fat deposition in the hepatocytes, producing necrosis of parenchymal cells [38]. Moreover, several studies have revealed size-dependent toxicity of AuNPs using several different cultured cell phenotypes with smaller particles being the most toxic [39–41].

Agglomerate Size

Recently our laboratory examined the size-dependent effects of three different silver particles (20, 40, and 80 nm) on human keratinocytes (unpublished data). Conflicting from past results, the smallest primary size silver (20 nm) particles were less toxic to cells than the larger NPs were. The 40-nm particles caused the most cell death, generated significant amounts of ROS, and changed the expression of more genes involved in stress and toxicity. However, upon further investigation, dynamic light scattering (DLS) data revealed that the 40-nm particles agglomerated the least in the media, making the final complex the smallest size material presented to cultured cells as an NM dose (Table 1). When accounting for stable agglomerated Ag NPs in the toxicology analysis, the solution agglomerates again exhibited size-dependent toxicity with the smallest agglomerate generating the most ROS and thereby being the most toxic to cells. Another study by Moulton et al. prepared Ag NPs through green chemical synthesis, which created a primary particle size of ~10 nm. Those 10-nm Ag NPs were nontoxic to human keratinocytes, despite numerous past studies using other Ag NP synthesis methods that produced 10-nm NP that were extremely cytotoxic to cells [27–29, 42]. Again, different agglomeration states were determined to be the reason for the differences. The green synthesized Ag NPs agglomerated to comparatively large micron-sized particles in media and were consequently excluded from the cells. In another study, Braydich-Stolle found that two different-sized TiO_2 particles, 10 and 100 nm, both agglomerated to the same size of approximately 1,800 nm in cell culture media [43]. This similar agglomeration caused an analogous same decrease in cell proliferation at 10 and 25 μg/ml doses. However, this cytotoxic effect disappeared at concentrations at or higher than 50 μg/ml dose, which produced exposure concentration levels of varied agglomeration sizes. Therefore, this report was one of the first to suggest that NMs primary and agglomerated sizes combine for toxicity and require consideration when performing size-dependent NM studies.

Nanomaterial Shape

Several studies show that the shape of the NM has an effect on cellular interactions. A study comparing Au nanosphere (NS) and Au (nanorod) NR with similar diameters (20 nm) revealed that Au NR reduced cellular viability by ~45 % at 25 μg/mL, whereas treatment with Au NS did not have any effect on cell viability, even at the highest dose of 100 μg/mL [44]. Additionally, the Au NR generated a significant increase in ROS and caused a depolarization of the mitochondrial membrane potential, while the Au NS did not produce similar deleterious effects. Furthermore, the AuNR demonstrated

Table 1 *Summary of silver characterization*. The 40-nm particles agglomerated the least in the media as compared to the 20 and 80 nm

			Z-average particle diameter (nm)	
Nanomaterial	Diameter size (mu)	Actual diameter size by TEM (nm)	Dispersed in water	Dispersed in serum free media
Silver spheres	20	20.3 ± 1.9	27.0 ± 10.8	1,130 ± 451
	40	42.0 ± 3.5	42.0 ± 5.0	278 ± 169
	80	79.8 ± 5.1	79.0 ± 25.2	881 ± 453

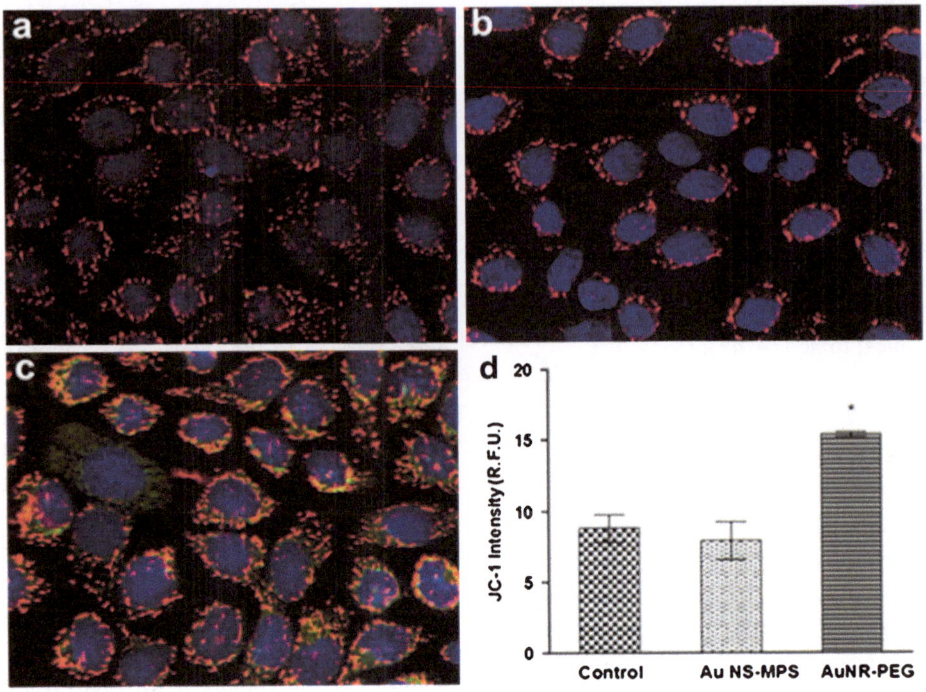

Fig. 2 *Evaluation of mitochondrial membrane potential (MMP) following gold nanomaterial exposure*. This assay uses the dye JC-1, which enters healthy mitochondria, aggregates, and fluoresces red. Upon disruption of the mitochondrial membrane, the dye disperses and fluoresces green throughout the cell, indicating apoptotic cells. (**a**) Control cells (untreated), (**b**) AuNS-MPS, (**c**) AuNR-PEG. There was significant amount of MMP lost after exposure to the AuNR-PEG. (**d**) Graph depicting the data shown in the images (*denotes significance in comparison to control values $p < 0.05$)

considerable upregulation of genes involved in cellular damage and stress (Fig. 2). A recent in vivo study by George et al. found that silver nanoplates were more toxic to fish gill epithelial cells and zebrafish embryos when compared to Ag spheres and Ag wires [45]. Additionally, a separate study found that while Ag nanowires were not toxic to a lung coculture model at low doses, they did cause a disruption to the mitochondrial membrane potential and an inflammatory response in test system [46]. Taken together, these data suggest that nanomaterial shape plays another key role in material NP-cell interactions.

Core Composition

As the importance of NM size emerged in toxicity responses and significant advancement occurred in the area of reproducible NM synthesis, researchers began to compare different NMs with the same sizes but different core compositions. Several studies have demonstrated that several NMs possess a strong cytotoxic potential, based solely on their core composition. A review by Schrand et al. provides a summary of the studies that explored core composition in the toxicity of NMs [47]. For example, a study by Hussain et al. demonstrated that AgNPs were more toxic to BRL 3A cells than MoO_3, Al, Fe_3O_4, and TiO_2, while Wang et al. determined that Cu and Mn NPs were more toxic than Al to PC12 cells [28, 48]. Jeng and Swanson revealed that ZnONPs were more toxic to N2A cells than TiO_2, Fe_3O_4, Al_2O_3, and CrO_3, and Sayes et al. showed that ZnO NPs were the most toxic to rat epithelial cells, rat primary alveolar macrophages, and cocultures of both, compared to equal dose amounts of Fe and SiO_2 NPs [49, 50]. These studies provide evidence that silver, copper, zinc, manganese, aluminum, iron, and chromium NPs as metal or their oxide forms are toxic to various cell types. More recent studies revealed that the higher toxic NPs are frequently those with the greater kinetic rate of ionic dissociation in solution, resulting in a strong induction of metal ion ROS induction and cellular distress [28, 48–52].

Surface Chemistry

NM surface chemistry is extremely important in predicting and dictating how a cell will respond to NM exposure. Several very general classes of surface moieties exist, including surfactants, biomolecules, chemical agents, and ligands to modify surface charge. For example, cetyltrimethylammonium bromide (CTAB) is a cationic surfactant used in the synthesis of AuNRs, which acts as a capping agent to control the size and the shape of AuNRs during solution synthesis. Several studies have found that direct use of CTAB caused significant cytotoxicity, and its removal resulted in a destabilization of AuNR suspension [13–16]. Chemical modification by direct functionalizing of the AuNR surface to create a thiol bond or exchanging the CTAB with a less toxic ligand such as polyethylene glycol (PEG) can eliminate CTAB as a toxic surface coating [13–16, 53–60]. Moreover, Grabinski et al. showed that both gold nanorod (GNR)-PEG and GNR-mercaptohexadecanoic acid (MHDA) exhibited minimal effects on cell proliferation and that the GNR-PEG induced less significant and unique changes in gene expression related to stress and toxicity than GNR-MHDA induced [60]. This study demonstrated that while AuNRs are not toxic, NM coating has a profound effect on cellular responses.

Another key component that falls under the surface chemistry category is surface charge. Recent studies suggest that the surface charge of the NP plays a critical role in the initial binding to the cell membrane and subsequent internalization into cells that provides a strong link to this physicochemical state to the resultant cytotoxicity. For example, according to Orr and colleagues, positively charged particles bind to cytoskeletal structures within the actin machinery, traveling along filopodia and microvilli-like structures on the cell body, but the same particles bearing an overall net negative charge did not bind to these cellular structures [61]. Gratton et al. showed that positively charged cylindrical nanoparticles were taken up into HeLa cells after 1 h to a much greater extent than their corresponding anionic particles [62]. Conversely, Wilhelm et al. showed that anionic maghemite nanoparticles demonstrated a high level of internalization by interacting strongly and nonspecifically with the cell membrane [63]. Goodman et al. discovered that the electrical charge of the nanoparticle plays a key role in determining toxicity, with cationic Au NPs displaying moderate toxicity, while their anionic counterparts exhibit no toxic effects [64].

Interestingly, another study demonstrated that surface charge of Au NPs mediates the mechanism of toxicity [65] finding that 1.5-nm Au NPs with different surface charges (positive, negative, and neutral) disrupted cell morphology and a dose-dependent toxicity found in charged Au NPs displayed low-dose (10 μg/ml) toxicity compared to neutral Au NPs (25 μg/ml). Significant mitochondrial stress (decreases in MMP and intracellular Ca^{++} levels) was found following charged Au NP exposure, but not with neutral Au NPs. Lastly, differences in regulation of DNA damage-related gene expression suggested a differential cell death mechanism found based on whether or not Au NPs were charged or neutral (Fig. 3).

Surface Functionalization

Several studies have shown that altering the surface chemistry of nanoparticles has been effective in preventing toxicity from the core NM chemicals [44, 54, 66–69]. However, some studies suggest that the stability of these coatings may degrade and fluctuate, depending upon time of composition and residency media and cell or tissue regional residency. For example, spermatogonial stem cells treated with Ag NPs containing hydrocarbon (HC) and polysaccharide (PS) surface coatings initially demonstrated differential biocompatibility, with HC-coated Ag being toxic to the cells and the PS-coated Ag

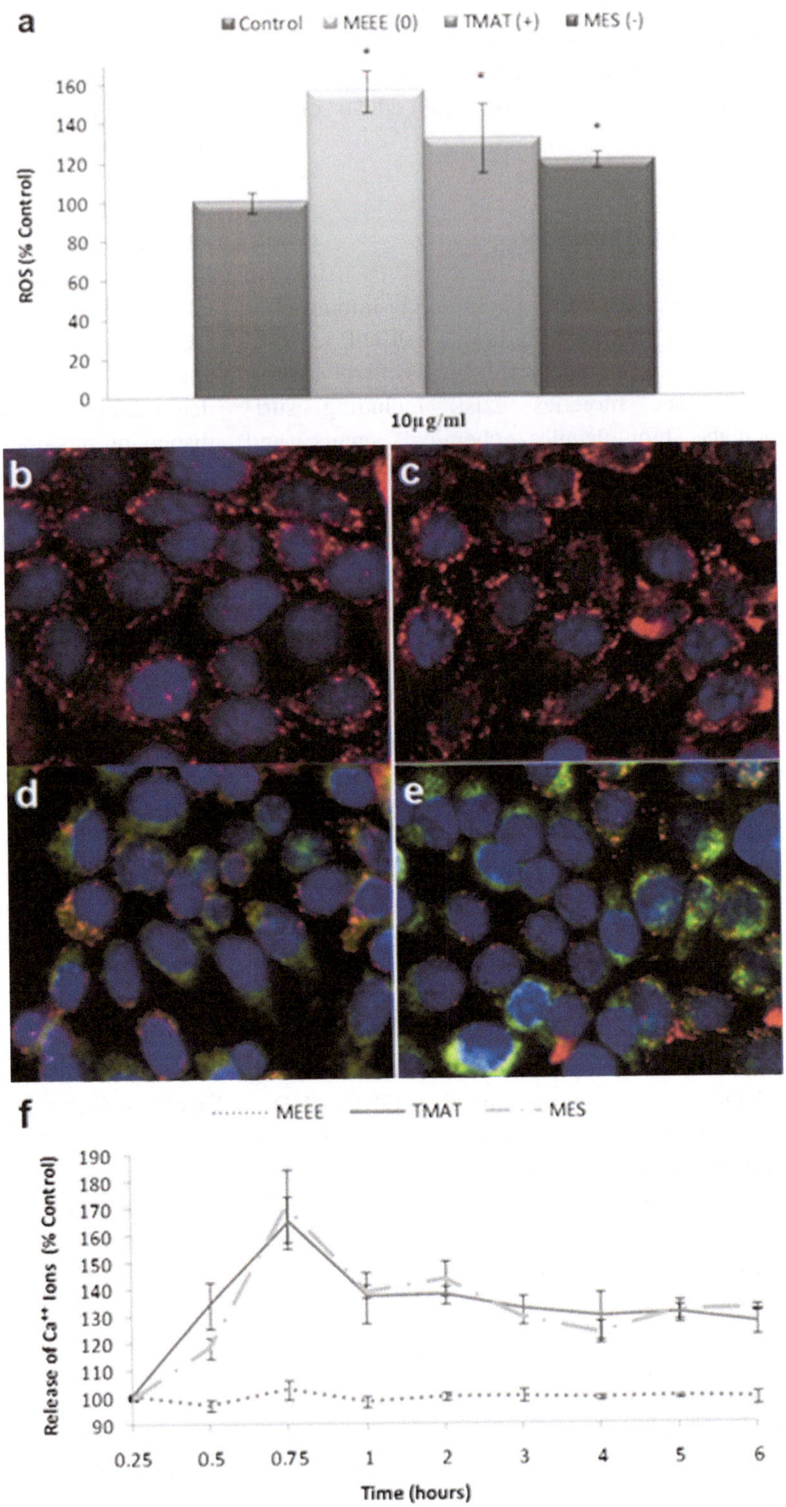

Fig. 3 *Evaluation of mitochondrial stress following gold nanoparticle exposure.* (**a**) The AuNPs generated significant amounts of ROS, regardless of charge (*denotes significance in comparison to control values $p < 0.05$). (**b–e**) Loss of mitochondrial membrane potential after exposure to 25 μg/ml AuNPs. (**b**) Control (**c**) MEEE (0), (**d**) TMAT (+), (**e**) MES (−). There was significant amount of MMP lost after exposure to both the charged AuNPs. (**f**) Release of calcium ions into cytosol after exposure to 25 μg/ml AuNPs. There was a significant amount of Ca^{++} ions released into the cytosol after exposure to both the charged particles but not the neutral particles

being nontoxic [70]. However, after 3 days of exposure, the PS Ag biocompatibility was lost and cell viability lowered to the same levels as cells treated with the HC Ag. Similarly, Schrand et al. demonstrated that the same PS-coated Ag was more biocompatible in neuroblastoma cells. However, after treatment with nerve growth factor to induce proliferation of the cells, both types of Ag prevented the NGF from inducing a growth response. Although the mechanism is yet to be discovered, this result may likely be an NP membrane receptor blocking effect, where the highest NP concentrations are found [9]. Studies by Freese et al. and Uboldi et al. showed that sodium citrate AuNPs were toxic to alveolar type II-like human epithelial cells, A549, NCIH441, and endothelial cells and that increased amounts of sodium citrate on the AuNP surface increased uptake levels in endothelial cells [71, 72]. These studies demonstrate the need for additional research on the role NM surface coatings play in cellular response and on the importance of NM surface composition and chemistries that can change over time of exposure. When engineering biocompatibility NMs, one must consider the surface chemistry, the coating, and the desorption dynamics to best investigate and understand a likely plethora of biological effects.

Current Challenges

So far, this chapter has presented some original data and topics from the literature involving different physical and chemical properties of NM and that NP exposure can produce diverse affects to a biocellular environment. One of the major challenges that nanotoxicology faces today is that many of these original biocompatibility tests were performed under very high, unrealistic exposure NM doses. Current investigations are now exploring lower, more realistic dosages of NMs, which are equivalent to an individual daily exposure use based upon work or an operation basis. Some of these studies have demonstrated that the lack of frank cell death does not automatically imply that exposure is completely without affect. A recent study demonstrated that although aluminum and aluminum oxide nanoparticles were not toxic to a lung coculture model, the aluminum NPs prevented this culture from generating a normal immune response when infected with bacteria. This apparent nontoxic NM exposure produced the diminution or loss of immune cell function requiring further exploration [73]. Another study showed that low-level, nontoxic exposures to gold, silver, and iron oxide caused a disruption in the EGF signaling, though each NP type appeared to act through independent mechanisms [74]. Li et al. showed that while 20-nm Au NPs were not cytotoxic to lung fibroblasts, they did produce significant amounts of oxidative DNA damage and downregulated the expression of DNA damage and cell cycle genes [75]. Additionally, other studies have demonstrated that NM can induce immune responses and can cause inflammation without causing significant cell death [76, 77]. Lastly, recent investigations have explored the combinatorial bioeffects of NMs at low dosages in conjunction with a magnetic field and demonstrated that co-exposure scenarios [78] influence the resultant bioeffects. Continued exploration into new bioeffects research remains critical for providing realistic exposure evaluation for human use exposure scenarios. Only after further significant study will data allow establishment of regulatory exposure limits and ascertain the safety levels for NM inclusion into consumer and military applications.

Conclusions

Nanotechnology has emerged as one of the fastest-growing interdisciplinary fields in science today and promises to continue to revolutionize industry, medicine, and consumer products with its translation into additional areas such as information technology, food safety, agriculture, and energy. However, the very distinctive and exceptionally novel physicochemical properties found with NMs may also cause some unwanted and unintended side effects. Therefore, it is critical to address the potential environmental, health, and safety impacts of these innovative materials before their full incorporation into any technology, product, or application. Scientists are starting to address the most critical issues. However, because NMs are being incorporated into new products daily, difficulties remain to match the pace of this exponential growth and deployment of new products. The preliminary studies described in this chapter are provided as an initial primer to understand and address the complexity of the problem. The current data provided shows this diversity of toxicity responses and potential untoward effects imparted by NM exposure. We hope that this information will excite interest in others and show that there is still much more work required to understand how emerging designs of synthesized NMs can and will influence cellular responses.

References

1. Dutta S, Ganguly B (2012) Characterization of ZnO nanoparticles grown in presence of Folic acid template. J Nanobiotechnol 10(1):1–10
2. Ellsworth DK, Verhulst D, Spitler TM, Sabacky BJ (2000) Titanium nanoparticles move to the marketplace. Chem Innov 30(12):30–35
3. Mahfuz H, Clements F, Rangari V, Dhanak V (2009) Enhanced stab resistance of armor composites with functionalized silica nanoparticles. J Appl Phys 105(6):1–7
4. Hauqe N, Khalel RR, Parvez N, Yadav S, Hwisa N, Al-Sharif Ms, Awen BZ, Molvi K (2010) Nanotechnology in Cancer Therapy: Review. J Chem Pharm Res 2(5):161–168
5. Tsai C-S, Yu T-B, Chen C-T (2005) Gold nanoparticle-based competitive colorimetric assay for detection of protein-protein interactions. Chem Commun (34):4273–4275
6. Bhattacharya R, Patra CR, Earl A, Wang S, Katarya A, Lu L et al (2007) Attaching folic acid on gold nanoparticles using noncovalent interaction via different polyethylene glycol backbones and targeting of cancer cells. Nanomed Nanotechnol Biol Med 3(3):224–238
7. Ma H (ed) (1991) Colloidal gold: principles, methods, and applications. Academic, San Diego
8. Dubas ST, Pimpan V (2008) Humic acid assisted synthesis of silver nanoparticles and its application to herbicide detection. Mater Lett 62 (17–18):2661–2663
9. Schrand AM, Braydich-Stolle LK, Schlager JJ, Dai L, Hussain SM (2008) Can silver nanoparticles be useful as potential biological labels? Nanotechnology 19(23):235104, Epub 2008/06/11
10. Black KCL, Yi J, Rivera JG, Zelasko-Leon DC, Messersmith PB (2013) Polydopamine-enabled surface functionalization of gold nanorods for cancer cell-targeted imaging and photothermal therapy. Nanomedicine (Lond) 8(1):17–28
11. Maurer EI, Sharma M, Schlager JJ, Hussain SM (2014) Systematic analysis of silver nanoparticle ionic dissolution by tangential flow filtration: toxicological implications. Nanotoxicology 8(7):718–727, epub 1 Aug 2013
12. Carney RP, Kim JY, Qian H, Jin R, Mehenni H, Stellacci F et al (2011) Determination of nanoparticle size distribution together with density or molecular weight by 2D analytical ultracentrifugation. Nat Commun 2:335
13. Cortesi R, Esposito E, Menegatti E, Gambari R, Nastruzzi C (1996) Effect of cationic liposome composition on in vitro cytotoxicity and protective effect on carried DNA. Int J Pharm 139 (1–2):69–78
14. Mirska D, Schirmer K, Funari SS, Langner A, Dobner B, Brezesinski G (2005) Biophysical and biochemical properties of a binary lipid mixture for DNA transfection. Colloids Surf B Biointerfaces 40 (1):51–59
15. Connor EE, Mwamuka J, Gole A, Murphy CJ, Wyatt MD (2005) Gold nanoparticles are taken up by human cells but do not cause acute cytotoxicity. Small 1(3):325–327, Epub 2006/12/29
16. Takahashi H, Niidome Y, Niidome T, Kaneko K, Kawasaki H, Yamada S (2005) Modification of gold nanorods using phosphatidylcholine to reduce cytotoxicity. Langmuir 22(1):2–5
17. Nel A, Xia T, Madler L, Li N (2006) Toxic potential of materials at the nanolevel. Science 311 (5761):622–627, Epub 2006/02/04
18. Suganeswari M, Shering A, Bharathi MP, JayaSutha J (2011) Nano particles: a novel system in current century. Int J Pharm Biol Arch 2(3):847–854
19. Jiang W, Kim BY, Rutka JT, Chan WC (2008) Nanoparticle-mediated cellular response is size-dependent. Nat Nanotechnol 3(3):145–150, Epub 2008/07/26
20. Coulter JA, Jain S, Butterworth KT, Taggart LE, Dickson GR, McMahon SJ et al (2012) Cell type-dependent uptake, localization, and cytotoxicity of 1.9 nm gold nanoparticles. Int J Nanomedicine 7:2673–2685
21. Silver S, Phung L, Silver G (2006) Silver as biocides in burn and wound dressings and bacterial resistance to silver compounds. J Ind Microbiol Biotechnol 33(7):627–634
22. Chen X, Schluesener HJ (2008) Nanosilver: a nanoproduct in medical application. Toxicol Lett 176(1):1–12
23. Eby DM, Schaeublin NM, Farrington KE, Hussain SM, Johnson GR (2009) Lysozyme catalyzes the formation of antimicrobial silver nanoparticles. ACS Nano 3(4):984–994, Epub 2009/04/07
24. Dibrov P, Dzioba J, Gosink KK, Häse CC (2002) Chemiosmotic mechanism of antimicrobial activity of Ag(+) in Vibrio cholerae. Antimicrob Agents Chemother 46(8):2668–2670
25. Feng QL, Wu J, Chen GQ, Cui FZ, Kim TN, Kim JO (2000) A mechanistic study of the antibacterial effect

of silver ions on Escherichia coli and Staphylococcus aureus. J Biomed Mater Res 52(4):662–668

26. Lok C-N, Ho C-M, Chen R, He Q-Y, Yu W-Y, Sun H et al (2007) Silver nanoparticles: partial oxidation and antibacterial activities. J Biol Inorg Chem 12 (4):527–534
27. Braydich-Stolle L, Hussain S, Schlager JJ, Hofmann MC (2005) In vitro cytotoxicity of nanoparticles in mammalian germline stem cells. Toxicol Sci 88 (2):412–419, Epub 2005/07/15
28. Hussain SM, Hess KL, Gearhart JM, Geiss KT, Schlager JJ (2005) In vitro toxicity of nanoparticles in BRL 3A rat liver cells. Toxicol In Vitro 19 (7):975–983, Epub 2005/08/30
29. Carlson C, Hussain SM, Schrand A, Braydich-Stolle LK, Hess KL, Jones RL, Schlager JJ (2008) Unique cellular interaction of silver nanoparticles: size-dependent generation of reactive oxygen species. J Phys Chem B 112:13608–13619
30. Speshock JL, Murdock RC, Braydich-Stolle LK, Schrand AM, Hussain SM (2010) Interaction of silver nanoparticles with Tacaribe virus. J Nanobiotechnol 8:19, Epub 2010/08/20
31. Riboh JC, Haes AJ, McFarland AD, Ranjit Yonzon C, Van Duyne RP (2003) A nanoscale optical biosensor: real-time immunoassay in physiological buffer enabled by improved nanoparticle adhesion. J Phys Chem B 107(8):1772–1780
32. Ghosh P, Han G, De M, Kim CK, Rotello VM (2008) Gold nanoparticles in delivery applications. Adv Drug Deliv Rev 60(11):1307–1315
33. Pissuwan D, Valenzuela SM, Cortie MB (2006) Therapeutic possibilities of plasmonically heated gold nanoparticles. Trends Biotechnol 24(2):62–67
34. Wang Y, Xie X, Wang X, Ku G, Gill KL, O'Neal DP et al (2004) Photoacoustic tomography of a nanoshell contrast agent in the in vivo rat brain. Nano Lett 4 (9):1689–1692
35. Yelin D, Oron D, Thiberge S, Moses E, Silberberg Y (2003) Multiphoton plasmon-resonance microscopy. Opt Express 11(12):1385–1391
36. Shukla S, Priscilla A, Banerjee M, Bhonde RR, Ghatak J, Satyam PV et al (2005) Porous gold nanospheres by controlled transmetalation reaction: a novel material for application in cell imaging. Chem Mater 17(20):5000–5005
37. Murphy CJ, Gole AM, Stone JW, Sisco PN, Alkilany AM, Goldsmith EC et al (2008) Gold nanoparticles in biology: beyond toxicity to cellular imaging. Acc Chem Res 41(12):1721–1730
38. Abdelhalim MA, Jarrar BM (2012) Histological alterations in the liver of rats induced by different gold nanoparticle sizes, doses and exposure duration. J Nanobiotechnol 10:5, Epub 2012/01/27
39. Pan Y, Neuss S, Leifert A, Fischler M, Wen F, Simon U et al (2007) Size-dependent cytotoxicity of gold nanoparticles. Small 3(11):1941–1949, Epub 2007/10/30
40. Pernodet N, Fang X, Sun Y, Bakhtina A, Ramakrishnan A, Sokolov J et al (2006) Adverse effects of citrate/gold nanoparticles on human dermal fibroblasts. Small 2(6):766–773, Epub 2006/12/29
41. Coradeghini R, Gioria S, Garcia CP, Nativo P, Franchini F, Gilliland D et al (2013) Size-dependent toxicity and cell interaction mechanisms of gold nanoparticles on mouse fibroblasts. Toxicol Lett 217 (3):205–216, Epub 2012/12/19
42. Moulton MC, Braydich-Stolle LK, Nadagouda MN, Kunzelman S, Hussain SM, Varma RS (2010) Synthesis, characterization and biocompatibility of "green" synthesized silver nanoparticles using tea polyphenols. Nanoscale 2(5):763–770, Epub 2010/07/22
43. Braydich-Stolle LK, Schaeublin NM, Murdock RC, Jiang J, Biswas P, Schlager JJ et al (2008) Crystal structure mediates mode of cell death in TiO2 nanotoxicity. J Nanoparticle Res 11(6):1361–1374
44. Schaeublin NM, Braydich-Stolle LK, Maurer EI, Park K, MacCuspie RI, Afrooz AR et al (2012) Does shape matter? Bioeffects of gold nanomaterials in a human skin cell model. Langmuir 28(6):3248–3258, Epub 2012/01/17
45. George S, Lin S, Ji Z, Thomas CR, Li L, Mecklenburg M et al (2012) Surface defects on plate-shaped silver nanoparticles contribute to its hazard potential in a fish gill cell line and zebrafish embryos. ACS Nano 6(5):3745–3759
46. Schaeublin NM, MK, Roberts JR, Hussain SM (2013) Silver nanowires induced inflammation in an in vitro human alveolar lung model. Unpublished data. Manuscript is in progress
47. Schrand AM, Rahman MF, Hussain SM, Schlager JJ, Smith DA, Syed AF (2010) Metal-based nanoparticles and their toxicity assessment. Wiley Interdiscip Rev Nanomed Nanobiotechnol 2(5):544–568, Epub 2010/08/04
48. Wang J, Rahman MF, Duhart HM, Newport GD, Patterson TA, Murdock RC et al (2009) Expression changes of dopaminergic system-related genes in PC12 cells induced by manganese, silver, or copper nanoparticles. Neurotoxicology 30(6):926–933, Epub 2009/09/29
49. Jeng HA, Swanson J (2006) Toxicity of metal oxide nanoparticles in mammalian cells. J Environ Sci Health A 41(12):2699–2711
50. Sayes CM, Reed KL, Warheit DB (2007) Assessing toxicity of fine and nanoparticles: comparing in vitro measurements to in vivo pulmonary toxicity profiles. Toxicol Sci 97:163–180
51. Chang Y-N, Zhang M, Xia L, Zhang J, Xing G (2012) The toxic effects and mechanisms of CuO and ZnO nanoparticles. Materials 5(12):2850–2871
52. Xia T, Kovochich M, Liong M, Madler L, Gilbert B, Shi H et al (2008) Comparison of the mechanism of toxicity of zinc oxide and cerium oxide nanoparticles based on dissolution and oxidative stress properties. ACS Nano 2(10):2121–2134, Epub 2009/02/12
53. Alkilany AM, Nagaria PK, Hexel CR, Shaw TJ, Murphy CJ, Wyatt MD (2009) Cellular uptake and cytotoxicity of gold nanorods: molecular origin of cytotoxicity and surface effects. Small 5(6):701–708

54. Hauck TS, Ghazani AA, Chan WC (2008) Assessing the effect of surface chemistry on gold nanorod uptake, toxicity, and gene expression in mammalian cells. Small 4(1):153–159, Epub 2007/12/18
55. Leonov AP, Zheng J, Clogston JD, Stern ST, Patri AK, Wei A (2008) Detoxification of gold nanorods by treatment with polystyrenesulfonate. ACS Nano 2(12):2481–2488
56. Wang S, Lu W, Tovmachenko O, Rai US, Yu H, Ray PC (2008) Challenge in understanding size and shape dependent toxicity of gold nanomaterials in human skin keratinocytes. Chem Phys Lett 463 (1–3):145–149
57. Parab HJ, Chen HM, Lai T-C, Huang JH, Chen PH, Liu R-S et al (2009) Biosensing, cytotoxicity, and cellular uptake studies of surface-modified gold nanorods. J Phys Chem C 113(18):7574–7578
58. Boca SC, Astilean S (2010) Detoxification of gold nanorods by conjugation with thiolated poly(ethylene glycol) and their assessment as SERS-active carriers of Raman tags. Nanotechnology 21(23):235601, Epub 2010/05/14
59. Niidome T, Yamagata M, Okamoto Y, Akiyama Y, Takahashi H, Kawano T et al (2006) PEG-modified gold nanorods with a stealth character for in vivo applications. J Control Release 114(3):343–347, Epub 2006/08/01
60. Grabinski C, Schaeublin N, Wijaya A, D'Couto H, Baxamusa SH, Hamad-Schifferli K et al (2011) Effect of gold nanorod surface chemistry on cellular response. ACS Nano 5(4):2870–2879
61. Orr G, Panther DJ, Phillips JL, Tarasevich BJ, Dohnalkova A, Hu D et al (2007) Submicrometer and nanoscale inorganic particles exploit the actin machinery to be propelled along microvilli-like structures into alveolar cells. ACS Nano 1(5):463–475, Epub 2007/12/01
62. Gratton SE, Ropp PA, Pohlhaus PD, Luft JC, Madden VJ, Napier ME et al (2008) The effect of particle design on cellular internalization pathways. Proc Natl Acad Sci U S A 105(33):11613–11618, Epub 2008/08/14
63. Wilhelm C, Billotey C, Roger J, Pons JN, Bacri JC, Gazeau F (2003) Intracellular uptake of anionic superparamagnetic nanoparticles as a function of their surface coating. Biomaterials 24(6):1001–1011
64. Goodman CM, McCusker CD, Yilmaz T, Rotello VM (2004) Toxicity of gold nanoparticles functionalized with cationic and anionic side chains. Bioconjug Chem 15(4):897–900
65. Schaeublin NM, Braydich-Stolle LK, Schrand AM, Miller JM, Hutchison J, Schlager JJ et al (2011) Surface charge of gold nanoparticles mediates mechanism of toxicity. Nanoscale 3(2):410–420, Epub 2011/01/14
66. Gole A, Murphy CJ (2005) Polyelectrolyte-coated gold nanorods: synthesis. Characterization and immobilization. Chem Mater 17(6):1325–1330
67. Qiu Y, Liu Y, Wang L, Xu L, Bai R, Ji Y et al (2010) Surface chemistry and aspect ratio mediated cellular uptake of Au nanorods. Biomaterials 31 (30):7606–7619
68. Zhang Y, Xu D, Li W, Yu J, Chen Y (2012) Effect of size, shape, and surface modification on cytotoxicity of gold nanoparticles to human HEp-2 and canine MDCK cells. J Nanomater 2012:7
69. Sun Y-N, Wang C-D, Zhang X-M, Ren L, Tian X-H (2011) Shape dependence of gold nanoparticles on in vivo acute toxicological effects and biodistribution. J Nanosci Nanotechnol 11(2):1210–1216
70. Braydich-Stolle LK, Lucas B, Schrand A, Murdock RC, Lee T, Schlager JJ et al (2010) Silver nanoparticles disrupt GDNF/Fyn kinase signaling in spermatogonial stem cells. Toxicol Sci 116 (2):577–589, Epub 2010/05/22
71. Freese C, Uboldi C, Gibson MI, Unger RE, Weksler BB, Romero IA et al (2012) Uptake and cytotoxicity of citrate-coated gold nanospheres: comparative studies on human endothelial and epithelial cells. Part Fibre Toxicol 9:23, Epub 2012/07/05
72. Uboldi C, Bonacchi D, Lorenzi G, Hermanns MI, Pohl C, Baldi G et al (2009) Gold nanoparticles induce cytotoxicity in the alveolar type-II cell lines A549 and NCIH441. Part Fibre Toxicol 6(1):18
73. Braydich-Stolle LK, Speshock JL, Castle A, Smith M, Murdock RC, Hussain SM (2010) Nanosized aluminum altered immune function. ACS Nano 4(7):3661–3670
74. Comfort KK, Maurer EI, Braydich-Stolle LK, Hussain SM (2011) Interference of silver, gold, and iron oxide nanoparticles on epidermal growth factor signal transduction in epithelial cells. ACS Nano 5(12):10000–10008
75. Li JJ, Zou L, Hartono D, Ong CN, Bay BH, Lanry Yung LY (2008) Gold nanoparticles induce oxidative damage in lung fibroblasts in vitro. Adv Mater 20 (1):138–142
76. Hutter E, Boridy S, Labrecque S, Lalancette-Hébert M, Kriz J, Winnik FM et al (2010) Microglial response to gold nanoparticles. ACS Nano 4(5):2595–2606
77. Dobrovolskaia MA, McNeil SE (2007) Immunological properties of engineered nanomaterials. Nat Nanotechnol 2(8):469–478
78. Comfort KK, Maurer E, Hussain SM (2013) The biological impact of concurrent exposure to metallic nanoparticles and a static magnetic field. Bioelectromagentics 34(7):500–511

HIF-1 and EGLN1 Under Hypobaric Hypoxia: Regulation of Master Regulator Paradigm

Aastha Mishra and M.A. Qadar Pasha

Abstract

High-altitude (HA) populations living for several thousands of years at elevations up to 5,000 m present many adaptive phenotypic changes; conversely, the lowland populations respond differentially like acclimatizing to the environment or predisposing to HA disorders. Hypobaric hypoxia environment at HA results into reduced blood arterial O_2 saturation in the body. It stimulates an array of physiological responses enabling the body to function optimally under this environment. Several of the physiological responses are regulated by hypoxia inducible factor-1 (HIF-1), a master transcription factor of O_2-sensing pathway. It regulates transcription of those genes that are required for either increasing O_2 availability or mediating responses to O_2 deprivation such as reduction of ATP turnover rate of the body. Under normoxic state, however, HIF-I is repressed by HIF-prolyl hydroxylase 2 (EGLN1), which in actuality hydroxylates HIF1α to facilitate its degradation. Furthermore, the adaptive phenotypes are the result of natural selection of genetic traits that counteract the effects of environmentally induced changes. Identification of variants in these genes may help in elucidating molecular mechanisms by which these two molecules function under hypobaric hypoxia. These elucidated mechanisms can be translated for practical applications to enhance the health management at HA. It may also help in designing new therapeutic targets, thereby transforming the basic knowledge into practical applications.

Introduction

High-altitude (HA), characterized by decreased atmospheric pressure (hypobaria) resulting in low partial pressure of oxygen (hypoxia), represents one of the extreme environments on Earth. Increase in altitude proportionately reduces the number of oxygen (O_2) molecules per breath, which in turn reduces the amount of O_2 availability to the blood and tissues in the body [1]. Thus, the hypobaric hypoxia environment at HA results into reduced blood arterial O_2 saturation (SaO_2) in the body [2]. SaO_2 is a percentage of the amount of hemoglobin

A. Mishra • M.A.Q. Pasha (✉)
Functional Genomics Unit, CSIR-Institute of Genomics and Integrative Biology, Delhi, India
e-mail: qpasha@igib.res.in

S.B. Singh et al. (eds.), *Translational Research in Environmental and Occupational Stress*,
DOI 10.1007/978-81-322-1928-6_8, © Springer India 2014

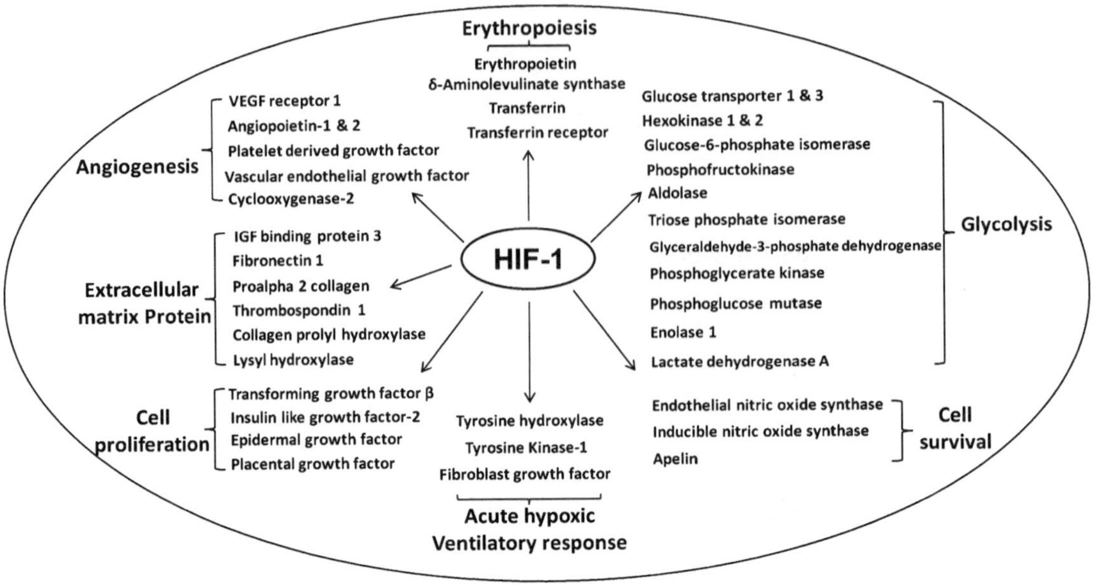

Fig. 1 Transcriptional activation of genes bearing hypoxia response element (HRE) by HIF-1 transcription factor. HIF-1 induces activation of all those genes that bear HRE in their promoters. Hence, HRE-bearing genes are also known as HIF-1-regulated genes. The HIF-1-regulated genes belong to the pathways like erythropoiesis, angiogenesis, glycolysis, extracellular matrix remodeling, and cell survival and proliferation to maintain vascular and adaptive homeostasis at HA

molecules which are oxygenated (oxyhemoglobin) in arterial blood [3]. Alleviated SaO_2 level directly affects O_2 transport cascade of the body and is used as a measure of stress to the underlying HA environment [4], which stimulates an array of physiological responses that enable the body to function optimally in the low O_2 environment [5]. In addition, variation in the percent of SaO_2 is an evidence of interindividual variations in ventilation and transfer of O_2 from air to the blood [4]. Hence, among the several characteristic features of the HA environment like extreme cold, low humidity, and solar and ionizing radiations, it is the hypobaric hypoxia that has the main effect on human beings and is the chief driving force for major acclimatization and adaptive processes at HA.

Oxygen-sensing pathway plays a cardinal role in maintaining the adaptive homeostasis by transcribing those genes that are required for either increasing O_2 availability of the body or mediating responses to O_2 deprivation such as reduction of ATP turnover rate [6]. Transcription of these genes is carried out by hypoxia-inducible factor-1 (HIF-1), a master transcription factor (TF) of hypoxia. HIF-1 induces several genes, to name few, glucose transporter-1, lactate dehydrogenase A, erythropoietin, transferrin, vascular endothelial growth factor, insulin-like growth factor-2, endothelin-1, and endothelial nitric oxide synthase [7, 8]. These HIF-1-regulated genes belong to pathways like glycolysis, red blood cell formation and maturation, angiogenesis, vascular homeostasis, and cell proliferation and survival that promote cellular adaptation to low O_2 availability (Fig. 1). Hence, HIF-1 becomes the central regulatory molecule affecting the physiology at HA. However, HIF-1 itself is regulated by yet another molecule EGLN1 (HIF prolyl hydroxylase 2) of O_2-sensing pathway [9]. EGLN1 translates change in O_2 signals and regulates the protein stability of HIF1α subunit of HIF-1. During normal O_2 supplies, it hydroxylates HIF1α to facilitate its degradation [10]. Conversely, the condition of low O_2 supply results in the loss of EGLN1 activity. This prevents hydroxylation reaction of HIF1α subunit and gives HIF-1 an opportunity to accumulate and function optimally under this environment [11]. Elucidation of the molecular mechanisms by which these two molecules, HIF-1 and EGLN1, function under hypobaric hypoxia can be translated for practical applications to enhance the health management at HA. It will also help in devising new therapeutic targets, thereby transforming the basic knowledge into practical applications.

Adaptive and Maladaptive Responses to Hypobaric Hypoxia

HA has always fascinated humans. As a result, around 140 millions are permanently settled. More than 40 million lowlanders are traveling to altitudes for various reasons like professional commitments, recreation, and adventure. In recent times, HA has also been utilized for training athletes and sports personnel to increase their hyperventilation capacity and endurance [12]. Moreover, the numerous clinical, physiological, and experimental lessons learned from altitude studies can be applicable to cardiovascular and respiratory diseases in which hypoxia is explicitly involved and that share pathophysiological features similar to HA disorders. We have, in recent times, witnessed several evidences of adaptation in HA natives [13–20]. This HA adaptation is responsible for incorporation of many characteristic features like blunted hypoxic pulmonary vasoconstriction response, thin-walled pulmonary vascular structure, increased blood flow, higher O_2 saturation during exercise, and elevated vasodilatation [21–25]. HA, thus, offers a natural laboratory set up to study fitness-enhancing traits that have conferred biological advantages to the organisms at HA.

Compared to the HA natives, the lowland population when exposed to a similar environment reacts differentially. There are subjects who perform physical activities without showing signs of discomfort; on the contrary there are subjects who experience various levels of physical exertion and develop HA-related disorders [26]. These interindividual variations in lowlanders depend on the extent of acclimatization to HA, which enables the body to function optimally in the low O_2 environment [27]. Both adaptation and acclimatization allow organisms to cope up with the environmental stress. While adaptation refers to the changes at genetic level acquired over a very long period of time by the process of natural selection, acclimatization generally refers to the physiological changes [28]. However, often many individuals due to the absence of the beneficial alleles in them remain unacclimatized and thus remain susceptible to various HA maladaptations [29, 30]. These maladaptations represent a number of severe life-threatening disorders [31]. The major disorders that inflict humans at HA are acute mountain sickness (AMS) [32], subacute mountain sickness (SAMS) [33], high-altitude pulmonary edema (HAPE) [34], high-altitude cerebral edema (HACE) [35], high-altitude pulmonary hypertension (HAPH) [36], and Monge's disease or chronic mountain sickness (CMS) [37]. While AMS, SAMS, HACE, and HAPE affect sojourners, the HAPH and CMS affect HA natives [26].

Adaptation and acclimatization to HA are driven by the alterations in the O_2-sensing pathway. Therefore, HIF-1 and EGLN1, the central regulatory molecules of O_2-sensing pathway, are implicated in the manifestation of these disorders, and elucidation of alterations in these molecules will help in our understanding of the pathophysiological mechanisms. Hence, the description of these two cardinal molecules, HIF-1 and EGLN1, finds relevance under hypobaric hypoxic stress.

HIF-1

HIF-1 is made up of two differentially regulated subunits, the alpha subunit (HIF1α, HIF2α, and HIF3α) and aryl hydrocarbon receptor nuclear translocator (ARNT) or β subunit. All the three isoforms of alpha subunits are O_2 regulated but differ in their expression pattern [7]. HIF1α is the most predominant and ubiquitously expressed form. HIF2α, also termed as endothelial PAS domain protein 1 (EPAS1), shares a number of similarities with HIF1α but is predominantly expressed in endothelium, carotid bodies, liver parenchyma, lung type II pneumocytes, and kidney epithelial cells [7]. The third isoform, HIF3α or inhibitory PAS domain protein (IPAS), is expressed at high levels in the thymus, cerebellar Purkinje cells, and corneal epithelium of the eye [7]. As the name suggests, inhibitory PAS domain protein acts as a dominant negative regulator of transcription mediated by other two α subunits, HIF1α and HIF2α [7]. HIF1β or ARNT is expressed constitutively and is insensitive to changes in O_2 levels. It dimerizes with HIF1α to form functional HIF-1, which is primarily responsible for the hypoxic responses of the body [7, 38].

Both the binding partners of HIF-1 dimerize through their basic-helix-loop-helix–Per-ARNT-Sim (bHLH–PAS) domain present in each subunit [8]. This domain is responsible for all the DNA-protein interactions of HIF-1 [8]. In fact, the discovery of HIF-1 is attributed to the binding of bHLH–PAS domain to the hypoxia response element (HRE; 5′-TACGTGCT-3′) present in the erythropoietin (EPO) gene [7]. *EPO* is recognized as a hypoxia-inducible gene that encodes for the hormone responsible for red blood cell proliferation [39]. More red blood cells ensure better oxygenation of the body during O_2 deprivation. Only after the discovery of HRE element, it was realized that the HIF-1 is responsible for the induction of HRE-bearing genes following the hypoxic stimulus (Fig. 2) [5]. Besides the bHLH–PAS domain, the HIF1α subunit contains several other regulatory domains like C-terminal O_2-dependent degradation (ODD) domain and transactivation domain (CTAD). The ODD domain of HIF1α is responsible for the protein stability, whereas the CTAD mediates its transcriptional response [6]. In normoxic condition, the proline residues at positions 402 and 564 of ODD domain of HIF1α are hydroxylated by EGLN1 [10]. This post-translational modification of ODD domain helps in polyubiquitination and proteasomal degradation of HIF1α (Fig. 2) [11]. CTAD interacts with CBP/p300, the functional coactivators that are required for transcriptional response of HIF-1. HIF1α subunit inhibitor (HIF1AN), another hydroxylase belonging to the same family of EGLN1, hydroxylates asparagine residue at position 803 of CTAD of HIF1α. This modification blocks the interaction of CTAD with CBP/p300, thereby obstructing the transcriptional activity of HIF-signaling pathway (Fig. 2). Hence, the protein stability and the activity of α subunits depend on the post-translational modifications and respond to changes in O_2 tension. These responses allow cells to adapt to the decreased O_2 availability. Several studies have suggested and shown that increase in HIF-1 protein levels associates with adaptation in all living beings including humans and animals living permanently at these mighty mountains [40–43].

HIF-1 Under Hypobaric Hypoxia

The physiological changes that include hematological, respiratory, and cardiovascular responses towards environmental hypoxia are mediated by HIF-1 [5]. Induction of HIF-1-regulated genes followed by hypobaric hypoxia stimulus helps the cells in maintaining adaptive homeostasis and in fulfilling the energy demands of the body [6]. Sojourns at HA lead to a series of physiological adaptations, which are critical for overcoming the deleterious effects of hypoxia. One such adaptation is the ventilatory acclimatization to hypoxia (VAH). VAH is characterized by a progressive increase in baseline ventilation, which ensures adequate oxygen supply [44]. Carotid bodies are the principle sensory organs for detecting arterial blood O_2 levels and a substantial body of evidence suggests that carotid body chemosensory reflex is critical for VAH [45–47]. Reduced VAH is accountable to various HA-related disorders [48]. The erythropoietic response of HIF-1 is yet another environmentally induced response to increase red blood cell mass for enhancing O_2 delivery to the tissues and cells of the body [39]. Pulmonary vascular homeostasis is the third and extremely important function of HIF-1 [5]. Hypoxia constricts the human pulmonary vasculature and increases the pulmonary arterial pressure.

Nevertheless, the functions of HIF-1 have their negative aspects too. For example, AHVR

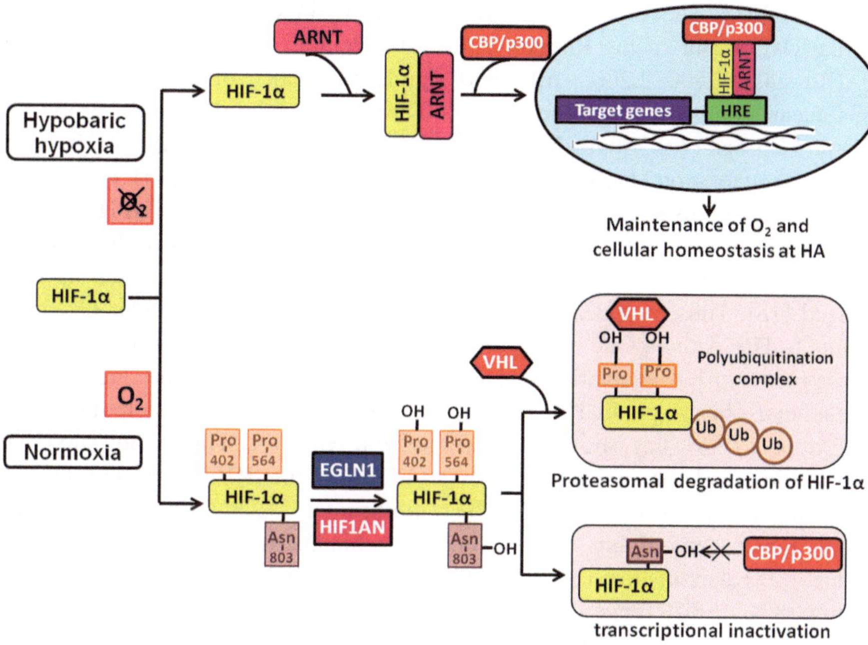

Fig. 2 HIF1α, the oxygen-controlled subunit, under hypobaric hypoxic and normoxic conditions. Under hypobaric hypoxia, EGLN1 and HIF1AN are inhibited to allow proper functioning of HIF-1 for maintenance of O_2 and cellular homeostasis. However, in the normoxic condition, the protein levels of HIF1α as well as its transactivational activity are inhibited by EGLN1 and HIF1AN, respectively. ARNT, aryl hydrocarbon nuclear receptor translocator; Asn, asparagine; EGLN1, HIF prolyl hydroxylase 2; HIF1α, hypoxia-inducible factor-1α; HIF-1AN, HIF1α subunit inhibitor; HRE, hypoxia response element; O_2, oxygen; OH, hydroxyl group; ub, ubiquitin proteins; VHL, von Hippel–Lindau tumor-suppressor protein; CBP/p300, CREB-binding protein/E1A binding protein p300; Pro, proline

marks the beginning of vascular remodeling that shunts the blood from poorly aerated regions of lungs to better ventilated region [49]. Although this helps in improving the ventilation to perfusion ratio during the process, it leads to structural changes in the lungs, hypoxic pulmonary vasoconstriction (HPV), and pulmonary hypertension (PH) due to vascular remodeling [50]. The latter is the hallmark of a number of HA-related disorders and becomes the major cause of morbidity at HA. It worsens the condition of CMS and HAPH in HA natives and HAPE in sojourners. PH, thus, becomes the major cause of deaths at HA [34, 36, 37]. Similarly, increase in red blood cell mass is undesirable and results in CMS in HA natives, a neuropsychological disorder with symptoms like memory loss, mental confusion, sleep disturbance, headache, anorexia, and fatigue [37]. Hence, HIF-1 mediates physiological responses to sustain hypobaric hypoxia but is also responsible for various environmentally induced maladaptive changes.

HA populations living for several thousands of years at elevations up to 5,000 m present many adaptive phenotypic changes. These phenotypes are the result of natural selection on genetically based trait variations that counteracts the effects of environmentally induced changes. For example, HA natives of Tibet (3,000–5,000 m) have hematological profile similar to what would be expected at sea level and are particularly resistant to developing CMS [21]. On the other hand Andeans (3,000–5,000 m) have better O_2 saturation in their body [22]. Likewise, Ethiopians, the indigenous human population residing on East African plateau (1,500–3,500 m), have Hb concentration and SaO_2 levels similar to the levels found at sea level [21]. The other HA native populations like Daghestani of the Republic of Daghestan (Russia, 1,500–2,200 m) and Eurasians of Caucasian mountains on the border between Asia and Europe (≤2,500 m) have also shown evidences of adaptation [17, 18]. Interestingly, all the HA populations have demonstrated the role of HIF pathway in O_2 pressure adaptation over a wider range of altitudes than previously recognized [18]. Thus, it appears that the genes pertaining to HIF pathway are undergoing the continuous process of HA adaptation and are conferring beneficial phenotype in HA natives [13–18]. *HIF2α* or *EPAS1* has shown natural selection in Tibetans, and the major alleles of these identified signals were also found to be associated with lower hemoglobin (Hb) level [14]. Yi X et al. [15] demonstrated an SNP of *EPAS1* showing 78 % of difference in its distribution in Tibetan highlanders and Han Chinese as a representation of the fastest allele frequency change observed at any human gene to date [15]. Association of these signals with Hb concentrations further suggests that these loci have functional importance in genetic adaptation to HA [14]. Similarly, *ARNT2* gave the signal for natural selection in Ethiopian highlanders when compared to their lowland counterparts, and the SNPs belonging to this gene were also associated with the Hb levels [20]. Studies using candidate gene approach reported the selection of wild-type alleles of exonic SNPs C1772T (rs11549465) and G1790A (rs11549467) of *HIF1α* in Tibetans [51]. Similarly, another study found a novel dinucleotide polymorphism in intron 13 of this gene being more frequent in Sherpas as compared with Japanese lowlanders [52].

EGLN1

The O_2-controlled HIFα isoforms are regulated by the members of ubiquitous Fe(II) and 2-oxoglutarate-dependent oxygenase superfamily [10]. The three alpha subunits of HIF-1 are regulated by O_2 through the enzymatic hydroxylation of specific amino acid residues in these proteins [11]. In mammals, three isoforms of prolyl hydroxylases have been identified that catalyze the post-translational modification of proline residues of α subunits of HIF-1 [53]. These hydroxylases are HIF prolyl hydroxylase 1 (HPH1 or EGLN3, PHD3), HIF prolyl hydroxylase 2 (HPH2 or EGLN1 or PHD2), and HIF prolyl hydroxylase 3 (HPH3 or EGLN2 or PHD1). Several studies on RNA interference of the three prolyl hydroxylases have established EGLN1 as the main cellular O_2 sensor and the one responsible for proline hydroxylation of HIF1α subunit [53]. The degradation of HIF-1 in normoxic condition is credited to the binding of von Hippel–Lindau tumor-suppressor protein (VHL) to the HIF1α subunit that directs the multiprotein ubiquitin ligase complex for polyubiquitination and proteasomal degradation [10]. The binding of VHL to HIF1α requires its post-translational hydroxylation by EGLN1, which requires O_2 along with iron, ascorbate, and 2-oxoglutarate as obligate cofactors [53]. The hydroxylation process transfers one O_2 atom from O_2 to the proline residue and the other reacts with 2-oxoglutarate. Fe(II) bound by two histidine residues and one aspartic acid forms the active site of the EGLN1, and ascorbate helps to maintain Fe(II) state of Fe required for maintaining full activity of the enzyme [53]. Hence, under low O_2 condition, the activity of hydroxylases is hampered resulting in stabilization and accumulation of HIF-1 complex.

EGLN1 Under Hypobaric Hypoxia

EGLN1 has emerged as the only molecule, whose signals have been found in the genome-wide association studies of almost every population of HA around the world [16–20]. The positively selected haplotypes of *EGLN1* in HA populations have also been associated with decreased Hb phenotypes [16]. The selection of *EGLN1* in Tibetans, Andeans, Eurasians, and Daghestanis suggests the common pathways of adaptation to HA indifferent populations [15–20]. Through candidate gene approach, Aastha Mishra et al. [26] has also shown selection of rs1538664, rs479200, rs2486729, rs2790879, rs480902, rs2486736, and rs973252 from the intronic region of this gene with respect to adaptation in Ladakhi highlanders and maladaptation due to HAPE in sojourners [26]. Ladakhi population belonging to Tibetan origin resides at ≥3,500 m of Ladakh region of India. The assessment of *EGLN1* expression in this study was equally pertinent. It has revealed more than 4-fold upregulation of *EGLN1* in HAPE, which causes instability to HIF1α that in turn prohibits the downstream genes from maintaining the cellular O_2 homeostasis. Furthermore, the risk variants rs1538664A, rs479200T, rs2486729A, rs2790879G, rs480902C, rs2486736A, and rs973252G, which were over-represented in HAPE correlated with increased *EGLN1* expression and decreased SaO_2 level. These association studies suggested the role of these polymorphisms in regulating the phenotypic differences. The team had also conducted TFSEARCH analysis that identified the alterations in TF binding sites due to the presence of variants in it (Fig. 3). The analysis observed that the protective alleles of these polymorphisms had binding sites for TFs like CREB, VBP, NIT2, and AP-1 that ensured normal functioning of HIF and thereby cellular O_2 homeostasis. Contrary to protective alleles, the appearances of risk alleles in these TF binding sites created sites for TFs like C-MYB, GATA-1, GATA-3, DELTA E, and BCD that are associated with tumor growth, proto-oncogenicity, and inflammatory and oxidative stress response. Thus, these analyses indicated that the significantly associated SNPs belonging to intron 1 of the gene might be involved in the regulation of this gene. Another candidate gene approach study has identified a causal non-synonymous mutation after re-sequencing the entire gene [54]. This non-synonymous mutation (rs186996510) showed significant difference between Tibetans and lowland population and was also associated with low Hb levels. This exonic SNP, highly prevalent in Tibetans but extremely rare in lowland population, may be contributing to HA hypoxic adaptation.

It hence may be concluded that genes like *EPAS1* and *EGLN1* show strong signals of natural selections that correlate well with the relevant clinical and bimolecular levels at HA. Moreover, the investigation of HA disorders elucidated the contributions of risk variants of these genes in maladaptations. Identification of the genetic variants associating with HA adaptations and maladaptations and exploration of their functionality through various correlation studies, thus, have helped in understanding of the molecular mechanisms working towards hypoxia tolerance in the body.

Future Perspectives

HIF-1–EGLN1–VHL complex controls the responses of mammalian cells to O_2 deprivation. This complex has immense potential for pharmacological manipulations in HA-related disorders. Exploitation of HIF-1 in tumor biology has been highly rewarding and the same could be expected in HA biology too. However, the journey of translational research of O_2-sensing pathway in the environmental stress such as hypobaric hypoxia cannot be realized until all the mechanisms of these pathways at HA are transparent to us. A lot of functions of HIF-1 are still obscure and require detailed investigation. Along with them, the potential of selected beneficial alleles of *EGLN1* and *HIF-1* at HA can be exploited in gene therapy and through pharmacogenomics.

Acknowledgments The work presented here was supported by CSIR under the projects SIP0006, MLP1401, and EXP0016. We acknowledge the support and encouragement of the director, CSIR-IGIB, during the preparation of this manuscript.

Author Contributions Aastha Mishra researched and wrote the manuscript. Qadar Pasha conceptualized, researched, wrote, and edited the manuscript.

Conflicts of Interest We declare that we have no conflicts of interest.

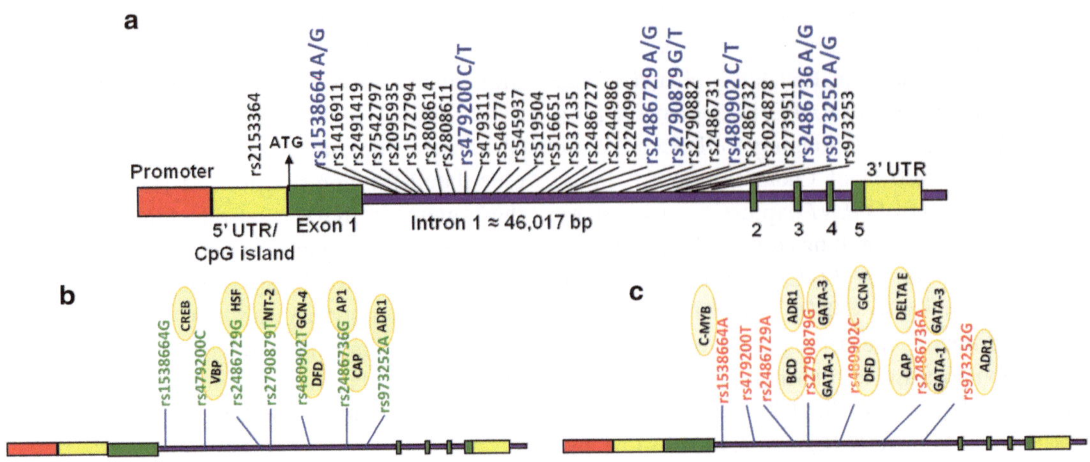

d

S. No.	Allele	TF-binding site	TF	Affinity score	Function	Reference
1a	rs1538664 G	-	NONE		-	
1b	rs1538664 A	AGTTGTC	C-MYB	85.6	Leukaemogenesis	60
2a	rs479200 C	TTACGTA	CREB	87.9	Lung responses to hypoxia; activation of tyrosine hydroxylase gene	55, 56
		GTTACGTAC	VBP	87.3	Proteasomal degradation of HIF-1α	57
2b	rs479200 T	-	NONE		-	-
3a	rs2486729 G	CGTAG	HSF	91.1	Cell protection from protein damaging stress	66
3b	rs2486729 A	-	NONE		-	-
4a	rs2790879 T	ATATTA	NIT2	88.8	Tumor suppressor candidate	58
4b	rs2790879 G	GAGATTAGG	GATA-3	89.1	Marker of breast cancer	63
		TGGAG	ADR1	87.7	Regulates glycerol metabolism in yeast	65
		GAGATTA	BCD	87.3	Proto-oncogene	64
		GGAGATTA	GATA-1	86.5	Haematopoeisis; cell growth and cancer	61
5a	rs480902 T	TGTGTCTTTGAGTC	GCN4	89.1	Protein and purine metabolism in yeast	68
		CATTAATGTGTCTTTG	DFD	87.4	Axial morphogenesis and patterning in drosophila	69
5b	rs480902 C	TGTGTCTCTGAGTC	GCN4	86.7	Protein and purine metabolism in yeast	68
		CATTAATGTGTCTCTG	DFD	85.8	Axial morphogenesis and patterning in drosophila	69
6a	rs2486736 G	CAGTCACCTA	AP-1	96.3	Stress response through mitogen-activated protein kinase	59
		CAGTCACCTA	AP-1	95.9		
		CAGTCACCTA	AP-1	95.6		
		CAGTCACCTA	AP-1	92.9		
		CTCAGTCA	CAP	94.4	Catabolite repression	67
6b	rs2486736 A	CAATCACCTA	GATA-1	91.4	Haematopoeisis; cell growth and cancer	61
		AATCACCTAC	DELTA E	85.6	Inflammation and energy metabolism regulation	62
		TCAATCA	CAP	86.1	Catabolite repression	67
		AATCACCTA	GATA-3	85.6	Marker of breast cancer	63
7a	rs973252 A	CCAAT	ADR1	92.3	Glycerol metabolism	61
7b	rs973252 G	CCAGT	ADR1	92.3		

Fig. 3 Schematic representation of polymorphisms of EGLN1 lying in the TF binding sites. (**a**) EGLN1 structure with the 30 polymorphisms, (**b**) The protective alleles of above mentioned seven polymorphisms lying in the TF binding sites of the respective TFs, (**c**) The risk alleles lying in the TF binding sites of the respective TFs and (**d**) the affinity score and function of the TFs that bind to the TF binding sites falling in the vicinity of the EGLN1 polymorphisms (Reproduced with permission, from Mishra et al. [26]. © The Biochemical Society)

References

1. West JB (2002) Highest permanent human habitation. High Alt Med Biol 3:401–407
2. Hultgren HN, Marticorena EA (1978) High altitude pulmonary edema. Epidemiologic observations in Peru. Chest 74:372–376
3. Hsia CC (1998) Respiratory function of hemoglobin. N Engl J Med 338:239–248
4. Qadar Pasha MA, Kocherlakota KS, Khan AP, Norboo T, Grover SK, Baig MA, Selvamurthy W, Bramhachari SK (2003) Arterial oxygen saturation under hypoxic environment of high-altitude associates with routine physical activities of natives. Curr Sci 85:502–506
5. Semenza GL (2009) Regulation of oxygen homeostasis by hypoxia-inducible factor 1. Physiology (Bethesda) 24:97–106
6. Smith TG, Robbins PA, Ratcliffe PJ (2008) The human side of hypoxia-inducible factor. Br J Haematol 141:325–334
7. Ke Q, Costa M (2006) Hypoxia-inducible factor-1 (HIF-1). Mol Pharmacol 70:1469–1480
8. Maxwell PH (2005) Hypoxia-inducible factor as a physiological regulator. Exp Physiol 90:791–797
9. Bruick RK (2003) Oxygen sensing in the hypoxic response pathway: regulation of the hypoxia-inducible transcription factor. Genes Dev 17:2614–2623
10. Berra E, Benizri E, GinouveÁs A, Volmat V, Roux D, PouysseÂgur J (2003) HIF prolyl-hydroxylase 2 is the key oxygen sensor setting low steady-state levels of HIF-1a in normoxia. EMBO J 22:4082–4090
11. To KKW, Huang LE (2005) Suppression of HIF-1α transcriptional activity by the HIF prolyl hydroxylase EGLN1. J Biol Chem 280:38102–38107
12. Saunders PU, Pyne DB, Gore CJ (2009) Endurance training at altitude. High Alt Med Biol 10:135–148
13. Bigham A, Bauchet M, Pinto D, Mao X, Akey JM, Mei R, Scherer SW, Julian CG, Wilson MJ, LópezHerráez D, Brutsaert T, Parra EJ, Moore LG, Shriver MD (2010) Identifying signatures of natural selection in tibetan and andean populations using dense genome scan data. PLoS Genet 6:1001116
14. Beall CM, Cavalleri GL, Deng L, Elston RC, Gao Y, Knight J, Li C, Li JC, Liang Y, McCormack M, Montgomery HE, Pan H, Robbins PA, Shianna KV, Tam SC, Tsering N, Veeramah KR, Wang W, Wangdui P, Weale ME, Xu Y, Xu Z, Yang L, Zaman MJ, Zeng C, Zhang L, Zhang X, Zhaxi P, Zheng YT (2010) Natural selection on EPAS1 (HIF2α) associated with low hemoglobin concentration in Tibetan highlanders. Proc Natl Acad Sci U S A 107:11459–11464
15. Yi X, Liang Y, Huerta-Sanchez E, Jin X, Cuo ZX, Pool JE, Xu X, Jiang H, Vinckenbosch N, Korneliussen TS, Zheng H, Liu T, He W, Li K, Luo R, Nie X, Wu H, Zhao M, Cao H, Zou J, Shan Y, Li S, Yang Q, Asan, Ni P, Tian G, Xu J, Liu X, Jiang T, Wu R, Zhou G, Tang M, Qin J, Wang T, Feng S, Li G, Huasang, Luosang J, Wang W, Chen F, Wang Y, Zheng X, Li Z, Bianba Z, Yang G, Wang X, Tang S, Gao G, Chen Y, Luo Z, Gusang L, Cao Z, Zhang Q, Ouyang W, Ren X, Liang H, Zheng H, Huang Y, Li J, Bolund L, Kristiansen K, Li Y, Zhang Y, Zhang X, Li R, Li S, Yang H, Nielsen R, Wang J, Wang J (2010) Sequencing of 50 human exomes reveals adaptation to high altitude. Science 329:75–78
16. Simonson TS, Yang Y, Huff CD, Yun H, Qin G, Witherspoon DJ, Bai Z, Lorenzo FR, Xing J, Jorde LB, Prchal JT, Ge R (2010) Genetic evidence for high-altitude adaptation in Tibet. Science 329:72–75
17. Pagani L, Ayub Q, MacArthur DG, Xue Y, Baillie JK, Chen Y, Kozarewa I, Turner DJ, Tofanelli S, Bulayeva K, Kidd K, Paoli G, Tyler-Smith C (2012) High altitude adaptation in Daghestani populations from the Caucasus. Hum Genet 131:423–433
18. Ji LD, Qiu YQ, Xu J, Irwin DM, Tam SC, Tang NL, Zhang YP (2012) Genetic adaptation of the hypoxia inducible factor pathway to oxygen pressure among Eurasian human populations. Mol Biol Evol 29:3359–3370
19. Aggarwal S, Negi S, Jha P, Singh PK, Stobdan T, Qadar Pasha MA, Ghosh S, Agrawal A, Indian Genome Variation Consortium, Prasher B, Mukerji M (2010) EGLN1 involvement in high-altitude adaptation revealed through genetic analysis of extreme constitution types defined in Ayurveda. Proc Natl Acad Sci U S A 107:18961–18966
20. Scheinfeldt LB, Soi S, Thompson S, Ranciaro A, Woldemeskel D, Beggs W, Lambert C, Jarvis JP, Abate D, Belay G, Tishkoff SA (2012) Genetic adaptation to high altitude in the Ethiopian highlands. Genome Biol 13:R1
21. Moore LG (2001) Human genetic adaptation to high altitude. High Alt Med Biol 2:257–279
22. Beall CM (2006) Andean, Tibetan and Ethiopian patterns of adaptation to high-altitude hypoxia. Integr Comp Biol 46:18–24
23. Beall CM, Laskowski D, Strohl KP, Soria R, Villena M, Vargas E, Alarcon AM, Gonzales C, Erzurum SC (2001) Pulmonary nitric oxide in mountain dwellers. Nature 414:411–412
24. Ahsan A, Charu R, Qadar Pasha MA, Norboo T, Afrin F, Baig MA (2004) eNOS allelic variants at the same locus associate with HAPE and adaptation. Thorax 59:1000–1002
25. Ali Z, Mishra A, Kumar R, Alam P, Pandey P, Ram R, Thinlas T, Mohammad G, Qadar Pasha MA (2012) Interactions among vascular-tone modulators contribute to high altitude pulmonary edema and augmented vasoreactivity in highlanders. Plos One 7:e44049
26. Mishra A, Mohammad G, Thinlas T, Qadar Pasha MA (2013) EGLN1 variants influence its expression and SaO2 levels to associate with high-altitude pulmonary edema and adaptation. Clin Sci 124:479–489
27. Gleadle JM, Ratcliffe PJ (1998) Hypoxia and the regulation of gene expression. Mol Med Today 4:122–129

28. Baker PT (1969) Human adaptation to high altitude. Science 163:1149–1156
29. Stobdan T, Karar J, Qadar Pasha MA (2008) High altitude adaptation: genetic perspectives. High Alt Med Biol 9:140–147
30. Qadar Pasha MA, Newman JH (2010) High-altitude disorders: pulmonary hypertension: pulmonary vascular disease: the global perspective. Chest 137:13S–19S
31. Barry PW, Pollard AJ (2003) Altitude illness. BMJ 326:915–919
32. Clarke C (2006) Acute mountain sickness: medical problems associated with acute and subacute exposure to hypobaric hypoxia. Postgrad Med J 82:748–753
33. Anand IS, Wu T (2004) Syndromes of subacute mountain sickness. High Alt Med Biol 5:156–170
34. Bartsch P, Mairbaurl H, Maggiorini M, Swenson ER (2005) Physiological aspects of high-altitude pulmonary edema. J Appl Physiol 98:1101–1110
35. Wilson MH, Newman S, Imray CH (2009) The cerebral effects of ascent to high altitudes. Lancet Neurol 8:175–191
36. Maggiorini M, Léon-Velarde F (2003) High-altitude pulmonary hypertension: a pathophysiological entity to different diseases. Eur Respir J 22:1019–1025
37. Monge C (1943) Chronic mountain sickness. Physiol Rev 23:166–184
38. Prabhakar NR, Semenza GL (2012) Adaptive and maladaptive cardiorespiratory responses to continuous and intermittent hypoxia mediated by hypoxia-inducible factors 1 and 2. Physiol Rev 92 (3):967–1003
39. Jelkmann W (2003) Erythropoietin. J Endocrinol Invest 26:832–837
40. Li HG, Ren YM, Guo SC, Cheng L, Wang DP, Yang J, Chang ZJ, Zhao XQ (2009) The protein level of hypoxia-inducible factor-1alpha is increased in the plateau pika (Ochotona curzoniae) inhabiting high altitudes. J Exp Zool A Ecol Genet Physiol 311:134–141
41. Lundby C, Pilegaard H, Andersen JL, van Hall G, Sander M, Calbet JA (2004) Acclimatization to 4100 m does not change capillary density or mRNA expression of potential angiogenesis regulatory factors in human skeletal muscle. J Exp Biol 207:3865–3871
42. Appenzeller O, Minko T, Pozharov V, Bonfichi M, Malcovati L, Gamboa J, Bernardi L (2003) Gene expression in the Andes; relevance to neurology at sea level. J Neurol Sci 207:37–41
43. Vogt M, Puntschart A, Geiser J, Zuleger C, Billeter R, Hoppeler H (2001) Molecular adaptations in human skeletal muscle to endurance training under simulated hypoxic conditions. J Appl Physiol 91:173–182
44. León-Velarde F, Richalet JP (2006) Respiratory control in residents at high altitude: physiology and pathophysiology. High Alt Med Biol 7:125–137
45. Prabhakar NR (2000) Oxygen sensing by the carotid body chemoreceptors. J Appl Physiol 88:2287–2295
46. Lahiri S, Rozanov C, Cherniack NS (2000) Altered structure and function of the carotid body at high altitude and associated chemoreflexes. High Alt Med Biol 1:63–74
47. Kumar P, Prabhakar NR (2012) Peripheral chemoreceptors: function and plasticity of the carotid body. Compr Physiol 2(1):141–219
48. Moore LG, Harrison GL, McCullough RE, McCullough RG, Micco AJ, Tucker A, Weil JV, Reeves JT (1986) Low acute hypoxic ventilatory response and hypoxic depression in acute altitude sickness. J Appl Physiol 60:1407–1412
49. Penaloza D, Stella JA (2007) The heart and pulmonary circulation at high altitudes healthy highlanders and chronic mountain sickness. Circulation 115:1132–1146
50. Naeije R, Brimioulle S (2001) Physiology in medicine: importance of hypoxic pulmonary vasoconstriction in maintaining arterial oxygenation during acute respiratory failure. Crit Care 5:67–71
51. Liu K, Sun X, Wang S, Hu B (2007) Association between polymorphisms of HIF-1alpha C1772T and G1790A and hypoxic acclimation in high altitude in Tibetans. Sheng Wu Yi Xue Gong Cheng Xue Za Zhi 24:654–658
52. Suzuki K, Kizaki T, Hitomi Y, Nukita M, Kimoto K, Miyazawa N, Kobayashi K, Ohnuki Y, Ohno H (2003) Genetic variation in hypoxia-inducible factor 1alpha and its possible association with high altitude adaptation in Sherpas. Med Hypotheses 61:385–389
53. Appelhoff RJ, Tian YM, Raval RR, Turley H, Harris AL, Pugh CW, Ratcliffe PJ, Gleadle JM (2004) Differential function of the prolyl hydroxylases PHD1, PHD2, and PHD3 in the regulation of hypoxia-inducible factor. J Biol Chem 279:38458–38465
54. Xiang K, Ouzhuluobu, Peng Y, Yang Z, Zhang X, Cui C, Zhang H, Li M, Zhang Y, Bianba, Gonggalanzi, Basang, Ciwangsangbu, Wu T, Chen H, Shi H, Qi X, Su B (2013) Identification of a tibetan-specific mutation in the hypoxic gene EGLN1 and its contribution to high-altitude adaptation. Mol Biol Evol 30:1889–1898
55. Leonard MO, Howell K, Madden SF, Costello CM, Higgins DG, Taylor CT, McLoughlin P (2008) Hypoxia selectively activates the CREB family of transcription factors in the in vivo lung. Am J Respir Crit Care Med 178:977–983
56. Kim KS, Lee MK, Carroll J, Joh TH (1993) Both the basal and inducible transcription of the tyrosine hydroxylase gene are dependent upon a cAMP response element. J Biol Chem 268:15689–15695
57. Tsuchiya H, Iseda T, Hino O (1996) Identification of a novel protein (VBP-1) binding to the von Hippel-Lindau (VHL) tumor suppressor gene product. Cancer Res 56:2881–2885
58. Lin CH, Chung MY, Chen WB, Chien CH (2007) Growth inhibitory effect of the human NIT2 gene and its allelic imbalance in cancers. FEBS J 274:2946–2956

59. Ameyar M, Wisniewska M, Weitzman JB (2003) A role for AP-1 in apoptosis: the case for and against. Biochimie 85:747–752
60. Ramsay RG, Gonda TJ (2008) MYB function in normal and cancer cells. Nat Rev Cancer 8:523–534
61. Ferreira R, Ohneda K, Yamamoto M, Philipsen S (2005) GATA1 function, a paradigm for transcription factors in hematopoiesis. Mol Cell Biol 25:1215–1227
62. Cardinaux JR, Allaman I, Magistretti PJ (2000) Pro-inflammatory cytokines induce the transcription factors C/EBPbeta and C/EBPdelta in astrocytes. Glia 29:91–97
63. Usary J, Llaca V, Karaca G, Presswala S, Karaca M, He X, Langerød A, Kåresen R, Oh DS, Dressler LG, Lønning PE, Strausberg RL, Chanock S, Børresen-Dale AL, Perou CM (2004) Mutation of GATA3 in human breast tumors. Oncogene 23:7669–7678
64. El Rouby S, Newcomb EW (1996) Identification of Bcd, a novel proto-oncogene expressed in B-cells. Oncogene 13:2623–2630
65. Cook WJ, Chase D, Audino DC, Denis CL (1994) Dissection of the ADR1 protein reveals multiple, functionally redundant activation domains interspersed with inhibitory regions: evidence for a repressor binding to the ADR1c region. Mol Cell Biol 14:629–640
66. Wu C (1995) Heat shock transcription factors: structure and regulation. Annu Rev Cell Dev Biol 11:441–469
67. Schultz SC, Shields GC, Steitz TA (1991) Crystal structure of a CAP-DNA complex: the DNA is bent by 90 degrees. Science 253:1001–1007
68. Hinnebusch AG (2005) Translational regulation of GCN4 and the general amino acid control of yeast. Annu Rev Microbiol 59:407–450
69. Tsuzuki S, Sekiguchi S, Hayakawa Y (2005) Regulation of growth-blocking peptide expression during embryogenesis of the cabbage armyworm. Biochem Biophys Res Commun 335:1078–1084

Electrochemical Biosensors for Hypoxia Markers

C. Karunakaran, T. Madasamy, M. Pandiaraj, Niroj K. Sethy, and Kalpana Bhargava

Abstract

An inadequate oxygen supply to cells/tissues causes hypoxia at high altitude, resulting in alteration in the levels of nitric oxide (NO) and its metabolites, *viz.* nitrite (NO_2^-) and nitrate (NO_3^-) and simultaneous increase in the formation of peroxynitrite which triggers the cell death by releasing cytochrome *c* (cyt *c*) from mitochondria. Therefore, measurements of these biologically important hypoxia biomarkers are imperative in human physiology as it provides valuable information regarding the personnel at high altitude. So, we have developed a cost-effective and portable electrochemical biosensor assay for the measurement of various hypoxia biomarkers in volume miniaturized samples using screen-printed electrodes (SPE). Modification of SPE surface with nanocomposites of polypyrrole and carbon nanotube/self-assembled layer on gold nanoparticle for biofunctionalization of specific biorecognization (enzymes/antibody) elements provides a selective and sensitive determination of various hypoxia biomarkers. Copper, zinc superoxide dismutase, and nitrate reductase-functionalized electrodes were used as biosensors for the determination of NO, NO_2^-, and NO_3^-. Supplementation of NO_3^- rich beetroot juice to human and several animal models enhanced the NO-like bioactivity. So, we have measured the total NO_2^- and NO_3^- levels in beetroot supplements and in human plasma before and after beetroot intake. Further, novel cyt *c* biosensor based on cyt *c* reductase was employed to measure cyt *c* release from the hypoxic-induced cell death and the results agreed well with the standard assay methods.

Introduction

Hypoxia is a pathological condition in which the entire human body or some parts of the body are deprived or suffer from lack of oxygen supply [1]. It results in a mismatch between oxygen supply and its demand at the cellular level [2]. This shortage of oxygen leads to cancer, cardiovascular

C. Karunakaran (✉) • T. Madasamy • M. Pandiaraj
Biomedical Research Lab, Department of Chemistry, VHNSN College (Autonomous), Virudhunagar 626 001, Tamil Nadu, India
e-mail: ckaru2000@gmail.com

N.K. Sethy • K. Bhargava
Peptide and Proteomics Division, DIPAS, DRDO, New Delhi 110 054, India

S.B. Singh et al. (eds.), *Translational Research in Environmental and Occupational Stress*,
DOI 10.1007/978-81-322-1928-6_9, © Springer India 2014

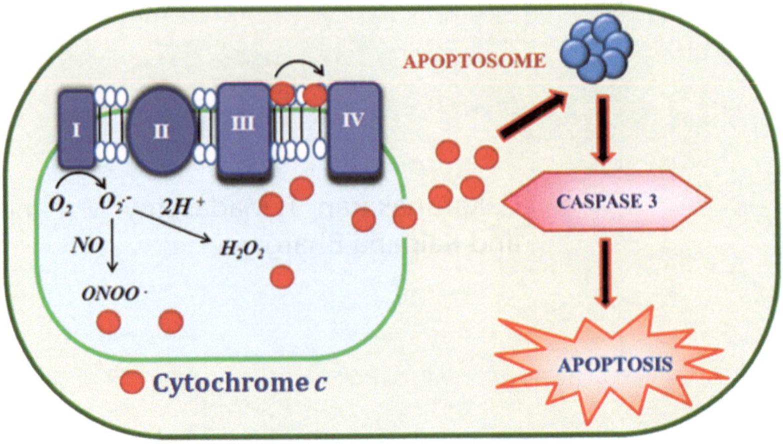

Fig. 1 Role of ROS/RNS and cytochrome c in hypoxia-induced cell death

dysfunction, and neurodegenerative diseases [3, 4]. Nitric oxide (NO) is a ubiquitous signaling molecule produced through the metabolism of L-arginine by nitric oxide synthases (NOS) [5]. Its availability is probably reduced at high altitude since its enzymatic production depends on the availability of oxygen, and exposure to hypoxia results in a paradoxical increase in the production of reactive oxygen species leading to NO inactivation [6]. Further, nitrite (NO_2^-) and nitrate (NO_3^-) are the stable end oxidative metabolites of NO and their concentrations are also altered in blood during hypoxia [7]. Subsequently, the concentration of peroxynitrite will be increased which triggers the cell death by releasing cytochrome *c* (cyt *c*) from mitochondria [8]. NO_2^- reduction to NO during hypoxia appears to contribute to physiological hypoxic signaling, vasodilation, modulation of mitochondrial respiration, and the cellular response to ischemic stress [9]. Therefore, NO, NO_2^-, NO_3^-, and cyt *c* are the important biomarkers of hypoxia and their measurements are aid in monitoring subjects at high altitude (Fig. 1).

Numerous strategies have been developed for the determination of NO, NO_2^-, and NO_3^- including Griess colorimetric assay [10, 11], fluorometry [12], and chemiluminescence methods [13]. Separation-based methods, *viz.* GC-MS [14, 15] and HPLC [16, 17], with variety of detection system have also been reported. Most of these techniques are indirect, are not suitable for *in vivo* measurement, and require sophisticated instruments. Enzyme-linked immunosorbent assay (ELISA), western blot, flow cytometry, and spectrophotometry are the commonly available biochemical techniques for the quantification of cyt *c* release [18–21]. Although these methods have high sensitivity, unfortunately, they are multi-step and time-consuming processes, require large, expensive instrument, require extensive pretreatment of the sample, and are also not suitable for rapid routine use. Recently, electrochemical biosensor techniques are proved to be powerful tools for the measurement of markers by providing practical advantages, such as operation simplicity, low expense of fabrication, and suitability for real-time detection. In addition, it provides fast response and more sensitive (particularly with the use of modified electrodes) and selective determination. Therefore, we developed electrochemical biosensor assays for these clinically important hypoxia biomarkers.

Further, the research has been extended to increase the NO bioavailability during hypoxia, possibly by administration of NO_3^- rich beetroot juice [22]. It promotes NO-like bioactivity and regulates biological activities like reduction of blood pressure, vasodilation, cytoprotection, cardioprotection, and protection from ischemia–reperfusion injury and promotes exercise capacity [7, 23]. Interestingly, vegetables like beetroot, spinach, and lettuce along with cured meat in our everyday diet are a major source of inorganic NO_3^- [24]. Recent investigations have also shown that higher intake

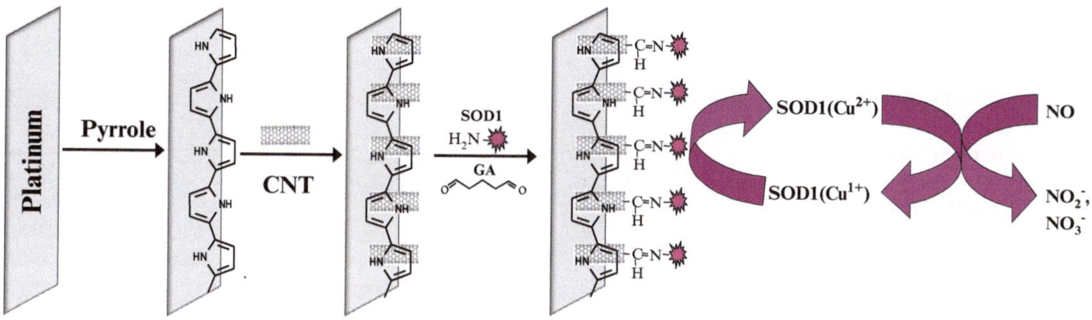

Scheme 1 Schematic representation of the construction of NO biosensor and the illustration of the electrochemical reactions during NO measurement

of nitrate-rich fruits and vegetables is associated with reduced risk of coronary and ischemic heart diseases [25]. In this exertion, we have also investigated the promotion of NO and its metabolites concentrations in blood using the developed electrochemical biosensors.

Copper, Zinc Superoxide Dismutase, and Nitrate Reductase-Based Biosensors for Nitric Oxide and Its Metabolites

Based on the nitric oxide and nitrite oxidase activities of copper, zinc superoxide dismutase (SOD1), an electrochemical biosensor for NO and NO_2^- was developed [26, 27]. Further, NO_3^- biosensor was developed by using nitrate reductase (NaR)-immobilized electrode [28].

Nitric Oxide Biosensor

Nitric oxide biosensor was developed by immobilizing SOD1 onto the carbon nanotubes (CNT)–polypyrrole (PPy) nanocomposite-modified Pt electrode. Scheme 1 represents the various steps involved in the construction of NO biosensor and the mechanism of measurement.

The electrochemical response of the constructed NO biosensor for NO measurement was investigated using cyclic voltammetry (CV). Figure 2 shows the electrochemical responses obtained for the SOD1–PPy–Pt and SOD1–CNT–PPy–Pt electrodes in the absence (curves a and b) and presence (curves c and d) of 100 μM NO at a scan rate of 50 mV s^{-1} in 0.1 M PBS (pH 7.0). In the absence of NO, a quasi-reversible peak at +0.06 V, characteristic of SOD1, was observed. However, after the addition of NO, SOD1–PPy–Pt and SOD1–CNT–PPy–Pt electrodes exhibited a new peak at +0.8 V and the current response increased anodically as the concentration raised. The observed anodic peak at +0.8 V is attributed to the oxidation of NO *via* a cyclic redox reaction of SOD1 active site Cu (I/II) moiety. It is clearly seen that the SOD1–CNT–PPy–Pt electrode (curve d) shows higher current than SOD1–PPy–Pt (curve c). It is perhaps due to the additional surface area provided by the CNT–PPy nanocomposite enhancing the immobilization of SOD1, as well as accelerating electron transfer between the underlying electrode and active site of SOD1.

Further, the electrochemical responses of the SOD1–CNT–PPy–Pt electrode in 0.1 M PBS containing various NO concentrations were investigated. The calibration curve was obtained by plotting the observed anodic peak currents vs. NO concentrations exhibiting a linear range of response over the concentration of NO from 0.1 μM to 1 mM ($r^2 = 0.999$, $n = 3$) with a detection limit of 100 nM and sensitivity of 1.1 μA μM^{-1}.

Virtual Instrumentation Based NO Analyzer

Commercially available NO analyzers are mainly based on the chemical reaction of NO with ozone (O_3) and the concentration of NO is measured with respect to the luminescent

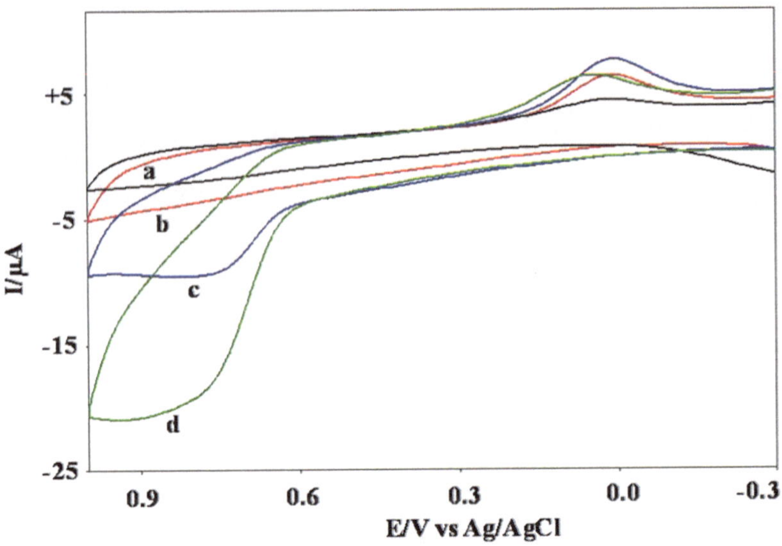

Fig. 2 Typical CV responses obtained for the SOD1–PPy–Pt and SOD1–CNT–PPy–Pt in the absence (curve *a* and *b*) and presence (curve *c* and *d*) of 100 μM NO solution in 0.1 M PBS (pH 7.0) at scan rate: 50 mV s^{-1} vs. Ag/AgCl

intensity [29]. These analyzers involve corrosive chemicals, high cost and also not a selective method for the determination of NO. Because of these limitations, we have developed here virtual electrochemical NO biosensor. Earlier, we have used the standard cyclic voltammetry (CV) instrument for the measurement of NO. However, this physical instrument involves high cost and the instrumentation systems are made up of predefined hardware components that are completely specific to their measurement function. So, in an effort to diminish the cost of the measuring unit, make it environment and user friendly, and also to achieve the specific determination of NO, we have demonstrated here a compact, flexible, and low-cost electrochemical NO analyzer. It was developed by integrating home-made potentiostat with data acquisition system (MyDAQ) and the control program was developed using graphical user interface software LabVIEW (Laboratory Virtual Instrumentation Engineering Workbench). Figure 3 illustrates the overall electrochemical setup (Panel A) and front panel of the measurement (Panel B). Further, the linear range, sensitivity, and detection limit of the NO biosensor were investigated using the virtual electrochemical analyzer and are comparable with those obtained using standard CV instrument. Using this NO analyzer, the concentrations of NO present in the exhaled breath and also from hydrogen peroxide-stimulated endothelial cells were estimated.

Nitrite Biosensor

In the literature, several reports described the measurement of NO_2^- based on its electrochemical reduction. However, these electroreductive reactions involved in the above reported methods gave NO as a product causing interference by its reaction with oxygen. Hence, the researchers focused on the electrocatalytic oxidation of NO_2^- for its selective measurement without any interference. Therefore, we have used here the nitrite oxidase activity of SOD1 [27]. Further, the active channel of SOD1 is narrow to pass NO_2^- [30]. So, SOD1 was used as a specific biorecognition element for the determination of NO_2^-. The electrochemical biosensor (SOD1–CNT–PPy–Pt) for the measurement of NO_2^- was fabricated as described in the section "Nitric Oxide Biosensor."

The electrochemical responses of the SOD1–CNT–PPy–Pt and CNT–PPy–Pt electrodes before and after the addition of 250 μM NO_2^- in 0.1 M PBS are displayed in Fig. 4. Upon addition of NO_2^-, SOD1

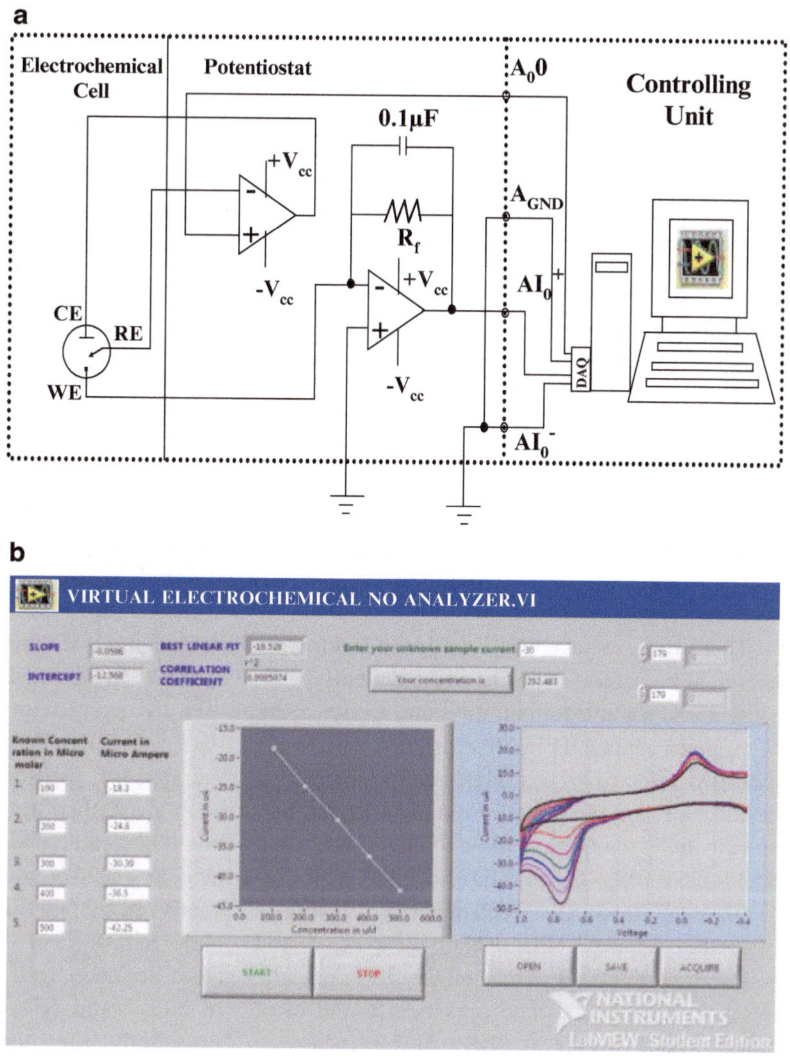

Fig. 3 Virtual electrochemical NO analyzer: (**a**) potentiostat circuit, (**b**) front panel of the measurement

functionalized CNT–PPy nanocomposite electrode exhibited a new irreversible anodic oxidation peak at +0.68 V (curve b) attributed to the anodic oxidation of NO_2^- to NO_3^-. This result clearly demonstrates the electrocatalytic anodic oxidation of NO_2^- mediated by SOD1 functionalized on CNT–PPy–Pt. The present NO_2^- biosensor shows the linear range of response over the concentration of NO_2^- from 0.1 μM to 750 μM and the sensitivity of 0.9 ± 0.003 μA μM^{-1} cm^{-2} with the detection limit of 0.1 ± 0.03 μM. Using this biosensor, the concentrations NO_2^- released from LPS-treated MCF-7 cells were estimated.

Nitrate Biosensor

Earlier, several enzymatic [31, 32] and nonenzymatic [33, 34] electrochemical biosensors have been reported for the determination of NO_3^-. Among these, the enzymatic biosensor method has been widely performed using NaR. NaR is a multidomain enzyme containing flavin adenine dinucleotide (FAD), two heme-Fe, and molybdopterin, which catalyzes the two electron reduction of NO_3^- to NO_2^- [35].

$$NO_3^- + \beta\text{-NAD(P)H} \overset{\text{NaR}}{\rightarrow} NO_2^- + H_2O + \beta\text{-NAD(P)} \quad (1)$$

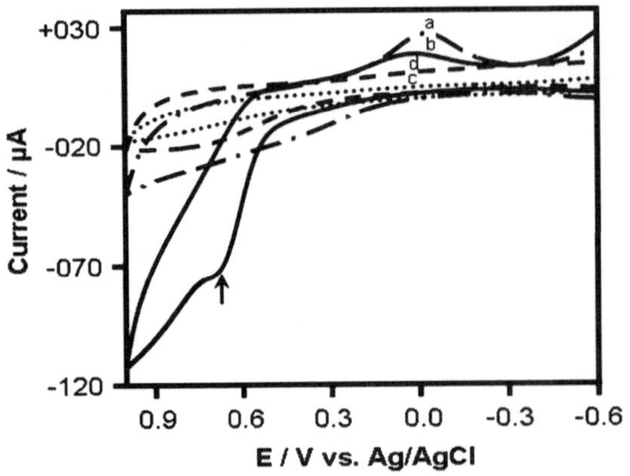

Fig. 4 Typical CV response of SOD1–CNT–PPy–Pt (*a* and *b*) and CNT–PPy–Pt electrodes (*c* and *d*) in the absence of (*a*) and (*c*) and (solid) in the presence of (*b*) and (*d*) of 250 μM NO_2^- solution in 0.1 M PBS (pH 7.0) containing 100 μM DTPA; scan rate: 50 mV s^{-1} vs. Ag/AgCl

Therefore, we have developed here an electrochemical biosensor for NO_3^- by covalently coupling NaR, a multidomain enzyme, to the SAM of cysteine on GNP in PPy matrix. This NaR modified electrode exhibited high sensitivity because of the facilitated electron transfer and enhanced loading of NaR at the SAM–GNP–PPy–Pt electrode.

Figure 5 shows the electrochemical response of the bare Pt, NaR–PPy–Pt, and NaR–SAM–GNP–PPy–Pt electrodes in the absence (a, b, and d) and presence (c and e) of 500 μM NO_3^- in 0.1 M PBS (pH 7.0) measured at the scan rate of 50 mV s^{-1} vs. Ag/AgCl. Before the addition of NO_3^-, no current changes were observed. Further, the current response at the bare Pt electrode is the same as without NO_3^- (data not shown). However, after the addition of NO_3^-, the current responses were significantly increased in both the NaR–PPy–Pt (curve c) and NaR–SAM–GNP–PPy–Pt (curve e) electrodes cathodically at the potential −0.76 V. It is due to the intrinsic specificity of the NaR which catalyzes the reduction of NO_3^- to NO_2^- *via* a cyclic redox reaction of its Mo (IV/VI) complex moiety. This biosensor showed the 1 μM to 1 mM linear range with a detection limit of 0.5 μM NO_3^- and sensitivity of 84.5 nA μM^{-1} cm^{-2}. Further, using the present NO_3^- biosensor, the concentration of NO_3^- present in the beetroot juice was estimated and validated with the standard Griess method. The observed results are shown in Table 1.

Simultaneous Determination of Nitrite and Nitrate Using SOD1 and NaR Coimmobilized Bienzymatic Biosensor

Measurement of various analytes using multienzymatic biosensor in a single experiment is a challenging research area. Therefore, we developed a novel bienzymatic biosensor for the simultaneous determination of NO_2^- and NO_3^- ions using SOD1 and NaR coimmobilized on CNT–PPy nanocomposite-modified platinum electrode (Scheme 2). Figure 6 shows the CVs obtained for the bienzymatic biosensor by increasing the concentration of NO_2^- from 500 nM to 300 μM and NO_3^- from 700 nM to 400 μM using scan rate of 50 mV s^{-1} in 0.1 M PBS (pH 7.0). The observed results exhibit the increase of well-distinguished anodic peak at +0.8 V ascribed to the electrochemical oxidation of NO_2^- catalyzed by SOD1 and the cathodic peak at −0.76 V attributed to the NO_3^- reduction catalyzed by NaR. These results indicate that the NaR–SOD1–CNT–PPy–Pt electrode can be successfully used for the simultaneous measurement of NO_2^- and NO_3^-. Further, the utility of the proposed bienzymatic biosensor for the biological samples was explored by using it for the simultaneous determination of NO_2^- and NO_3^- in human plasma [28].

The simultaneous measured values of NO_2^- and NO_3^- in plasma samples are given in Table 2. The mean ± standard deviation (SD) values of 510.3 ± 3.9 nM for NO_2^- and 16.76 ± 1.2 μM of NO_3^- were obtained.

Electrochemical Assay for NO and Its Metabolites Using Screen-Printed Modified Electrodes

In the previous sections, we have described the determinations of NO, NO_2^- using SOD1, and

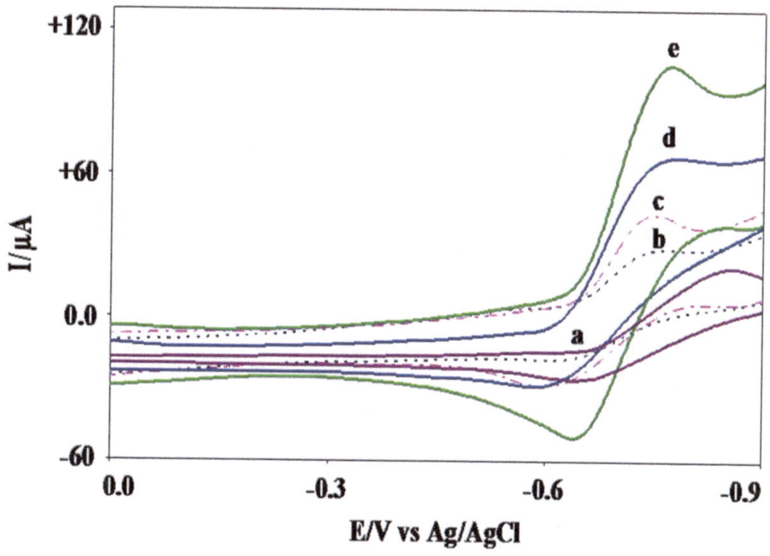

Fig. 5 Electrochemical response of bare Pt, NaR–PPy–Pt, and NaR–SAM–GNP–PPy–Pt electrodes in the absence (*a*, *b*, and *d*) and presence (*c* and *e*) of 500 μM NO_3^- in 0.1 M PBS at scan rate: 50 mV s^{-1} vs. Ag/AgCl

Table 1 Measurement of nitrate in beetroot juice using nitrate biosensor and validation with Griess method

No. of experiments	Nitrate conc. by biosensor (μM ± SD)	Nitrate conc. by Griess method (μM ± SD)
01	593 ± 13.5	590 ± 11.8
02	612 ± 15.3	610 ± 12.2
03	601 ± 15.0	600 ± 12.0

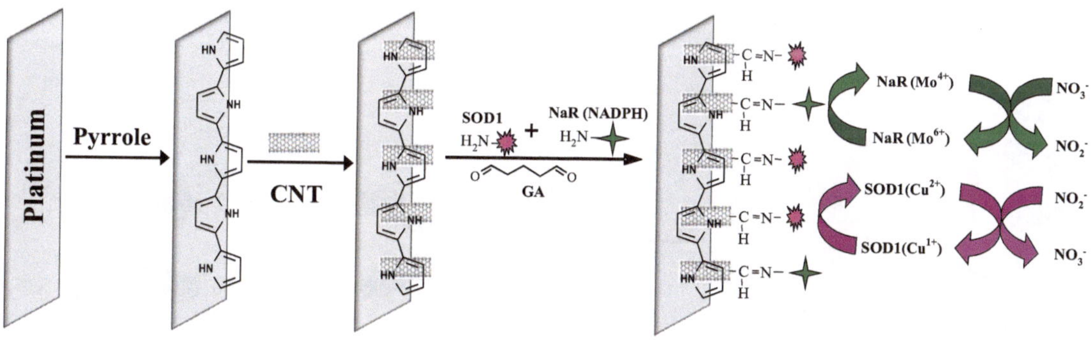

Scheme 2 Schematic representation of the construction of bienzymatic biosensor NaR–SOD1–CNT–PPy–Pt electrode and illustration of reactions takes place during the simultaneous determination of NO_2^- and NO_3^-

NO_3^- using NaR modified Pt electrodes. For all the measurements, 1–2 mL of the sample is required making it difficult to prepare the sample (blood plasma, serum, and cell cultures) for measurements. Therefore, in order to reduce the sample volume, we have developed an electrochemical assay for the collective measurement of NO and its metabolites in a single drop of the sample using SOD1 immobilized on CNT integrated screen-printed carbon electrode (SPCE) as shown in Scheme 3. It allows the mass production of electrochemical biosensors at low cost in comparison to other usual electrodes. The measurement is based on the electrochemical oxidation of NO and NO_2^- mediated by SOD1 *via* a cyclic redox reaction of its Cu (I/II) complex moiety (vide supra). The intrinsic concentrations of NO and NO_2^- were

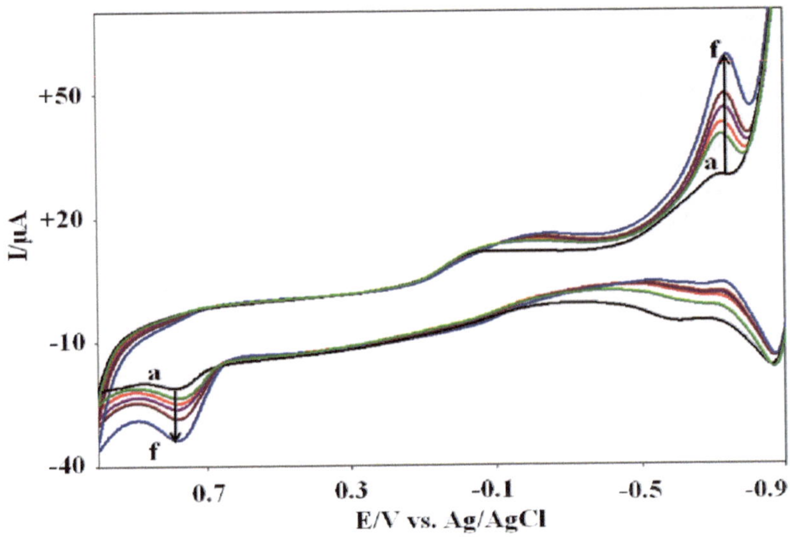

Fig. 6 Typical CV responses obtained for the NaR–SOD1–CNT–PPy–Pt electrode in (*a*) 500 nM NO_2^- + 700 nM NO_3^-, (*b*) 10 μM NO_2^- + 30 μM NO_3^-, (*c*) 30 μM NO_2^- + 50 μM NO_3^-, (*d*) 50 μM NO_2^- + 100 μM NO_3^-, (*e*) 100 μM NO_2^- + 200 μM NO_3^-, and (*f*) 300 μM NO_2^- + 400 μM NO_3^- solution at scan rate: 50 mV s^{-1} vs. Ag/AgCl

Table 2 Simultaneous measurement of NO_2^- and NO_3^- in human plasma using the bienzymatic biosensor

Sample no.	Conc. of NO_2^- (nM mL^{-1})	Conc. of NO_3^- (μM mL^{-1})
1	492 ± 3.0	20.80 ± 1.1
2	594 ± 3.0	19.75 ± 1.3
3	583 ± 5.0	19.26 ± 1.0
4	518 ± 4.3	14.40 ± 1.2
5	390 ± 2.4	9.40 ± 1.4
6	485 ± 6.0	16.95 ± 1.1

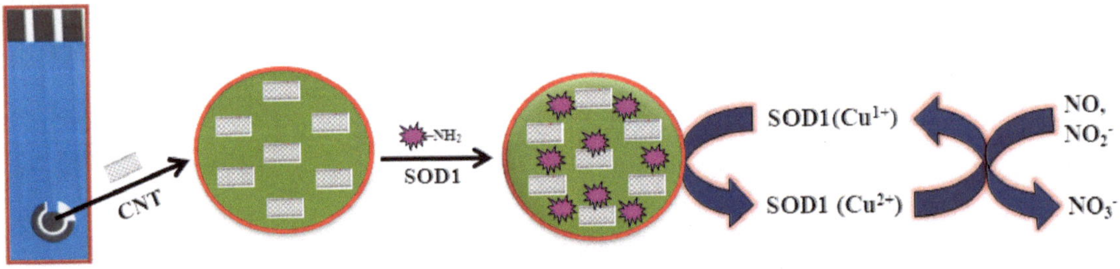

Scheme 3 Representation of the construction of SOD1–CNT–SPCE and the electrochemical mechanism for the determinations of NO and NO_2^-

measured in the sample, whereas NO_3^- was measured after its enzymatic conversion into NO_2^- using NaR and subtracting the $[NO_2^-]$ from the total NOx $[NO_3^- + NO_2^-]$. The possible interferences present in the biological samples were eliminated by coating with nafion (selective for NO) and cellulose acetate membranes (selective for NO_2^- and NO_3^-) [36].

Electrochemical Response to NO and NO_2^-

Figure 7 illustrates the electrochemical responses obtained for the SOD1–CNT–SPCE coated with nafion membrane in the presence of various NO concentrations at the scan rate of 50 mV S^{-1}. The observed anodic peak currents *vs.* NO concentrations were plotted as shown in inset

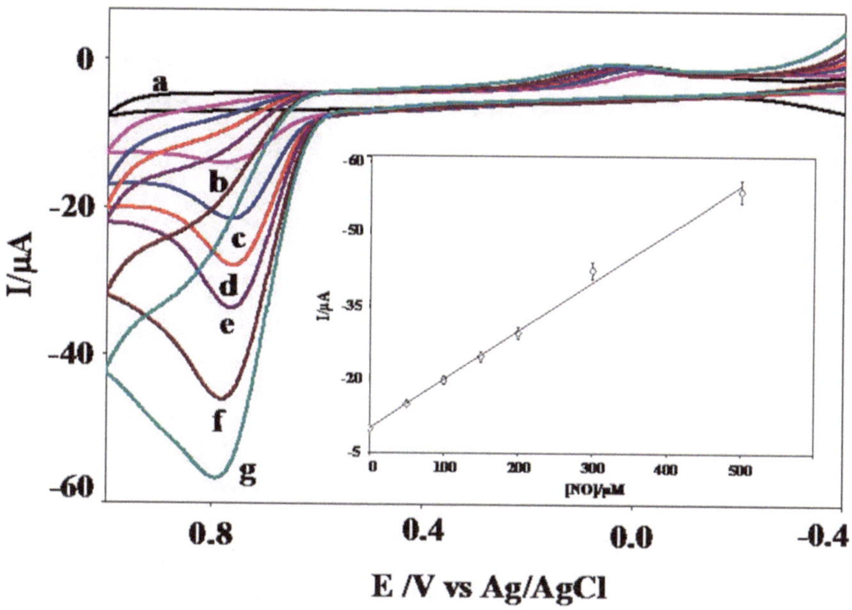

Fig. 7 Electrochemical responses obtained for the SOD1–CNT–SPCE in 0.1 M PBS containing (*a*) control, (*b*) 50 μM, (*c*) 100 μM, (*d*) 150 μM, (*e*) 200 μM, (*f*) 300 μM, and (*g*) 500 μM of NO solution at scan rate: 50 mV s^{-1} vs. Ag/AgCl. Linear calibration curve (inset of Fig. 7) $y = -0.0981 \times -10.096, r^2 = 0.9953$

of Fig. 7. The calibration curve thus obtained exhibits a linear range of response over the concentration of NO from 200 nM to 500 μM, but for clarity here we have shown from 50 μM to 500 μM ($r^2 = 0.9953$, $n = 3$) with a detection limit of 100 nM and sensitivity of 85.4 nA μM^{-1}. Further, the electrochemical responses of the same electrode coated with cellulose acetate membrane at the various concentrations of NO_2^- are shown in Fig. 8. The calibration curve obtained here also exhibits a linear range of response over the concentration of NO_2^- from 100 nM to 1 mM ($r^2 = 0.9973$, $n = 3$) with a detection limit of 100 nM and sensitivity of 96.4 nA μM^{-1}.

Measurement of Total NO_2^- and NO_3^- Levels in Human Plasma

The concentration of blood nitrite is reportedly altered in high-altitude subjects at hypoxia condition [37]. Recent study reported that the administration of NO_3^--rich beetroot juice to human and several animal models promotes NO-like bioactivity and regulates biological activities like reduction of blood pressure, vasodilation, cytoprotection, cardioprotection, and protection from ischemia–reperfusion injury [22, 23, 38]. Therefore, we have attempted to measure the total NO_2^- and NO_3^- levels in human plasma of four subjects before and after beetroot supplementation. The concentrations of NO_2^- were measured before and after enzymatic reduction of NO_3^- into NO_2^-. Further, the concentration of NO_3^- was deduced by subsequent subtraction of $[NO_2^-]$ from the total $[NO_2^- + NO_3^-]$:

$$[NO_3^-] = [NO_2^- + NO_3^-] - [NO_2^-]$$

One drop of the plasma sample was placed on the cellulose acetate membrane-coated SOD1–CNT–SPCE to measure total NOx. The corresponding current response was observed at +0.83 V. Further, another aliquot of the plasma sample was taken to measure total NOx present in the sample using colorimetric assay kit (Griess method), and the obtained results were tabulated as shown in Table 3. The obtained values using biosensor and Griess methods were critically compared and found to be in agreement with each other. It could be concluded from Table 3 that the administration of exogenous NO_3^- possibly increases the total

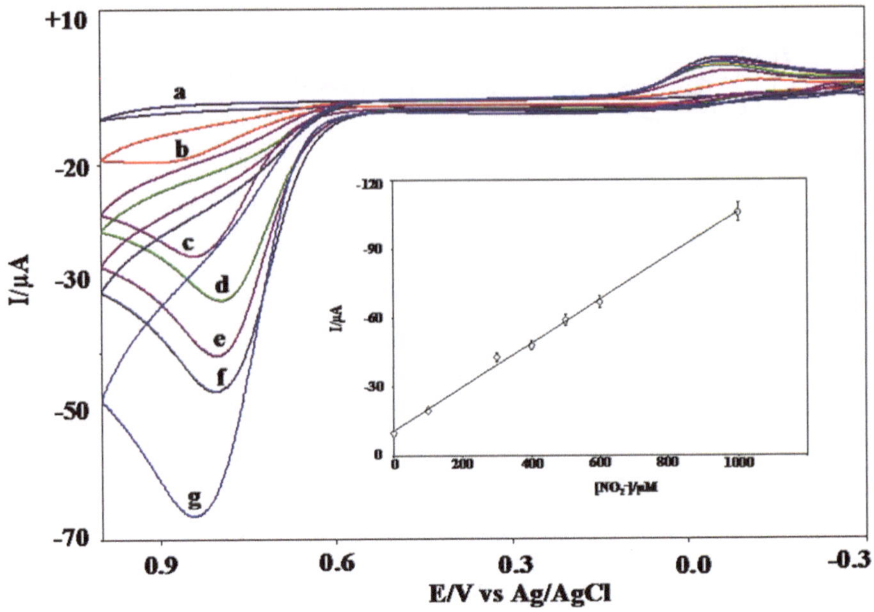

Fig. 8 Electrochemical responses obtained for the SOD1–CNT–SPCE in 0.1 M PBS containing (*a*) control, (*b*) 100 µM, (*c*) 300 µM, (*d*) 400 µM, (*e*) 500 µM, (*f*) 600 µM, and (*g*) 1,000 µM of NO_2^- solution at scan rate: 50 mV s^{-1} vs. Ag/AgCl. Linear calibration curve (*inset* of Fig. 8) $y = -0.0953 \times -10.91$, $r^2 = 0.9973$

Table 3 Measurement of total NO_2^- + NO_3^- levels in human plasma before and after beetroot supplementation

Individual	Total NO_2^- + NO_3^- conc. by Griess method (µM ± SD)		Total NO_2^- + NO_3^- conc. by Sensor (µM ± SD)	
	Before beetroot juice	After beetroot juice	Before beetroot juice	After beetroot juice
01	28.4 ± 1.8	29.7 ± 1.7	27.0 ± 0.9	28.8 ± 0.7
02	24.5 ± 1.1	32.1 ± 2.0	23.9 ± 1.3	31.7 ± 1.4
03	21.9 ± 1.3	26.1 ± 1.8	22.5 ± 0.8	24.8 ± 1.3
04	24.7 ± 1.4	28.9 ± 1.5	24.8 ± 1.1	29.2 ± 1.4

NO_2^- and NO_3^- concentrations in blood, thereby enhancing the NO–NO_2^-–NO_3^- pathway.

Cytochrome *c* Biosensor

Cyt *c* is a heme protein located in the mitochondrial intermembrane space. It plays a key role in biological respiratory chain, whose function is to transfer electrons between cytochrome *c* reductase (CcR) (complex III) and cytochrome *c* oxidase (CcO) (complex IV) [39, 40]. Mitochondria, besides their primary physiological function to generate ATP through oxidative phosphorylation, are also important source for the production of cellular ROS. Recent findings have shown that hypoxia can cause mitochondrial dysfunction, protein oxidation, and excessive cellular damage, all of which ultimately lead to the release of cyt *c* from mitochondria to cytoplasm [41]. This translocation of mitochondrial cyt *c* to cytosol is a critical event in the activation of intracellular signaling; it results in a cascade of caspase activation and leads to programmed cell death (apoptosis) [42]. Thus, the absolute quantification of cyt *c* release as a biomarker of apoptosis is of great importance in clinical diagnosis and therapeutic research.

Recently, amperometric biosensors for the direct determination of cyt *c* have been reported with good detection limits [43, 44]. However, they suffer from lack of selectivity in the measurement, especially in cells or biological samples, due to the fact that the interaction of the recognition elements (negative charge) with cyt *c* (positive charge) is purely based on electrostatic interactions. So, these methods are prone to interferences by other positively charged species present in the samples and hence are not applicable for the quantification of cyt *c* release in biological systems. Enzymatic biosensors for the determination of cyt *c* have also been investigated by incorporating cytochrome *c* oxidase (CcO) [45, 46]. However, the CcO-based biosensors are capable of determining only the reduced form of cyt *c* (Fe^{2+}) by mediating electron transfer between the cyt *c* (Fe^{2+}) and the electrode. But in apoptotic cells, only the oxidized form of cyt *c* (Fe^{3+}) triggers the time-dependent caspase activation and serves as a proapoptotic molecule [47]. Moreover, in permeabilized cell models, the cytosolic cyt *c* (Fe^{2+}) is rapidly oxidized (Fe^{3+}) by the mitochondrial CcO [48], thus making the CcO-based biosensors difficult to quantify the apoptotic form of cyt *c* (Fe^{3+}). Further, during immobilization, it is reported that the electron transfer is blocked in the active centers of CcO [49]. Consequently, the analytical applications of CcO-based biosensors are also limited.

Thus, there is a real need for simple, rapid, selective, and inexpensive methods for cyt *c* (Fe^{3+}) measurement for point-of-care and research applications. So, we developed an alternate method for the detection of mitochondrial cyt *c* release using CcR functionalized with nanocomposites decorated electrodes [50]. Two kinds of nanomaterial-based biosensor platforms were used: (a) CNT incorporated PPy matrix on Pt electrode and (b) SAM-functionalized GNP in PPy–Pt. The coupling of oxidation–reduction and deprotonation–protonation of ubiquinone (QH_2) with the CcR is the central mechanism in cyt *c* reduction [51]. So, we have used hydroquinone

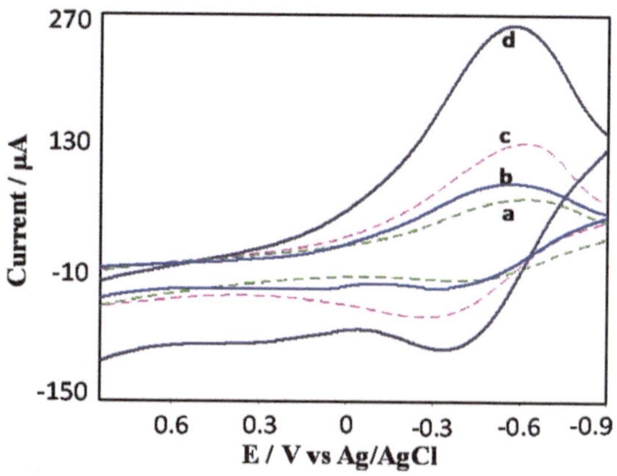

Fig. 9 CVs obtained for CcR–SAM–GNP–PPy–Pt (*dotted lines*) and CcR–CNT–PPy–Pt electrodes in the absence (*a* and *b*) and in the presence (*c* and *d*) of 500 μM of cyt c, respectively, in 0.1 M PBS containing 100 μM DTPA; scan rate: 50 mVs^{-1} vs. Ag/AgCl

(HQ) as a mediator for the reduction of cyt *c* by CcR. The current responses for the CcR–CNT–PPy–Pt biosensor were studied in 0.1 M PBS (pH 7.0) containing 100 μM cyt *c* in different concentrations of HQ. The current increased sharply with increasing HQ concentrations to a maximal response at 0.5 mM and then became steady with the further increase in HQ concentrations. Thus, the optimized 0.5 mM HQ was used for our experiments.

Figure 9 displays the CVs obtained for CcR–SAM–GNP–PPy–Pt (dotted lines) and CcR–CNT–PPy–Pt electrodes (solid lines) in 0.1 M PBS containing 0.5 mM HQ in the absence and presence of 500 μM cyt *c*. In both the cases, before the addition of cyt *c*, characteristic redox peaks (−0.45 and −0.34 V *vs.* Ag/AgCl) due to the immobilized CcR were observed. Upon addition of cyt *c*, the current increased cathodically at −0.45 V and also anodically at −0.35 V, which was attributed to the redox reaction of cyt *c* by the CcR. Under similar conditions, there were no changes in CV responses observed for the PPy–Pt, GNP–PPy–Pt, and CNT–PPy–Pt electrodes. However, for these electrodes (without CcR) in the presence of cyt *c*, new reversible redox peaks appeared at 0.05 and 0.07 V *vs.* Ag/AgCl, characteristics of cyt *c* [52], whereas the peaks due to cyt *c* were not exhibited by the CcR

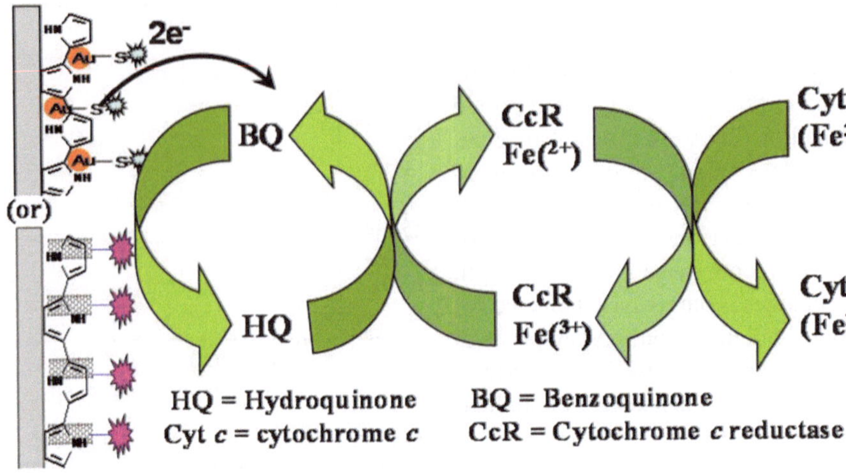

Scheme 4 Schematic illustration of the biochemical reaction occurring at the biosensor surface

immobilized electrodes. Therefore, the increases in current responses at −0.45 and −0.34 V are mainly due to the redox reaction of cyt *c* by the CcR but not due to the cyt *c* in solution. The present data thus confirms the immobilized CcR showing an enhanced biocatalytic activity towards the reduction of cyt *c* (Fe^{3+}). The schematic representation of the biochemical reaction of the CcR with cyt *c* occurring at the electrodes surface is shown in Scheme 4.

The typical CVs obtained for several concentrations of cyt *c* in 0.1 M PBS containing 0.5 mM HQ using these two cyt *c* biosensors at 50 mVs^{-1} are shown in Fig. 10. The current responses to cyt *c* obtained with the CcR–CNT–PPy–Pt biosensor were linear from 1 to 1,000 μM ($r^2 = 0.997$), with a detection limit of 0.5 ± 0.03 μM and sensitivity of 0.46 ± 0.003 $\mu A\ \mu M^{-1}\ cm^{-2}$.

Measurement of cyt *c* Release from Mitochondria

An important step in the mitochondrial pathway is the release of cyt *c* from mitochondria into cytosol. It has been earlier demonstrated that the cyt *c* translocation from the mitochondria into cytosol preceded doxorubicin (DOX) induced apoptosis in various cell and animal models [53, 54]. So, we have chosen DOX for the induction of apoptosis in human lung carcinoma A549 cells. Recent findings revealed that the oxidized form of cyt *c* (Fe^{3+}) mainly induced the caspase activation, thereby causing apoptosis over the reduced form of cyt *c* (Fe^{2+}) [48]. It clearly indicates that the measurement of only the oxidized form of cyt *c* (Fe^{3+}) in cytosol presumably serves as a marker for apoptotic process in cells. In this report, cyt *c* (Fe^{3+}) measurements were performed on the cytosolic fractions of DOX-treated and untreated human lung carcinoma apoptotic A549 cells using the CcR–CNT–PPy–Pt biosensor and western blot. After 24 h exposure of cells with 1 μM DOX, the cyt *c* concentration in cytosolic fractions of the cells (3.63 ± 0.02 μM) was increased when compared to that in untreated cells (2.4 ± 0.02 μM). Treatment for 48 h resulted in further increase in cytosolic cyt *c* (5.3 ± 0.018 μM) levels. These results are quite comparable with the cell viability studies and western blot analysis. Further, we have also developed a miniaturized label-free electrochemical immunosensor assay for the measurement of cyt *c* using antibody functionalized screen-printed electrodes [55].

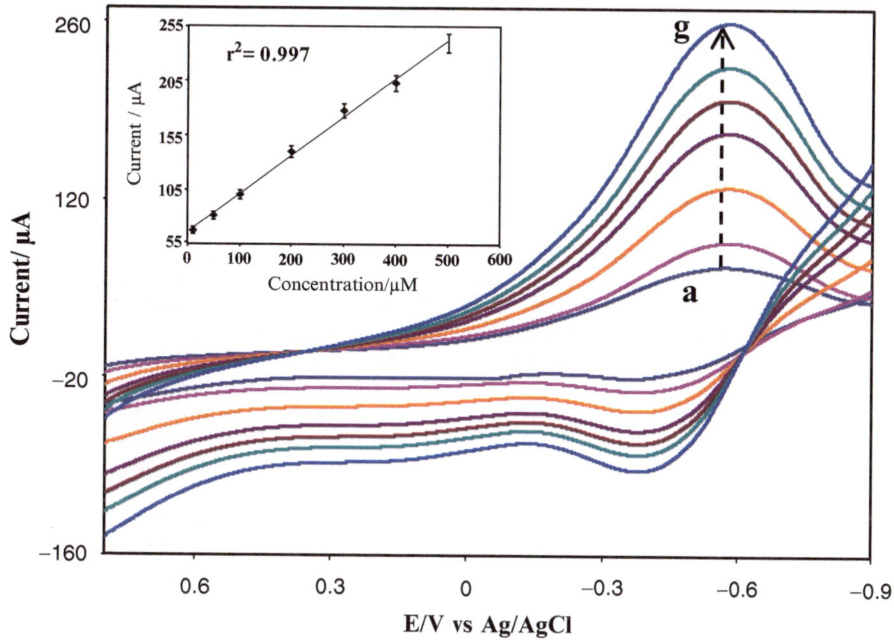

Fig. 10 Typical CV responses of the CcR–CNT–PPy–Pt electrode in 0.1 M PBS containing 0.5 mM HQ, without (*a*) and with 50, 100, 200, 300, 400, and 500 µM of cyt c (*b*–*g*) measured at scan rate of 50 mVs^{-1}. A linear calibration plot of cathodic peak currents against cyt *c* concentrations (inset of Fig. 10). Each point represents the mean (±0.03 SD) of three measurements

Conclusion

This chapter focused primarily on the significance of hypoxia biomarkers, *viz.* NO, NO_2^-, NO_3^-, and cyt *c* and explained briefly the development of biosensors for the measurement of these biomarkers by using SOD1, NaR, and CcR modified Pt/screen-printed electrodes. The nanocomposites, *viz.* CNT/GNP in conducting PPy, used as host matrices enhanced the immobilization of enzymes without affecting its biological activity and direct electron transfer between the active site of the enzyme and the underlying electrode resulting in high sensitivity, wider linear range, lower limit of detection, and fast response. The sample volume was further miniaturized into one drop (50 µL) for the measurement using screen-printed modified electrodes. The electroanalytical parameters thus obtained were remarkably improved over Pt modified electrode in 1–2 mL of the sample. These biosensors were successfully applied to measure the concentrations of cyt *c*, NO, and its metabolites in various biological samples. Further, low-cost microcontroller-based handheld biosensors for detecting low levels of hypoxia biomarkers with high sensitivity are in progress for monitoring the high-altitude subjects.

Acknowledgment Thanks are due to the support provided by the DIPAS-DRDO, DBT, New Delhi, and the Managing Board of Virudhunagar Hindu Nadar's Senthikumara Nadar College (Autonomous), Virudhunagar, Tamil Nadu, India.

References

1. Solaini G, Baracca A, Lenaz G, Sgarbi G (2010) Hypoxia and mitochondrial oxidative metabolism. Biochim Biophys Acta 1797:1171–1177
2. Lee K, Roth RA, Lapres JJ (2007) Hypoxia, drug therapy and toxicity. Pharmacol Ther 113:229–246
3. Dosek A, Ohno H, Acs Z, Taylor AW, Radak Z (2007) High altitude and oxidative stress. Respir Physiol Neurobiol 158:128–131

4. Fearon IM, Faux SP (2009) Oxidative stress and cardiovascular disease: novel tools give (free) radical insight. J Mol Cell Cardiol 47:372–381
5. Guix FX, Uribesalgo I, Coma M, Munoz FJ (2005) The physiology and pathophysiology of nitric oxide in the brain. Prog Neurobiol 76:126–152
6. Hemmingsson T, Horn A, Linnarsson D (2009) Measuring exhaled nitric oxide at high altitude. Respir Physiol Neurobiol 167:292–298
7. Lundberg JO, Weitzberg E, Gladwin MT (2008) The nitrate-nitrite-nitric oxide pathway in physiology and therapeutics. Nat Rev Drug Discov 7:156–167
8. Pacher P, Beckman JS, Liaudet L (2007) Nitric oxide and peroxynitrite in health and disease. Physiol Rev 87:315–424
9. Raat NJH, Noguchi AC, Liu VB, Raghavachari N, Liu D, Xu X, Shiva S, Munson PJ, Gladwin MT (2009) Dietary nitrate and nitrite modulate blood and organ nitrite and the cellular ischemic stress response. Free Radic Biol Med 47:510–517
10. Nataliya B, Andrei N (2005) A spectrophotometric assay for nitrate in an excess of nitrite. Nitric Oxide 13:93–97
11. Guevara I et al (1998) Determination of nitrite/nitrate in human biological material by the simple Griess reaction. Clin Chim Acta 274:177–188
12. Masserini RT, Fanning KA (2000) A sensor package for the simultaneous determination of nanomolar concentrations of nitrite, nitrate and ammonia in seawater by fluorescence detection. Mar Chem 68:323–333
13. Bates JN (1992) Nitric oxide measurement by chemiluminescence detection. Neuroprotoc 1:141–149
14. Tsikas D, Bohmer A, Mitschke A (2010) Gas chromatography–mass spectrometry analysis of nitrite in biological fluids without derivatization. Anal Chem 15:384–5390
15. Tsikas D, Suchy MT, Mitschke A, Beckmann B, Gutzki FM (2012) Measurement of nitrite in urine by gas chromatography – mass spectrometry. Methods Mol Biol 844:277–293
16. Jedlickova V, Paluch Z, Alusik S (2002) Determination of nitrate and nitrite by high-performance liquid chromatography in human plasma. J Chromatogr B 780:193–197
17. Jobgen WS, Jobgen SC, Li H, Meininger CJ, Wu G (2007) Analysis of nitrite and nitrate in biological samples using high-performance liquid chromatography. J Chromatogr B 851:71–82
18. Kotamraju S, Konorev EA, Joseph J, Kalyanaraman B (2000) Doxorubicin-induced apoptosis in endothelial cells and cardiomyocytes is ameliorated by nitrone spin traps and ebselen. J Biol Chem 275:33585–33592
19. Waterhouse NJ, Trapani JA (2003) A new quantitative assay for cytochrome c release in apoptotic cells. Cell Death Differ 10:853–855
20. Crouser ED, Gadd ME, Julian MW, Huff JE, Broekemeier KM, Robbins KA, Pfeiffer DR (2003) Quantification of cytochrome c release from rat liver mitochondria. Anal Biochem 317:67–75
21. Nakashima T, Sano M (2011) Single-walled carbon nanotube as 1D array of reacting sites: reaction kinetics of reduction of cytochrome c in tris buffer. J Phys Chem C 115:20931–20936
22. Webb AJ et al (2008) Acute blood pressure lowering, vasoprotective, and antiplatelet properties of dietary nitrate via bioconversion to nitrite. Hypertension 51:784–790
23. Larsen FJ, Weitzberg E, Lundberg JO, Ekblom B (2007) Effects of dietary nitrate on oxygen cost during exercise. Acta Physiol (Oxf) 191:59–66
24. Joshipura KJ et al (2001) The effect of fruit and vegetable intake on risk for coronary heart disease. Ann Intern Med 134:1106–1114
25. Crowe FL et al (2011) Fruit and vegetable intake and mortality from ischaemic heart disease:results from the European Prospective Investigation into Cancer and Nutrition (EPIC)-Heart study. Eur Heart J 32:1235–1243
26. Madasamy T, Pandiaraj M, Santosh K, Balamurugan M, Robson B, Arunvenkatesh K, Vairamani K, Kotamraju S, Karunakaran C (2012) Virtual electrochemical nitric oxide analyzer using copper, zinc superoxide dismutase immobilized on carbon nanotubes in polypyrrole matrix. Talanta 100:168–174
27. Rajesh S, Koteswararao A, Kalpana B, Ilavazhagan G, Kotamraju S, Karunakaran C (2010) Simultaneous electrochemical determination of superoxide anion radical and nitrite using Cu, ZnSOD immobilized on carbon nanotubes in polypyrrole matrix. Biosens Bioelectron 26:689–695
28. Madasamy T, Pandiaraj M, Balamurugan M, Kalpana B, Sethy NK, Karunakaran C (2014) Copper, zinc superoxide dismutase and nitrate reductase coimmobilized bienzymatic biosensor for the simultaneous determination of nitrite and nitrate. Biosens Bioelectron 52:209–215
29. Robinson JK, Bollinger MJ, Birks JW (1999) Luminol/H_2O_2 chemiluminescence detector for the analysis of nitric oxide in exhaled breath. Anal Chem 71:5131–5136
30. Karunakaran C, Zhang H, Crow JP, Antholine WE, Kalyanaraman B (2004) Direct probing of copper active site and free radical formed during bicarbonate-dependent peroxidase activity of bovine and human copper, zinc-superoxide dismutases. J Biol Chem 279:32534–32540
31. Kirstein D, Kirstein L, Scheller F, Borcherding H, Ronner J, Dieckmann S, Steinrucke P (1999) Amperometric nitrate biosensors on the basis of Pseudomonas stutzeri nitrate reductase. J Electroanal Chem 474:43–51
32. Cosnier S, Innocent C, Jouannea Y (1994) Amperometric detection of nitrate via nitrate reductase immobilized and electrically wired at the electrode surface. Anal Chem 66:3198–3201
33. Gamboa JCM, Pena RC, Paixao TRLC, Bertotti M (2009) A renewable copper electrode as an amperometric flow detector for nitrate determination in mineral water and soft drink samples. Talanta 80:581–585

34. Groot MTD, Koper MTM (2004) The influence of nitrate concentration and acidity on the electrocatalytic reduction of nitrate on platinum. J Electroanal Chem 562:81–94
35. Quan D, Shim JH, Kim JD, Park HS, Cha GS, Nam H (2005) Electrochemical determination of nitrate with nitrate reductase immobilized electrodes under ambient air. Anal Chem 77:4467–4473
36. Madasamy T, Pandiaraj M, Koteswararao AK, Rajesh S, Kalpana B, Sethy NK, Kotamraju S, Karunakaran C (2013) Gold nanoparticles with self-assembled cysteine monolayer coupled to nitrate reductase in polypyrrole matrix enhanced nitrate biosensor. Adv Chem Lett 1:2–9
37. Van Faassen EE et al (2009) Nitrite as a regulator of hypoxic signaling in mammalian physiology. Med Res Rev 29:683–741
38. Dezfulian C, Raat N, Shiva S, Gladwin MT (2007) Role of the anion nitrite in ischemia – reperfusion cytoprotection and therapeutics. Cardiovasc Res 75:327–338
39. Ow YP, Green DR, Hao Z, Mak TW (2008) Cytochrome c: functions beyond respiration. Nat Rev Mol Cell Biol 9:532–542
40. Huttemann M, Pecina P, Rainbolt M, Sanderson TH, Kagan VE, Samavati L, Doan JW, Lee I (2011) The multiple functions of cytochrome c and their regulation in life and death decisions of the mammalian cell: from respiration to apoptosis. Mitochondrion 11:369–381
41. Radhakrishnan J, Wang S, Ayoub IM, Kolarova JD, Levine RF, Gazmuri RJ (2007) Circulating levels of cytochrome c after resuscitation from cardiac arrest: a marker of mitochondrial injury and predictor of survival. Am J Physiol Heart Circ Physiol 292:767–775
42. Zhou F, Wu S, Wu B, Chen WR, Xing D (2011) Mitochondria-targeting single-walled carbon nanotubes for cancer photothermal therapy. Small 7:2727–2735
43. Zhao J, Zhu X, Li T, Li G (2008) Self-assembled multilayer of gold nanoparticles for amplified electrochemical detection of cytochrome c. Analyst 133:1242–1245
44. Liu Y, Wei W (2008) Detection of cytochrome c at biocompatible nanostructured Au-lipid bilayer-modified electrode. Anal Sci 24:1431–1436
45. Li J, Cheng G, Dong S (1996) Direct electron transfer to cytochrome c oxidase in self assembled monolayers on gold electrodes. J Electroanal Chem 416:97–104
46. Ashe D, Alleyne T, Iwuoha E (2007) Serum cytochrome c detection using a cytochrome c oxidase biosensor. Biotechnol Appl Biochem 46:185–189
47. Pasdois P, Parker JE, Griffiths EJ, Halestrap AP (2011) The role of oxidized cytochrome c in regulating mitochondrial reactive oxygen species production and its perturbation in ischaemia. Biochem J 436:493–505
48. Brown GC, Borutaite V (2008) Regulation of apoptosis by the redox state of cytochrome c. Biochim Biophys Acta 1777:877–881
49. Hrabakova J, Ataka K, Heberle J, Hildebrandt P, Murgida DH (2006) Long distance electron transfer in cytochrome c oxidase immobilized on electrodes. A surface enhanced resonance Raman spectroscopic study. Phys Chem Chem Phys 8:759–766
50. Pandiaraj M, Madasamy T, Paradesi Naidu G, Balamurugan M, Kotamraju S, Rao VK, Kalpana B, Karunakaran C (2013) Nanomaterial based biosensors for cytochrome *c* using cytochrome *c* reductase. Bioelectrochemistry 91:1–7
51. Mulkidjanian A (2005) Ubiquinol oxidation in the cytochrome bc1 complex: reaction mechanism and prevention of short- circuiting. Biochim Biophys Acta 1709:5–34
52. Puig MC, Berbel XM, Rouillon R, Blanchard CC, Marty J (2009) Development of a cytochrome c based screen-printed biosensor for the determination of the antioxidant capacity of orange juices. Bioelectrochemistry 76:76–80
53. Luipertz R, Watjen W, Kahl R, Chovolou Y (2010) Dose- and time-dependent effects of doxorubicin on cytotoxicity, cell cycle and apoptotic cell death in human colon cancer cells. Toxicology 271:115–121
54. Kalyanaraman B, Joseph J, Kalivendi S, Wang S, Konorov EA (2002) Doxorubicin-induced apoptosis: implications in cardiotoxicity. Mol Cell Biochem 234:119–124
55. Pandiaraj M, Niroj Sethy K, Kalpana B, Rao VK, Karunakaran C (2014) Designing label-free electrochemical immunosensors for cytochrome *c* using nanocomposites functionalized screen printed electrodes. Biosens Bioelectron 54:115–121

Determining Nutritional Requirements of Indian Soldiers: An Outcome of Translational Research

Som Nath Singh

Abstract

Soldiers are deployed in different climatic conditions such as hot deserts, hot humid jungles, snowbound high-altitude areas, underwater (submariners), etc. Adequate nutritional support is a prime requirement for maintaining highest level of physical fitness under different climatic conditions and operational situations. The energy expenditures of soldiers during training and different operations are much higher than civilian population. The average energy expenditure of different units of Indian Army under normal and specialized trainings is 2,900–4,500 kcal. Under extreme environmental conditions, the thirst and appetite responses also get affected which will have an impact on food and water intake. Decreased food intake due to high-altitude anorexia causes negative energy balance leading to loss of body mass in lowlanders visiting high altitudes. High-carbohydrate diets are beneficial under hypoxic environment of high altitudes during early days of acclimatization. Maintenance of adequate hydration and electrolyte level is very important for hot environments. The planning of ration for combat operations is a challenging task, as balance between nutrient requirement for optimum health and palatability needs to be ensured. Translational research has played a great role in formulating dietary recommendations and ration scales of armed forces and at the same time, various ready-to-eat products emerged as the innovative products of military nutrition research.

Introduction

The relationship between diet and health got scientific basis only since nineteenth century after Lavosier established that oxidation of nutrients provides energy and Magendie recognized the importance of proteins for survival. Nutritional deficiency diseases such as beriberi, pellagra, and scurvy were all known and described much before the food factors responsible for these illnesses were discovered. Nutrition is the area where translational research has been applied extensively in formulating healthy diet plans, taking inputs from animal

S.N. Singh (✉)
Nutrition Division, Defence Institute of Physiology and Allied Sciences, Lucknow Road, Timarpur, New Delhi 110054, India
e-mail: nutrition-dipas@hotmail.com

S.B. Singh et al. (eds.), *Translational Research in Environmental and Occupational Stress*, DOI 10.1007/978-81-322-1928-6_10, © Springer India 2014

studies and direct interventions on humans. Depletion and replenishment studies carried out on experimental models like dogs and livestock as well as on humans resulted in identification and role of different nutrients. Incidences of chronic nutritional deficiency diseases like protein energy malnutrition, beriberi, rickets, etc., mentioned in the nutrition textbooks are now rarely seen in the community and the credit goes to translational research in food planning. However, overnutrition and sedentary lifestyle-related degenerative diseases are increasing. The use of functional foods and nutraceuticals for health promotion is also increasing. The advancement in food technology has resulted in many ready-to-eat convenience food products initially developed for expeditions, and armed forces are being included in day-to-day menu.

The success of the military missions depends on overall health and training of soldiers and use of advanced war technology. Physical demands of combat combined with mental stress caused due to long working hours affect health and morale of soldiers. The situation may get compounded under extreme environmental conditions, even if it is for short duration. Adequate nutritional support before and during operation effectively reduces adverse effects of combat stress. Soldiers need to be "fighting fit" and that is only possible when they are well nourished, trained, and preventive measures against infectious disease have been taken [4]. In early days, armies largely depended on local food availability and faced starvation if there was scarcity in the area of campaign. The Napoleonic dictum, "an Army marches on its stomach," holds true even under modern war situations.

The study of energy expenditure by soldiers during different types of work was conducted during World War I (1914–1918) by Cathcart and Orr and the composition of potential ration components was analyzed by Plimmer. The outbreak of scurvy among Indian troops caused heavy mortality in 1916. British soldiers suffered heavily with beriberi in Mesopotamia (today's Iraq) and situation improved when they entered Palestine, where fresh fruits became available. Based on this observation, a provision of issuing certain items of food to soldiers along with cash allowance was made. After independence, the basic scales of ration during peace time and geographical areas were formulated and implemented after certain modifications based on studies at DIPAS [31].

Soldiers work under different climatic conditions such as plains, hot deserts, hot humid jungles, snowbound high-altitude areas, underwater (submariners), etc., and their nutritional needs are different [35]. Nutritional requirements of armed forces personnel and adequacy of ration scales have been determined after conducting studies under actual field conditions. Our institute has made significant contribution in the field of military nutrition since 1963, when, for the first time, the ration scale for high altitude was introduced [20]. Subsequently, various ration scales, viz. peace and field scales, high-altitude officers ration scale, submarine ration, different diets for Army Hospitals, ration scale for military and Sainik School boys, ration scale for Border Roads Organization, etc., were evaluated for their nutritional adequacy [16–19, 27, 39, 40]. The nutritional requirements of Indian Army and specialized trainings such as high-altitude warfare and commandoes have been reevaluated using detailed investigations [36, 37, 45].

Energy Requirements of Soldiers

The assessment of total energy expenditure (TEE) is required for determining energy requirements of an individual. TEE has three components: basal metabolic rate (BMR), activity-related energy expenditure (AEE), and thermic effect of food (TEF). BMR constitutes major part of the total energy expenditure in case of sedentary individuals, AEE is highly variable (based on this, people are grouped into sedentary, moderate, and heavy activity groups), and TEF is about 10 % of total energy expenditure, when individual is consuming mixed diet. The TEE of adults mainly depends upon their basal metabolic rate (BMR) and physical activity (PA) levels, which varies largely among the population [11].

Total energy expenditure by individual should be met by intake. Imbalance in energy expenditure may lead to over- or undernutrition. Earlier food intake at which body weights were maintained within normal range was considered as requirement. Intake data in the form of reference dietary intakes (RDI) are still used for determining safe intakes of nutrients for energy (average food intake + 2 standard deviations).

TEE assessment is difficult to perform on free-living healthy individuals. Soldiers have a busy working schedule, leaving less scope for their involvement in the experiment for a long time. However, units have well-defined work schedule due to which use of factorial method for assessing energy expenditure can be made. The energy cost of different activities is determined using oxygen consumption under experimental setup and time spent in activities is recorded to compute energy expenditure [9].

Application of doubly labeled water (DLW) technique to human subjects has been a major breakthrough in measuring free-living energy expenditure, which is now accepted as the gold standard [3, 34]. The DLW method involves the ingestion of a dose of stable isotopic water (^2H_2O and $H_2{}^{18}O$) and periodic collection of urine sample for washout kinetics of ^2H and ^{18}O. The difference in elimination rates of two isotopic species gives amount of CO_2 produced in metabolic activity which can be used to calculate TEE over a period of 15–21 days. The use of accelerometer-based activity monitoring devices is gaining popularity due to convenience and was validated against DLW technique [1, 38, 43].

The published data on energy expenditures by soldiers is mainly from studies carried out during military trainings conducted for 3–10 days. The energy expenditure is more during such trainings in comparison to normal days at units. Tharion et al. [44] have reviewed energy requirements of military personnel and reported energy expenditure of male soldiers is in the range of 3,109–7,131 kcal/day with a mean of 4,610 ± 650 kcal ($n = 424$). The energy expenditure of female soldiers is reported to be 2,850 ± 620 kcal/day (range 2,332–5,597, $n = 77$). The highest energy expenditure was reported in case of Norwegian ranger cadets (93.5 kcal/kg/day) followed by Zimbabwean combat commando soldiers (85.4 kcal/kg/day) and during crucible exercise of US Marine recruits (82 kcal/kg body weight) [26, 44]. The energy expenditures by Indian soldiers were studied by our team under different conditions of climate and training [36] and are given in Table 1. The energy expenditures of Indian soldiers during basic military training and commando training are comparable to US Army rangers. The main contributing factor to high energy expenditure is extended working hours of soldiers (16–20 h/day) and physically demanding training activities.

Table 1 Energy expenditure by Indian troops under different environments and training

Type of troops and environment	Total energy expenditure (kcal/day)
Army	
Sea level – combat and support	3,511 ± 601
Desert – combat and support	3,304 ± 593
Training centers – infantry	4,670 ± 345
Training center – support*	3,487 ± 728
Commando training*	4,498 ± 1,353
HA warfare training*	4,837 ± 322
HA (2,700–4,500 m) combat and support	3,880 ± 474
HA- (>4,500 m) – combat	4,270 ± 550
Air force	
Officers	3,615 ± 198
Ground crew	2,900 ± 320
Navy	
Ship crew	3,313 ± 578
Submariners	3,168 ± 282
MARCOs and divers	4,055 ± 465

Energy expenditure measured using oxygen consumption, and * indicates where DLW method was used. Refs. [36, 37, 45]

Requirements of Nutrients

The nutrient balance studies over a period of 3 months (taking data on food intake, energy expenditure, nutrient levels in body, excretion levels, and stable body composition) have been taken as the basis of formulation of ration scales for Indian soldiers. In a study conducted during

2001–2005 by DIPAS [36], the mean energy expenditure of the soldiers from all units studied at sea level was found to be 3,511 ± 601 kcal/day.

On this basis, required energy content of soldiers ration scale was set to 3,862 kcal/day (by making 10 % allowance for metabolic loss and plate and kitchen wastages) with a range of 3,385–4,437 kcal/day. For planning of energy content of ration, maximum 4 % wastage along with 6 % losses during digestion and absorption of food have been taken. The composition of basic ration scale for Indian soldiers is nutritionally balanced as energy contribution from carbohydrate, fat, and protein is 60.8, 26.9, and 12.3 %, respectively, with adequate levels of micronutrients. The variety of food items of five basic food groups, namely, (1) cereals, grains, and products, (2) pulses and legumes, (3) milk and meat products, (4) fruit and vegetables, and (5) fat, have been included in ration scales. The intake of different nutrients by soldiers and a comparison with RDA for civilians [13] in heavy activity group are made in Table 2 to give readers a view of adequacy of soldier's ration.

Table 2 Nutrient intake of soldiers at sea level and RDA for Indians (heavy activity group)

Energy and nutrients	Intake	RDA 2010 for heavy activity group#
Energy (kcal)	3,854 ± 344*	3,490
		(3,800 in RDA 1989)
Protein (g)	118 ± 41	60
Fat (g)	115 ± 62	40 [visible fat]
CHO (g)	586 ± 23	–
Vitamin A (μg)	575 ± 130	600
Vitamin C (mg)	65 ± 22	40
Thiamin (mg)	1.6 ± 0.3	1.7
Riboflavin (mg)	1.5 ± 0.3	2.1
Niacin (mg)	26.8 ± 6.4	21
Iron (mg)	34.7 ± 7.6	17
Calcium (mg)	1,474 ± 318	600
Phosphorus (mg)	2,386 ± 150	400
Sodium (mg)	8,189 ± 180	1,100–3,300
Potassium (mg)	2,679 ± 250	1,875–5,625
Zinc (mg)	13.58 ± 1.20	12
Copper (mg)	3.31 ± 0.50	2.2

The intakes of macronutrient are based on raw ration and micronutrient intake on basis of plate samples analyzed. Values are Mean ± SD. *Energy contribution from CHO, fat, and protein is 61 %, 27 %, and 12 %, respectively. #For computing energy requirement, weight of reference man is taken as 60 kg

Different dietary habits and food preferences and cooking practices across the country make diet planning a complex exercise. Wheat and rice are two major forms of staple diets in India. Majority of Indians are lacto-vegetarians and meat consumption is less than three times a week. Diet is also influenced by religious practices, especially in case of processing of meat. Most of the diets are well balanced in terms of macro- and micronutrients. Protein quality in terms of rate-limiting essential amino acids also gets balanced with use of cereals and pulses in most of food preparations. Bioavailability of iron, calcium, and zinc is low due to presence of high amounts of phytates and oxalates. Intake of iodized salt is taking care of iodine deficiency disorders, still presence of isothiocyanates as goitrogens in certain vegetables of Cruciferae family is matter of concern. Use of extrusion products like pasta and noodles is also gaining popularity in Indian diets and armed forces are not exception to it. Use of bread in place of chapaties is common practice in case of Navy, with more use of ready-to-eat products. Hydrogenated oil which was used earlier in Army ration scales has been totally replaced with different types of refined vegetable oils, based on nutritional studies related to cardiovascular health.

Effect of Environmental Extremes on Nutritional Requirements

The energy and nutrient requirements are generally more under environmental extremes and have been reviewed separately [35]. Extreme environments such as cold, heat, hypoxic environment, and microgravity change the requirement of both energy and micronutrients. Appetite and thirst perceptions are generally inappropriate under extreme environments that lead to inadequate food and water intake. The availability of food and water is also limited due to difficulties of transport and other logistic constraints and needs detailed planning. Proper nutrition should not be overlooked, as this is a critical component of effective work acclimatization under harsh conditions. Various studies on relationship of diet and extreme environment are the outcome of military research or expeditions to mountains and polar regions. High-altitude exposure increases energy expenditure, ranging from 6.9 % to 25 %, due to increase in basal metabolic rate during initial days of acclimatization [24]. Increased energy expenditure is also due to the heavier load carried by the troops, as cold protective garments and efforts in walking in snowbound hilly terrain. Carbohydrates provide higher yield of energy per mole of oxygen, therefore are beneficial at high altitude [25].

Negative nitrogen balance is reported at high altitude and is mainly due to decreased food intake due to high-altitude anorexia [49, 50]. To overcome the problem of anorexia, a variety of food components are included in special ration and efforts are being made to develop appetizers. There is no change in fat digestibility at altitude of <4,500 m [29, 41]. However, at extreme altitude, there are reports that fat absorption gets impaired [24].

Antioxidant nutrients such as vitamin E, C, and A (β-carotene), as well as selenium, copper, zinc, and manganese may be required in greater amounts in cold and high-altitude environments to prevent oxidative stress. These antioxidants act in concert to combat the oxidative stress arising from different sources. During rough weather, when supply of fresh fruits and vegetables becomes limited at high altitude, supplementation of vitamin C is recommended due to its antioxidant role [14, 42].

Both cold and heat exposures increase energy requirements [35]. Under cold environments, food intake is not a problem and tolerance of proteins is better in comparison to hot environments. Proteins have more thermic effect and at the same time, more water is required to eliminate nitrogenous products of amino acid metabolism.

Adequate fluid replenishment is the primary consideration under hot environment. Drinking adequate amount of water at regular intervals prevents dehydration and heat illness and maintains work performance [2, 32, 33]. Heat acclimatization relatively has no effect on water requirements. Thirst is a poor indicator of hydration status. Intense thirst is usually noticed at 5–6 % body weight loss due to hypohydration. At this much loss of body water, both physical and mental performances get impaired. Energy expenditure in hot environments is increased because of additional work of ventilation and increased sweat gland activity. There is rise of ~10 % in energy requirement at 38 °C. It was found that NaCl requirement increases due to loss in sweat; 15–16 g of salt normally taken in diet is quite adequate for acclimatized people [21]. With acclimatization for 3 days, sodium losses in sweat get reduced; however, body is not able to conserve potassium. Therefore, supplementation of potassium in drinks may enhance the process of acclimatization [22, 23].

Combat Rations

Under military missions, the energy expenditure increases significantly depending on environment. The maintenance of normal food supply is also difficult and soldiers have to depend on individualized ration carried along with them. Therefore, lightweight, calorie dense, meal ready to eat (MRE) with minimum calories are developed and evaluated [4, 6–8, 10, 12, 15, 28, 46–48]. Research on minimum calorie and nutrient requirement of individuals that can sustain metabolic functions during the emergency situations will help in better planning and development of combat and survival rations. The existing knowledge on calorie restriction and starvation is mainly based on studies on obese and fasting individuals with moderate or minimal physical activities and is not fully applicable to physically active soldiers. Semi-starvation deteriorates both physical and cognitive performances, whereas calorie restriction is reported to have several positive health benefits including decrease in postprandial oxidative stress. Under conditions of calorie restriction, certain amount of adaptation in BMR also takes place to minimize energy expenditure.

There are reports that an intake of 2,000–2,200 kcal per day can sustain work efficiency of well-nourished active individual with energy expenditure of 3,800 kcal/day for 7–15 days [28]. Earlier study on Indian troops with an energy deficit of about 1,500 kcal or on a 50 % of normal intake has not shown any abnormality in biochemical markers and physical efficiency for a period of 10–15 days [30]. Underconsumption of MRE due to several reasons is also reported [4, 5].

Future Perspectives

There is a need to develop combat rations which can sustain physical as well as cognitive performance of soldiers during initial 3–7 days of high-intensity combat missions. The first question needs to be answered is, what should be the energy content and ratio of macro- and micronutrients for sustained performance? Research is also required to determine the type and optimum level of macronutrients in rations, e.g., complex versus simple carbohydrates, proteins with specific essential amino acid composition, or fatty acid composition. The optimum levels of protective nutrients, nutraceuticals, and other bioactive performance enhancers for assault rations need to be investigated under actual field conditions. With the engagement of soldiers in managing low-intensity conflict in certain areas of the country, some special challenges have emerged for food technologists. One such challenge is the masking of odor of ready-to-eat food, which gives clue about the soldier's presence to opponent. At the same time, the aroma and flavors are necessary components to provide satisfaction of eating the food. With advancement in food technology, it is now possible to provide consumer-balanced diet. With the emergence of new field of nutrigenomics, which will help in understanding of individual variations in nutrient requirements, there is a possibility of formulating individualized nutraceuticals/foods for improving performance and well-being of soldiers in near future.

References

1. Bassett DR Jr, Ainsworth BE, Swartz AM, Strath SJ, O'Brien WL, King GA (2000) Validity of four motion sensors in measuring moderate intensity physical activity. Med Sci Sports Exerc 32:S471–S480
2. Clarkson PM (1993) The effect of exercise and heat on vitamin requirements. In: Marriot BM (ed) Nutritional needs in hot environments. National Academy Press, Washington, DC
3. Cole TJ, Coward WA, Elia M et al (1990) The doubly labeled water method for measuring energy expenditure. Technical recommendations for use in humans. In: Printice AM (ed) A consensus report by the IDECG Working Group. IAEA, Vienna
4. Committee on Military Nutrition Research, Food and Nutrition Board (2005) Nutrient composition of rations for short-term, high – intensity combat operations. The National Academies Press, Washington, DC
5. Committee on Military Nutrition Research, Institute of Medicine (2005) In: Marriott BM (ed) Not eating enough: overcoming under consumption of military operational rations. National Academies Press, Washington, DC
6. Consolazio CF, Moutoush LO, Johnson HL, Krzywicki HJ, Isaac GJ, Witt NF (1968) Metabolic aspects of calorie restriction: hypohydration effects on body weight and blood parameters. Am J Clin Nutr 21:793–802
7. Consolazio CF, Moutoush LO, Johnson HL, Krzywicki HJ, Isaac GJ, Witt NF (1968) Metabolic aspects of calorie restriction: nitrogen and mineral balances and vitamin excretion. Am J Clin Nutr 21:803–812
8. Davenport RE, Spaide JK, Hodges RE (1971) An evaluation of various survival rations. Am J Clin Nutr 24:513–523
9. FAO/WHO/UNU (2004) Human energy requirements. Report of the joint FAO/WHO/UNU Expert Consultation. FAO technical report series 1, Rome
10. Friedl KE, Moore RJ, Hoyt RW, Marchitelli LS, Martinez-Lopez LE, Askew EW (2000) Endocrine markers of semistarvation in healthy lean men in multistressor environment. J Appl Physiol 88:1820–1830
11. Geissler CA, Dzumbira TM, Noor MI (1986) Validation of a field technique for the measurement of energy expenditure: factorial method versus continuous respirometry. Am J Clin Nutr 44:596–602
12. Heilbronn LK, Smith SR, Martin CK, Anton SD, Ravussin E (2004) Alternate – day fasting in nonobese subjects: effect on body weight, body composition, and energy metabolism. Am J Clin Nutr 81:69–73
13. Indian Council of Medical Research (2010) Nutrient requirements and recommended dietary allowances for Indians. National Institute of Nutrition, Hyderabad

14. Ilavazhagan G, Sridharan K, Sharma SK, Bansal et al (1996) Role of vitamin C and E under high altitude and combat stress. Report no. DIPAS/26/96
15. Lichton IJ, Miyamura JB, McNutt SW (1988) Nutritional evaluation of soldiers subsisting on meal, ready – to- eat operational rations for an extended period: body measurements, hydration, and blood nutrients. Am J Clin Nutr 48:30–37
16. Malhotra MS, Chandra U et al (1972) Food and nutritional requirements of submarine crew. Report no. DIPAS/6/72
17. Malhotra MS, Chandra U et al (1971) Rationalization of army ration scales for peace and field areas. Phase II- Peace Scales. Report no. DIPAS/4/71
18. Malhotra MS, Chandra U, Sridharan K et al (1977) Rationalization of army ration scale for officers at high altitude. DIPAS/2/77
19. Malhotra MS, Rai RM, Sridharan K, Bhaskaran A (1970) Rationalization of army ration scales for peace and field areas. Phase I- Field Scales. Report no. DIPAS/6/70
20. Malhotra MS, Ramaswami SS, Sengupta J, Venketaswami Y et al (1967) Adequacy of the ration scale at high altitude. DIPAS/4/67
21. Malhotra MS, Sharma BK, Sivaraman R (1976) Requirement of sodium chloride during summer in tropics. J Appl Physiol 14:823–828
22. Malhotra MS, Sridharan K, Venkataswamy Y (1976) Potassium losses in sweat under heat stress. Aviat Space Environ Med 47:503–504
23. Malhotra MS, Sridharan K, Venkataswamy Y, Rai RM, Pichen G, Radhakrishnan U, Grover SK (1981) Effect of restricted potassium intake on its excretion and on physiological responses during heat stress. Eur J Appl Physiol 47:169–179
24. Marriot BM, Carlson SJ (1996) Nutritional needs in cold and high altitude environments. National Academy Press, Washington, DC
25. McClelland GB, Hochachka P, Weber J-M (1998) Carbohydrate utilization during exercise after high altitude acclimation: a new perspective. Proc Natl Acad Sci U S A 95:10288–10293
26. Mudambo KSMT, Scrimgeour CM, Rennie MJ (1997) Adequacy of food rations in soldiers during exercise in hot, day time conditions assessed by doubly labeled water and energy balance methods. Eur J Appl Physiol 76:346–351
27. Nayar HS, Solanki KS, Kherji HB, Sridharan K et al (1979) Rationalization of patient's diets and scales. Report no. DIPAS/01/79
28. Nindl BC, Barnes BR, Alemany JA, Frykman PN, Shippee RL, Friedl KE (2007) Physiological consequences of US Army Ranger training. Med Sci Sports Exer 39:1380–1387
29. Rai RM, Malhotra MS, Dimri GP, Sampathkumar T (1975) Utilization of different quantities of fat at high altitude. Am J Clin Nutr 28:242–245
30. Rai RM, Sridharan K et al (1983) Effect of repeated strenuous exercise under low energy intake on physical performance. Report no. DIPAS/06/83
31. Rao V (2012) Armies, war and their food. Cambridge University Press India Pvt. Ltd., New Delhi
32. Sawka MN, Neuter PP (1989) Interaction of water bioavailability, thermoregulation and exercise performance. In: Marriott BM, Rasemant C (eds) Fluid replacement and heat stress. National Academy Press, Washington, DC
33. Sawka MN, Francesconi RP, Young AJ, Pandolf KB (1984) Influence of hydration level and body fluids on exercise performance in the heat. JAMA 252:1165–1169
34. Schoeller DA, Van Santen E (1982) Measurement of energy expenditure in humans by doubly labeled water method. J Appl Physiol 53:955–959
35. Selvamurthy W, Singh SN (2003) Nutritional requirements for human adaptation at extreme environments. Proc Indian Natl Science Acad B 69:485–506
36. Singh SN, Singh SB et al (2006) Assessment of nutritional requirements of armed forces personnel at various conditions of climate and training. Report no. DIPAS/02/06
37. Singh SN, Vats P et al (2007) Assessment of nutritional requirements of trainees attending Ghatak Courses at Infantry School, Belgaum Report no. DIPAS/06/07
38. Singh SN, Vats P, Shukla V, Kurpad AV (2010) A comparison of free living energy expenditure determinations of physically active Indians using different methods and validation against doubly labeled water. In: Bhaskaran M et al (eds) Non nuclear applications of heavy water and deuterium. Macmillan Press India Ltd., New Delhi
39. Sridharan K, Ranganathan S, Mukherjee AK, Singh SN et al (1999) Evaluation of nutritional status of boys studying in Military School, Dholpur. Report no. DIPAS/11/99
40. Sridharan K, Nayar HS, Rai RM et al (1979) Nutritional improvement of ration scales of various tradesmen and pioneers in Border Road Organization. Report no. DIPAS/09/79
41. Sridharan K, Malhotra MS, Upadhyay TN, Grover SK, Dua GL (1982) Changes in gastro – intestinal functions in humans at an altitude of 3,500 m. Eur J Appl Physiol 50:145–154
42. Sridharan K, Ranganathan S, Mukherjee AK, Kumria MML, Vats P, Sharma RP, Gauttam RK (1996) Study of effects of tinned food on troops at high altitude with special reference to vitamin status. Report no. DIPAS/30/96
43. St-Onge M, Mignault D, Allison DB, Rabasa-Lhoret R (2007) Evaluation of a portable device to measure daily energy expenditure in free-living adults. Am J Clin Nutr 85:742–749
44. Tharion WJ, Lieberman HR, Montain SJ, Young AJ, Baker-Fulko CJ, DeLany JP, Hoyt RW (2005) Energy requirements of military personnel. Appetite 44:47–65
45. Vats P, Singh SB, Singh SN et al (2007) Evaluation of nutritional requirement of trainees at High Altitude Warfare School. Report no. DIPAS/07/07

46. Viswantahan KR, Prasad NN, Ramanuja MN, Narayanan VA (1991) Evaluation of low energy pack ration by short-term feeding to soldiers. Def Sci J 41:263–275
47. Weigle DS, Breen PA, Mattys CC, Callahan HS, Meeuws KE, Burden VR, Purnell JQ (2005) A high-protein diet induces sustained reduction in appetite, ad libitum caloric intake, and body weight despite compensatory changes in diurnal plasma leptin ad ghrelin concentrations. Am J Clin Nutr 82:41–48
48. Wenkam NS, Fox M, Thiele NN, Lichton IJ (1989) Energy and nutrient intakes of soldiers consuming MRE operational rations: physiological correlates. J Am Diet Assoc 89:407–409
49. Westerterp KR (2001) Energy and water balance at high altitude. News Physiol Sci 16:134–137
50. Westerterp KR, Kayser B, Wouters L, Le Trong J-L, Richalet JP (1994) Energy balance at high altitude of 6,542 m. J Appl Physiol 77:862–866

Improvements in Adjuvants for New-Generation Vaccines

Lilly Ganju and Divya Singh

Abstract

Over the last decade, extensive research for development of new vaccine adjuvants is being carried out. Present generation vaccines, particularly those based on recombinant proteins and DNA, are not only less reactogenic and also less immunogenic. Therefore, there is an urgent need for the development of new and improved vaccine adjuvants. Many novel adjuvants have been cleared for license, and many are in late stages of clinical trials. Recent investigations in innate immunity have offered new insights into immunostimulatory actions of adjuvants and have facilitated a more rational selection of adjuvants. Despite the impressive response of approved adjuvants in generating immunity against pathogens, there remains a need for improved adjuvants that enhance strong T-cell immunity and protective antibody response. The discovery of more potent adjuvants will also allow engineering of vaccines against infections that do not naturally elicit protective immunity. A logical approach to the development of new and more effective vaccine adjuvants requires a better understanding of the action of adjuvant-antigen formulations. Here, we discuss these advances and the need for better adjuvant development.

Introduction

The invention of smallpox vaccine by Edward Jenner, in the year 1800 AD, brought a revolution in health-care techniques and engendered new fields of vaccinology and immunology. Today many vaccines are available in the market for diverse immune interventions. The fact is that many vaccines are developed in the laboratory, but unfortunately never see the light of the day as they have apparently no immunogenecity at all, when tested alone. This could virtually happen for any antigen, but it is particularly true for small peptides that serve as antigens [5]

In the process of recruiting the immune system to produce specific or nonspecific immunity, the antigen by itself may not be adequate as a stimulatory agent to trigger an immune response. This is the condition when immunologists look

L. Ganju (✉) • D. Singh
Immunomodulation Division, DIPAS, DRDO, Timarpur, New Delhi, India
e-mail: lganju@rediffmail.com

S.B. Singh et al. (eds.), *Translational Research in Environmental and Occupational Stress*, DOI 10.1007/978-81-322-1928-6_11,

for an adjuvant, or an adjuvant strategy, in order to elicit a more pronounced immune response. The term adjuvant is derived from the Latin *adjuvāre*, meaning "to help" – first described by Ramon, while working with natural compounds, as "substances in combination with a specific antigen produced more robust and specific immune response than the antigen alone" [108]. An adjuvant can be defined as a substance or agent that serves as an immunological vehicle to enhance the innate immune response, or stimulate the adaptive immune response, to generate the desired levels of immunity with minimal toxicity or long-lasting immune effects on its own [129].

The Need for Adjuvants

There is a dearth of potent adjuvants for various vaccines. The demand for the development of new adjuvants and strategies that are safe and effective leads to continuing extensive research in this field, and as a result, different types of materials are being experimented with. This highly heterogeneous group of agents is categorized for rational selection of an adjuvant, with their classification based on chemical action, and mechanism of intervention.

Foreign invaders, e.g., bacteria, viruses, and parasites, contain antigens which a host immune system recognizes and mounts an immune response to. They also contain inbuilt "adjuvants" which trigger the immune system in a manner resulting in directing its response along an effective pathway. Thus, a purified antigen, derived from a bacterium, a virus, or a cancer cell, has a reduced ability to stimulate the immune system in the desired manner. Synthetic subunit vaccines are expensive to produce; but addition of an adjuvant will result in lesser quantity of antigen required, with fewer injections, thereby saving on the cost of vaccination. This will also help in developing better combination vaccines, the number of distinct vaccines, as well as the amount of vaccination antigen required will be lesser. This in turn may also reduce the competition of antigens and carrier-specific epitope suppression.

Table 1 Types of adjuvants, based on their composition

Type	Description
Mineral salt based	
Alum	Aluminum hydroxide; aluminum phosphate
Calcium phosphate	$CaHPO_4$, commercially available adjuvant
Bacteria derived	
CpG Oligodeoxynucleotides (ODNs)	Synthetic ODN containing unmethylated CpG motifs [12, 70]
Monophosphoryl Lipid A (MPL A)	Derived from LPS
AS04	TLR4-agonist MPL, with aluminum salt [23]
Emulsions	
MF59	Oil-in-water (o/w) emulsion [86]
AS03	α-tocopherol and squalene in an (o/w) emulsion[87]
Freund's incomplete adjuvant (FIA)	Water-in-oil emulsion
Montanide	Water-in-oil emulsion
Adjuvant 65, Lipovant	Water-in-oil emulsion
Carrier adjuvants	
Liposomes	Synthetic spheres of lipid layers encapsulating antigens [98]
Immunostimulatory complexes (ISCOMS)	Virus-like particles of 30–40 nm and dodecahedric structure, composed by Quil A, lipids, and cholesterol
Cytokines	
GM-CSF, IL-2, IL-12, Type 1 interferon, IFN-γ	Potentiate the immune response to vaccination
Carbohydrate adjuvants	
QS21	Saponin derived from a mixture of soluble triterpene glycosides purified from the soap bark tree (*Quillaja saponaria*)
Advax™ adjuvant	A crystallized natural plant-derived polysaccharide [94]
Acemannan	A natural polysaccharide extracted as a mucilaginous gel of the *Aloe barbadensis* [98]

For purified antigens, the importance of addition of effective adjuvants to optimize the immune response, is thus emphasized.

On the basis of quantum of adjuvanticity, a huge number of candidate adjuvants have been categorized (Table 1). However, such classifications have inherent limitations though, as a consequence of highly diverse individual variation. Till date, there is no reliable algorithm available which can reliably predict the interaction of an adjuvant, permitting the selection of an antigen based on its physicochemical or immunological properties [109]. Very often, commercial aspects, viz. ease of production, cost, toxicity, availability, etc., become difficult deciding factors. The modal factors in the choice of an adjuvant are the ability to cause additive effects, or lower the quantity of antigens needed, or qualitatively reduce the side effects and enhance the safety of the adjuvant. A balance is struck on the principles of "risk benefit analysis." For routine childhood vaccines, mandatory safety is the biggest concern, while adjuvanticity may be restricted because these vaccines are needed to develop lifelong immunity against specific diseases in a normal healthy body where any risk of side effects is unacceptable; whereas for a person with an infection like HIV or with a disease like cancer, therapeutic vaccines with a certain level of toxicity are acceptable, based upon the accruing benefits.

Importance of Adjuvant Mediation in Vaccines

Vaccines confer protection mainly through humoral immunity [101]. But new-generation vaccines, including recombinant, conjugated or attenuated, polysaccharide subunit vaccines, etc., are developed with the intention to provide a stable and sustained protection. Despite the enhanced effect of such vaccines, substantial groups of people are not able to achieve adequate protection. Addition of an effective adjuvant or a switch from aluminum hydroxide (alum) – the most widely used adjuvant – helps enormously in boosting the immune response in such groups. Adjuvants have been used with a variety of vaccines to elicit safe, early, high, and long-lasting immune response. This has been clearly observed in aluminum adjuvants as compared to un-adjuvanted preparations [6, 52].

For the last decade, adjuvants have received much attention due to their ability to selectively improve the humoral immune response and stimulate the cellular immune response [7, 17, 47, 118]. Reports suggest this second role of adjuvants become increasingly important: guiding to the most effective adaptive response against each specific pathogen [68, 105]. Recent findings suggest that characterizing the immunostimulatory properties of an adjuvant will help in understanding and predicting the translational potential of a constrained or non-immunogenic vaccine into an efficacious one [104]. Besides generating strong antibody response (humoral response), they also act as immunomodulators by influencing the type and character of the antibody generated and help in seroconversion rates in the general population (including nonresponders, elderly, infants, and diseased) and generation of memory response [36, 76, 127] and rapid initial response, which is crucial during pandemic outbreaks of infections [37, 61, 69].

The choice of the adjuvant is of critical importance for the isotype and subclass of IgG, pattern of the cytokine production and recruitment of T cells. At times, it also modifies the antigen moieties in a controlled manner. Humoral response leads to the generation of IgG antibody subtypes, and increase their avidity and affinity. T cells recognize the antigen which is presented to it by antigen-presenting cells (APCs), through major histocompatibility complex (MHC) [89]. At this juncture where antigen recognition by T cell may get affected, adjuvants play an important role in modulating the response [47, 80], leading to elicit both T-helper cells and cytotoxic T lymphocytes (CTLs).

Adjuvants help in modulating the immune response, with regard to MHC class I or class II [16, 112]. MHC class I response is usually generated by an intracellular pathogen, e.g., virus, leading to cytotoxicity [97]. This type of response is not usually obscured by proteins or peptide antigens which elicit MHC class II response. Adjuvants like ISCOMS and QS-21 can elicit CTL with protein [97, 112] and peptides [7, 51].

Adjuvants also contribute in modulating the immune response to different T-helper cells as Th1 or Th2. This is of prime importance for generating a response against different types of microbes or organisms. For intracellular organisms, e.g., protozoa, viruses, invasive bacteria, and parasitic infections, Th1 type of immune response is accompanied by secretion of IL-2 and IFN-γ leading to cell-mediated immune response and production of relatively high levels of IgG2a antibodies. The Th1 type response activates CTLs which induce death of cells infected with intracellular pathogens and natural killer cells (NK), playing a major role in apoptosis in tumors and virus-infected cells. Th2 type immune response is modulated by IL-4 and IL-10 and is stimulated by extracellular organisms, e.g., helminthes or toxins, protein antigens, and inactivated pathogens. Th2 response leads to antibody generation like IgG1 and IgE [47]. Aluminum adjuvants are known to stimulate a Th2-type response [7, 17, 52, 53]. Recent reports indicate that the combination of one or more adjuvants helps in a dramatic shift from Th2-type to Th1-type response [92, 129]. It has also been suggested that the addition of one adjuvant to another may modulate the adaptive response of the immune system [88].

Safety Issues

The triggering of the immune response is required only up to a level of immunogenicity. Certain adjuvants have complex immunomodulatory activities [73, 74], and one of them is the enhancement of IgE production [10] which could turn out to be an unfavorable adjuvant effect. At times, triggering of immune response in the host can result in an unintended attack on the host itself to which the host may react to save itself, failing which it may end up in developing adverse side effects or deterioration. Therefore balancing the required and desirable effects of an adjuvant becomes the most important safety issue. The incorporation of an adjuvant in a vaccine has to be critically reviewed and observed so as to guide the immune response in the most desirable direction.

Since the safety of adjuvants is the biggest concern, particularly in neonatal and pediatric vaccines, criteria have been set out to ensure the safety of adjuvanted vaccines [26]. Guidelines have been laid down by European Societies for development of adjuvants defining the criteria of an ideal adjuvant [29]. It is important to observe the initiation, quality and magnitude of the immune response which is influenced by many nonspecific factors: important ones being the type and dose of the antigen, route of immunization, number of boosters, and shelf life of the vaccine as well as of the adjuvant. The type of the vaccine immensely affects the strategy of adjuvant administration, e.g., oral delivery, transcutaneous, or intranasal route [35, 43, 48, 84, 111]. In addition to safety with regard to local reactions, systemic reactions (general toxicity and pyrogenicity), hypersensitivity, autoimmune reactions, carcinogenicity, teratogenicity, etc. need to be evaluated at the time of development and production. Adjuvant should be chosen based on the type of immune response desired and formulated with the antigen in such a way that both are optimally distributed and presented to the relevant lymphatic tissue. The most commonly used adjuvant, alum, has apparently been reported to have adverse reactions, and some of these events could be because of apparent antigenicity of alum itself [30].

Properties for Safety

Adjuvants should be chemically defined so as to get the consistent reproducibility while manufacturing. The preparation should elicit a protective immune response with weak and conjugated antigens, with a minimum of antigen dosage and a reduced number of boosters. It should be effective in neonatal infants, in elderly and immunocompromised (diseased) individuals. Adjuvants should be stable, with a comparatively long shelf life and be biodegradable and nonimmunogenic by itself. It should be able to work in a synergistic manner when mixed with other adjuvants in a formulation. These are the factors that go in the making of an effective adjuvant – eliciting an immune response which is also safe.

Mechanisms of Action

Research during the past two decades has offered new insight into the mechanisms of action of vaccine-adjuvant formulations. Evidence suggests that adjuvants employ different mechanisms to elicit different immune responses: (1) sustained release of antigen at the site of injection, which is also called the depot effect; (2) recruitment of immune cells at the site of injection; (3) increased antigen uptake and presentation to APC; (4) activation and maturation of APCs, which leads to increased MHC class II and co-stimulatory molecules expression and migration to the draining lymph nodes; (5) upregulation of cytokines and chemokines; and (6) activation of inflammasomes [18, 33, 58].

Till now, it was understood that vaccine adjuvants stimulate only adaptive immune response, but recent advances have identified the basic role of the innate immune response programming the protective immune response in a better way [104]. It is now clear how the innate immune response stimulates the adaptive immune response.

Decades of research have been spent in search of good adjuvants that can enhance the immune response to the desired level without any side effect, but very few adjuvants have been licensed for human use. These include alum (aluminum salts), ASO4 (combination of monophosphoryl lipid A, i.e., TLR4 ligand, adsorbed to alum), and water-in-oil emulsion (MF59) [15, 83]. The dearth of adjuvants for human use probably can be fulfilled by understanding their in-depth mechanism of action.

The design of an adjuvant is the key that stimulates the class of immune response, such as antigen-specific helper T-cell subset, cytotoxic T cells, or long-term sustenance and memory T cells/B cells. These pathways of action depend on how successfully antigens mediate their immunogenicity through dendritic cells (DC) and pattern recognition receptors (PRRs). The major role of these PRRs is to recognize a large group of microorganisms through their molecular structures.

In the last decade, various new families of PRRs have been identified, including TLRs, nucleotide oligomerization domain (NOD)-like receptors (NLRs), C-type lectin-like receptors (CLRs), and Retinoic acid-inducible gene 1 (RIG-1)-like receptors (RLRs). Sometimes particulate adjuvants are not recognized by specific PRRs, but they still induce adaptive immune responses [9]. The "Danger Hypothesis" [85] proposed that danger signals from damaged cells can also trigger activation of the immune system, apart from self-/nonself-discrimination against infection. Damaged cells at the injection site release molecules associated such as uric acid, nucleotides, adenosine triphosphate (ATP), reactive oxygen intermediates, and cytokines [117]. These noninfectious damage signals are named as damage-associated molecular patterns (DAMPs) to distinguish them from pathogen-associated molecular patterns (PAMPs) [9].

Alum – composed of precipitates of aluminum hydroxide and aluminum phosphate, to which the antigen gets adsorbed – has been widely used in most of the human vaccines for the last 90 years. Despite the fact that it has many undesirable side effects, it is used as a benchmark [51, 125, 126]. The mechanism of action – how on emulsification with the antigen, alum enhances the Th2-biased response – remained a "dirty little secret" until recent past [64]. It was believed that alum generates a depot effect where the emulsion retains the antigen at the site of injection and releases it slowly, which leads to sustained antigen presentation. A little bit of the dirty secret was revealed when it was reported that alum induces antibody responses independent of TLR signaling [40]. Very recently, it was also reported that removal of the injection site two hours after antigen and alum administration had no effect on humoral- or cell-mediated immunity [62]. Furthermore, it also acts on cells essential for clonal expansion and production [65]. It is also reported that alum works through NLRP-3 inflammasomes [27, 71, 77] though it is still a matter of controversy (Fig. 1). MF-59, an oil-in-water emulsion, has been licensed for the influenza vaccine: whereas other adjuvants work by promoting the recruitment of APCs, MF-59 triggers CD11 b + cells, induces cytokines, cytokine receptor, and adhesion molecules. The microarray analysis has revealed that adjuvants work at gene level as MF-59, CpG DNA, and alum do induce a set of 168 genes at the site of immunization. Though the role of the inflammasome in adjuvant activity of MF-59 has also been evaluated [28, 115], yet the PRR activation remains unknown [88]. Similarly, another novel adjuvant, poly-[di-(sodium-carboxylato-ethyl-phenoxy)-phosphazene] (PCEP), induced stronger expression of adjuvant core response genes compared to CpG at the site of injection. Locally, PCEP-triggered production of proinflammatory cytokines and chemokines, like CCL2 [8].

Though innate immunity plays an important role in regulation of T-cell responses, evidence also emphasizes its importance in regulating the quality, function, magnitude, and persistence of the antibody response [100]. The quality of the antibody which protects against infection matters most despite high antibody titers. Therefore, again, an understanding of the mechanisms of innate immunity regulation for the most appropriate quality of antibody response will help in designing better adjuvants.

Another key factor is the persistence of antibody response. Some vaccines, such as carbohydrate or weak protein vaccines for children and elderly people, induce short-lived immunity. This has especially been demonstrated in some viral vaccines [110]. It is critical to enhance the persistence of such responses by internally stimulating memory cells, which mediate and facilitate in antibody-dependent cell-mediated cytotoxicity and induce an antibody response that lasts for several decades [16, 123]. A combination of adjuvants and vectors may be useful sometimes – in developing a synergistic specific response. This became clearly noticeable when a combination of antigen plus a TLR4 ligand, along with TLR7 ligand, induced synergistic antigen-specific neutralizing antibodies [67]. This mechanism could be possible due to the triggering of TLRs on a B cell and signaling via MyD88 and TRIF in DCs. These findings indicate that innate immunity regulates the magnitude, quality, and persistence of antibody response and generate memory B-cell development and persistence of plasma cells. Besides, there are certain molecules, such as CD40, CD95, BCL-6, IL-21, ICOS, BAFF, etc., known to enhance the number and survival of plasma cells [78]. Studies are in progress to use a combination of these molecules for enhanced results of memory response. There are certain adjuvants which work on this newly discovered trend in the mechanism of action via TLRs, the best example is ASO4 which works as a ligand through TLR4 because of monophosphoryl lipid A (MPL) and signals through TRIF [81]. Interestingly, LPS – from where MPL is derived – signals via both MyD88 and TRIF, which results in enhanced proinflammatory cytokines and toxicity. It is somewhat unanticipated to know that alum induces Th2 type of immune response, whereas ASO4, which has part of alum, induces Th1 type of immune response [23]. TLRs generally induce DCs to boost Th1 response, although it is also seen that mild TLR ligands induce Th2 or T-regulatory cell responses [105] and other Th17 responses [75]. Thus the design and selection of an effective and appropriate adjuvant also depends on and is influenced by the type of CD4 cells generated or required for protection.

We can infer that the precise mechanism of enhancing the protective response with adjuvant and at the same time antibodies protecting against pathogens, seems to prefer working via the innate immune system. An understanding of the basic levels and pathways of action of particular DC or specific PRRs will be helpful in designing adjuvants for the appropriate class of immune response. At higher levels, the immune response is also orchestrated by cell-to-cell interaction, e.g., between basophils and other immune cells and DCs [55, 93, 113, 124], confirming the action of adjuvants with multiple cell types; whereas at the highest level, finally cytokines and chemokines instruct the DCs to start functioning. It is also important to understand the mechanisms of different routes through which adjuvant may be administered – which must be able to reprogram the tissue-specific responses of DCs. The insight into the process of decision-making by DCs depends upon the various transcription factors, which remain to be fully understood.

Types of Adjuvants

Adjuvants represent different classes of compounds across the spectrum of products derived from microbes, oil emulsions, liposomes, microparticles, natural, and synthetic substances. These products function diversely and characterize different mechanisms of action [33, 54, 96]. Studies demonstrate two types of adjuvants based on their functional group – TLR-dependent and TLR-independent [40, 90]. TLR-dependent adjuvants act directly through DCs, causing them to migrate to the T-cell area of lymph nodes, in the process inducing the upregulation of MHC II, cytokines, chemokines, and other co-stimulatory molecules [33, 58], whereas TLRs independently induce inflammatory response, increase T-cell activation, and generate B-cell abundancy [40, 130].

Alum, a conventional adjuvant, provokes a strong Th2 response but ineffective against pathogens that require cell-mediated immune response. Furthermore, intraperitoneal administration of alum recruits monocytes at the site which uptakes the antigen and migrates to the draining lymph nodes and then differentiate into inflammatory DCs [71]. There is increase in Nalp3 inflammasome activation and production of IL1ß and IL-33 [77] which eventually leads to increased activation of immune cells (Fig. 1).

Jules Freund in the 1930s proved that an antigen contained in a water-in-oil emulsion markedly enhanced immune response [34, 57]. Known as Complete Freund's adjuvant (CFA), it is unusable for humans – with whole-killed mycobacteria in it. However, incomplete Freund's adjuvant (IFA), lacking mycobacteria, has been used as a potent adjuvant for a diverse number of vaccines [2, 19, 121]. Both Freund's adjuvants induce strong Th1 and Th17 responses, depending upon the mycobacterial components requiring host MyD88 or TRIF.

Dependent signaling pathways, independent of inflammasome [116, 122], vary with the experimental model [27, 40]. Emulsification of antigens with surfactant or paraffin oil alone without mycobacterium substantially boosts the antibody response. Although the mechanism of action is still unknown, one study suggested the requirement of NOD2 receptor [86]. It is a

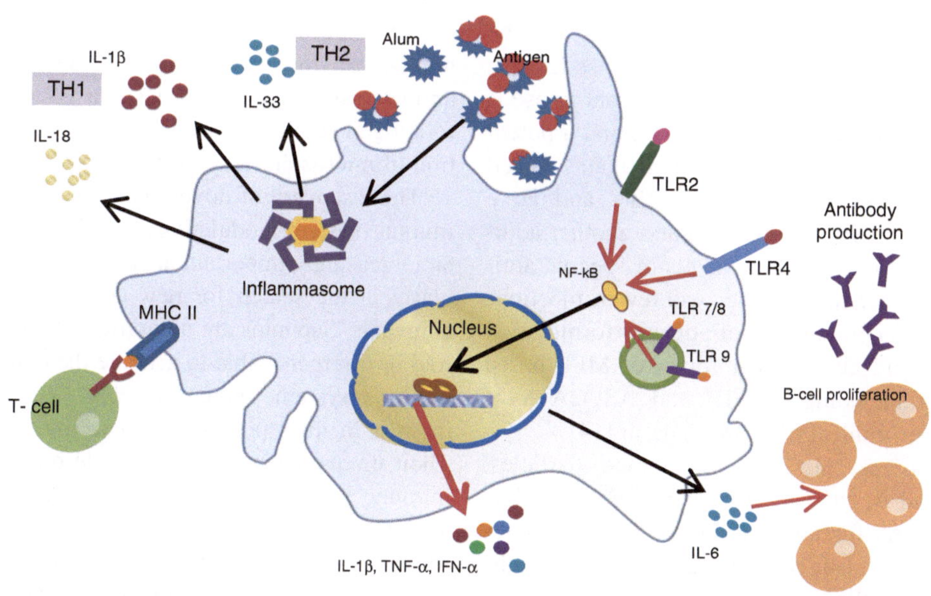

Fig. 1 Mechanism of induction of immune response by adjuvants

well-known fact that these emulsions upon injection cause cellular damage so it is speculated that necrotic cell death may be contributing to their adjuvant activity. Therefore, it becomes very important to select an appropriate adjuvant required for the protection.

MF59 a squalene-based oil-in-water emulsion has got approval for use with influenza vaccine. Squalene is an intermediate compound in the human steroid biosynthetic pathway and is precursor to cholesterol. Unlike alum, it induces increased CD8 response rather than CD4 T-cell response. Because of its ability to induce higher levels of hemagglutinin-inhibiting antibodies, it shows its potential to be used for influenza vaccines during pandemics [102, 107] along with similar ASO3 (GSK Biologicals), a 10% water-in-oil emulsion-based adjuvant, and both have been licensed for vaccines against pandemic influenza and widely used for 2009 influenza pandemic [88]. On comparison with other adjuvants like CPG and alum, MF59 showed the most efficient adjuvant activity by inducing maximum number of genes and rapid influx of CD11b + cells. Furthermore, it is a potent inducer of gene encoding, cytokine receptor, and cytokine adhesion molecules involved in leukocyte migration. Post-administration, it is internalized by APCs which migrate to the lymph nodes, eliciting efficient response [25].

Another oil/water emulsion adjuvant in use is MPL, a nontoxic derivative of LPS and a potent stimulator of Th1 response. It activates T-cell effector responses as part of HBV and HPV vaccines [13, 99]. In this category, another adjuvant giving rise to higher levels of specific antibody response with efficacy and fewer injections is ASO4, a combination of aqueous formulation of MPL and alum. Other forms of MPL-based adjuvant including ASO1B and ASO2A have undergone clinical trials for malaria [14].

Liposomes, the lipid membrane particles, LPS, lipid A, etc. can serve as vehicles or delivery systems for antigens [1, 3, 4]. Liposomes are usually made of biodegradable materials, e.g., phospholipids, and originally developed as carriers for drugs and biologically active substances. However, they have some limitations of causing sensitivity and instability. In addition, their high cost of manufacturing and scale up production limits their use. Instead, a superior liposome adjuvant referred as virosomes are prepared by inserting virus fusion protein into a liposome bilayer [45, 103]. These virosomes have proved to be potent adjuvants without any inflammatory reaction. They are membrane bound and serve to amplify fusogenic activity which facilitate the uptake by APC and induce a natural antigen-processing pathway. And this may be the reason why virosome-based vaccines stand out due to their excellent safety profile [44].

ISCOMS are particulate adjuvants composed of saponins purified from bark of the South American tree, *Quillaja saponaria*, formulated with cholesterol and phospholipids. This adjuvant doesn't act through PRRs, rather enhance antigen uptake and prolong retention by DCs in the draining lymph nodes, where DC activation leads to strong antibody and T-cell response [79]. They do not get biased about Th1 or Th2 cell response; at the same time, they induce CD8+ and CD4+ T-cell responses, both in human and animal vaccines [21].

Certain microspheres have also been evaluated as adjuvant with better targeting of antigen to APCs on mucosal surface or by reducing the number of doses by controlling the release of antigen. But there are still some issues related to antigen stability during encapsulation, hydration of microspheres, and their storage.

The discovery of novel plant compounds with immune system-modulating activities has become an increasingly important area of research, particularly in the search for new-generation vaccine adjuvants. Saponins are natural glycosides of steroid or triterpene, able to activate the mammalian immune system, which has led to significant interest in their potential as vaccine adjuvants. Their unique capacity to stimulate both the Th1 immune response and the production of CTLs against exogenous antigens makes them ideal for use in subunit vaccines and vaccines directed against intracellular pathogens as well as for therapeutic cancer vaccines. The most widely used saponin-based adjuvants are Quil A and its derivatives QS-21. In some human vaccine designs, improving immunogenicity of a synthetic malaria peptide vaccine SPf66 was achieved by using SPf66/QS-21 formulation [66]. Adjuvants derived from *Ginseng saponins* (ginsenosides) – the active substances in the root of *Panax ginseng* – on bacterial antigens have shown a marked enhancing effect on vaccinating pigs against *Erysipelothrix rhusiopathiae* infection. The saponins from the root of *P. grandiflorum* increased a specific antibody and cellular response against ovalbumin in mice and could lead to the development of promising balanced Th1- and Th2- directing immunological adjuvants [131]. Advax™ adjuvant derived from inulin was evaluated with influenza vaccine to enhance immunogenicity and protection in mice [94].

The development of many other additional adjuvants is the result of the shortcomings of alum. Often, combinations of adjuvants in one formulation have yielded better results in a synergistic manner. Adjuvants like Montanides, MDP, etc. have been tried with various vaccines. An increasing number of experimental adjuvants are in developmental stages, such as squalene and phosphate adjuvants, QS21 [42, 128].

Realizing the need for a novel adjuvant, Defense Institute of Physiology and Allied Sciences (DIPAS), DRDO, India, has developed an adjuvant, DIP-HIP, derived from a high altitude medicinal plant, Seabuckthorn (*Hippophae rhamnoides* L. (SBT)). SBT, a unique and valuable plant from the family Elaeagnaceae, has recently gained worldwide attention, mainly for its medicinal and nutritional potentials [11, 38, 63, 95, 132]. In a comparative study, animals administered with DIP-HIP in formulation with different types of antigens – recombinant, conjugated, or native proteins, etc., evaluating an extensive range of different adjuvant systems – significantly and consistently enhanced the safe, stable, and efficacious antigen-specific response with minimum amount of antigen and reduced boosters with a long-term antibody sustenance. DIP-HIP was found to be consistently superior to alternative adjuvants when adjudged collectively by bioactive qualities, manufacturability, syringeability, stability, safety, and immunopotency criteria. This adjuvant can be produced as "point-of-use" by simple physical mixing procedure to generate an emulsion that is stable and homogenous. Alternatively, DIP-HIP has been produced in different batches stable at 2–8 °C for at least 3 years. The shelf life of DIP-HIP in the extract form and as formulation with antigens at 4 °C is quite reasonable.

A relative contribution of Th1/Th2 type of immune response was indicated by substantially enhanced titers of IgG1 and IgG2a antibody subtypes. The cytokine profile of IFN-γ and IL-4 correlated well with the Th1 and Th2 types of immune responses which are also supported by higher DTH response, indicating thereby the overall magnitude of humoral- and cell-mediated immune response generated by DIP-HIP and its ability to amplify both the arms of immune system. Interestingly, using different strains and species of animals, DIP-HIP responded equally well. Local tolerance evaluated by immunization through different routes like intramuscular or intraperitoneal using scoring systems, both macroscopically and histologically, did not show any variation nor caused any muscular damage, granulomatous reaction, or dystrophy. There was no hemolysis caused on treatment of both humans and animal erythrocytes with DIP-HIP. The herbal adjuvant developed by DIPAS is safe, effective, and comparable with commercially available adjuvants like CFA and alum. The product is being tested for commercial antigens to qualify for the license. The extract is in crude form and is being fractionated into various components using supercritical CO_2 extraction procedure followed by HPTLC and HPLC analyses. The bioactive fractions are being analyzed for their adjuvant activity.

Model Development

The most important part of adjuvant development is selection of a good model that can predict immune responses in humans. There are many limitations in developing a model relevant to adjuvant development and evaluation. The most commonly used tools are activation of APCs *in vitro* and immunity in experimental animal models which could be inbred strains of mice, including knockout and transgenic mice. Despite many similarities, the immune responses of mice and humans differ in many aspects [41, 105] which lead to a failure of promising formulations in clinical confirmation trials. Different animal strains and species even behave differently to various adjuvants [56, 82]. In mouse studies, the route of immunization used is intraperitoneal or intravenous injections, whereas in humans, it is subcutaneous or intramuscular. It is understood that different routes affect activation of specific dendritic subsets. Sometimes, the choice of antigen itself can also affect the outcome of the study. Similarly, *in vitro* studies using blood cells have limitations in evaluating the adjuvant. Cell cultures are not reliable for studying the vaccine-adjuvant formulations that rely on inflammatory cytokine responses from noncirculating tissue cells. An adjuvant acts in part by altering the distribution of antigen or its presentation at the injection site, which may not give confirmatory results in cell culture. These differences pose an obstacle in rapid translation of findings from mice to humans or cell culture to humans. Some time ago, guinea pigs and rabbits were suggested to be better models for evaluating the adjuvanticity and toxicity of adjuvant formulations [46, 119].

To avoid these shortcomings, the possible approaches could be to use humanized mice or develop mice that have similar cell-specific expressions or finally use nonhuman primates. But all these approaches are very costly for maintaining such animals and cell culture conditions, which often limit the interest of development of a good adjuvant. Therefore, if one has to follow the traditional way of evaluating the adjuvanticity, then it is recommended that the evaluation may be performed in at least two strains of mice with different haplotypes or guinea pigs and/or rats for studying the immunological effect of adjuvant and its effective adjuvanticity.

Therefore, the present approach of systems biology has been considered of utmost importance to focus on the study of vaccine adjuvant formulations in humans [39, 106]. A consortium has been set up to analyze the human immune responses to vaccine formulations and infections by high-throughput approaches. This approach will benefit in systemically understanding and characterizing the immune response in humans to accelerate the studies in the field of human biology and vaccinology.

The age group distribution of the world population gives a deep insight into the vaccine adjuvant-induced immunity problems in humans, e.g., in the very young and very old, the immunocompromised, or the diseased. Therefore, whenever a new adjuvant is developed, there is a need to reexamine and reevaluate the clinical trials in which multiple parameters of both innate and adaptive immune responses can be evaluated by cutting-edge technologies [39, 105, 106]. Thus, systems biology is a needed and powerful modeling approach.

Novel Adjuvants and Approaches

Three-dimensional virtual screening, whereby a large number of small molecules are docked into the three-dimensional model of a protein receptor, is an important tool in the field of drug discovery and optimization. The identification of potential lead compounds from databases of small molecules significantly reduces the time spent on experimental screening and is therefore now an integral part of drug design. Within vaccinology, structure-based virtual screening is an approach of unprecedented power and scope.

Using virtual screening, CCR4 antagonists have been identified that act as adjuvants for both cellular and humoral immune responses. These molecules were tested *in vivo* with vaccines in mice; enhanced immunogenicity was observed with SP50. The enhancing effects observed in these experiments are particularly striking given that the vaccine vectors employed are known to be intrinsically immunogenic [20].

Cyclic-diguanylate (c-di-GMP), a bacterial signaling molecule, possesses protective immunostimulatory activity. In a study, c-di-GMP was evaluated as a vaccine adjuvant and compared to LPS, CpG oligonucleotides, and a conventional aluminum salt-based adjuvant. It elicited a more potent activator of both humoral- and Th1-like immune responses as evidenced by the robust IgG2a antibody response and the strong IFN-γ, TNF-α, and IP-10 responses in mice and *in vitro* in nonhuman primate peripheral blood mononuclear cells. Further, compared to LPS or CpG, c-di-GMP demonstrated a more pronounced ability to induce germinal center formation, a hallmark of long-term memory, in immunized mice. Together, these data add to the growing body of evidence supporting the utility of c-di-GMP as an adjuvant in vaccination for sustained and robust immune responses and provide a rationale for further evaluation in appropriate models of immunization [24].

Receptor-targeted small molecule adjuvants (SMA) are among the most under-explored type of immunomodulatory adjuvants. Examples include imidazoquinolines (Imiquimod and Resiquimod) which target toll-like receptors (TLRs) (specifically TLR-7 and TLR-8) and were developed as nucleoside analogues for antiviral or antitumor therapy; Bestatin, a tumor adjuvant acting as an inhibitor of aminopeptidase N [CD13]; and Levamisole and Bupivacaine (DNA vaccine adjuvants). Other examples of non-macromolecular adjuvants include monophosphoryl-lipid A, muramyl dipeptide, QS21, PLG, Seppic ISA-51, and CpG oligonucleotides. Optimized CpG oligonucleotides, which target TLR-9, are now entering late phase trials as adjuvants for the poorly immunogenic hepatitis B vaccine [22].

Evaluation of Novel Adjuvants and Delivery Systems

The use of repository adjuvants like mineral salts is accompanied by the formation of an inflammatory focus at the site of injection which may lead to the synthesis of proinflammatory cytokines and stimulation of innate immunity important for the initial steps of the immune response.

Quality evaluation of a vaccine-adjuvant formulation therefore covers aspects such as demonstration of the compatibility of the adjuvant(s) with the antigenic component(s) present in the vaccine, proof of an adequate and consistent association of the antigen with the adjuvant, demonstration that no significant de-association takes place in the course of the shelf life, degree of association throughout the shelf life, and effect of the adjuvant on the ability to assay components, biochemical purity, and pyrogenicity. As an example of association, adsorption is specific for aluminum hydroxide gels, aluminum phosphate gels, calcium phosphate gels, and ISCOMS, while ionic interaction occurs with charged dimethyldioctadecylammonium (DDA) micelles. For emulsions or liposomes, the mechanism is encapsulation. With saponin derivatives or other extracts, interactions with antigens are lipophilic/hydrophilic or ionic [29].

Adjuvants alone are evaluated for their local tolerance (inflammation, consideration of route of administration), induction of hypersensitivity and anaphylaxis (IgE antibody), antigen species (rabbit), pyrogenicity, validated *in vitro* models, systemic toxicity to tissues/organs (histopathology), pharmacodynamic studies (adsorption/elimination from tissues), and reproductive toxicity (reflecting intended schedule of use).

Regulatory Issues

The response generated by antigen-adjuvant formulations is vaccine specific and multifactorial. Thus, data cannot even be interpolated to another distinct antigen, a similar antigen with the same adjuvant or even the same antigen-adjuvant formulation administered via a different route. Therefore, the regulatory guidelines for issuing a license for one formulation would be inapplicable for other formulations [114]. This makes the task for the license-issuing authorities more difficult and vulnerable to wrong application of guidelines. However, the Center for Biologics Evaluation and Research (CBER) of the US FDA has developed a new regulatory approach to vaccine adjuvants and adjuvanted preventive vaccines for infectious disease indications. The Committee for Medicinal Products for Human Use, of the European Medicines Evaluation Agency (EMEA), has also published a guide on similar issues [29].

Safety and tolerability are two critical considerations impacting early introduction of new adjuvants for human use and pose the greatest barrier to regulatory approvals for new adjuvants. Use of novel adjuvants demands extensive preclinical studies, including local reactogenicity and systemic toxicity testing, because vaccines are administered to healthy individuals including infants and children, and there are potential safety concerns. WHO has also suggested nonclinical evaluations to help in proceeding with the clinical development of new adjuvant-antigen formulations (WHO Guidelines on Nonclinical Evaluation of Vaccines, 2003). Applying for a license approval for any new adjuvant or vaccine development falls in the category of "Investigational New Drug Application," and it becomes mandatory to furnish a full report on the parameters evaluated – including its toxicology, pharmacology, biochemistry, and comparison with other similar adjuvanted and non-adjuvanted vaccines – and any incidental or clinical data obtained on human use with full justification – which will add value to the existing adjuvant, etc. All the clinical data should be fully justified.

Sometimes, the pressure of a looming pandemic also helps in streamlining the licensure mechanisms, which happened in the case of avian flu H5N1 influenza vaccine [32]. Sometimes, separate guidelines are set for frequent pandemic vaccines such as pandemic influenza vaccines, which advice clinical development approaches to help in expediting the licensing procedures. For more benefit of the society, for value addition of these pandemic vaccines, certain approaches have been demonstrated and briefly described in two FDA guidance documents [49, 50].

Presently, studies on specific projects on new adjuvant and their dose optimization are given much attention. Many funding agencies are releasing substantial amount of funds in developing the new adjuvants. The matter is being taken seriously as a global issue, more so for basic research on pandemic diseases and early product development of novel adjuvants, including both immunopotentiators and delivery systems. Research should be focused on the rational design of an adjuvant, founded upon a clear understanding of its mode of action on innate immune system, thus allowing for the enhancement of beneficial aspects and reduction of toxic side effects.

With the goal of new adjuvant discovery and development, multiple contracts for "Innate Immune Receptors and Adjuvant Discovery" have been awarded, including Development of Vaccines, Adjuvants, Immunotherapeutics, and Diagnostics for Biodefense, under "Adjuvant Development Program." This program also includes the designing of special units to assure availability of experienced individuals for rapid evaluation of new vaccines and novel adjuvant and use of high-throughput library screens. Other programs have encouraged both public and private sectors to participate in vaccine and adjuvant research in high-priority areas to help combat a wide variety of complex diseases. Such programs develop the strategies for dose reduction [31, 72, 91] using various appropriate technologies for enhancing the immunogenicity and effectiveness of vaccine-adjuvant formulations in collaboration with various industries and academia for use in the developing world and mass campaigns and in disseminating knowledge in this field to the scientific community [59, 60, 120].

Future

Advances in the newer technologies like recombinant DNA, genomics, proteomics, metabolomics, etc. have speeded up the synthesis of newer vaccines, but at the same time demand a need for improved and well-characterized adjuvants and formulations. There is more emphasis on vaccines for neonates and immunocompromized individuals for a better understanding of mechanism of action of adjuvant and consequent immune response.

Several barriers have to overcome to meet the requirement of development of newer adjuvants. The first and foremost concern is the side effects and toxicity, particularly with the pediatric vaccines. In addition, the regulatory issues for approval of adjuvant have substantially raised the concerns. Adjuvants do not get FDA clearance as stand-alone compounds/products, but as a formulation of a registered vaccine adjuvant – antigen formulation. But this requires huge cost and effort for trying various combinations of untested antigens and adjuvants. Moreover, new antigen-adjuvant formulations become proprietary until the adjuvant evaluated is registered. All these regulations limit the development of new adjuvants and thus lead to slow development of novel adjuvants. The strategies to accelerate the development of adjuvants would be to create an organization for standardized evaluation of vaccine-adjuvant formulations capable of inducing safe, strong, stable, and long-lasting humoral and cellular immune responses, with a reduced number of protective boosters, for humans.

While investigating the mechanisms, contribution of TLR pathway has been significant in understanding how and why adjuvant used during vaccinations are important in augmenting adaptive immune responses to specific vaccine antigen. However, with the upcoming knowledge and the research findings, it is gathered that TLR activation is not necessarily required for enhancing the immunogenicity by adjuvant. Therefore, it can be concluded that there could be other receptors besides TLRs that have not yet been characterized, thus opening the door to future research.

Finally, development of candidate adjuvant should focus more on establishing easy, modular, reproducible, stable, and transferable standard operating procedures and fingerprint records for processing and production, which eventually will allow sustainable formulations with long shelf lives without major changes in large-scale manufacturing and sourcing.

Conclusion

Learning how to enhance the persistence of immune responses with relevant adjuvants is critical. Recent in-depth research related to the possible mechanisms of action of few adjuvants has significantly progressed our understanding of the immunology and biochemistry of these adjuvants. However, does the new research explain the mechanism of action of clinically approved adjuvant? Probably the answer is "no," as the appropriate experiments performed particularly in humans do not give the clear indications, be it individual variation or behavior of adjuvant in humans. Moreover, advances in technologies have accelerated the development of newer vaccines which has increased the need for improved adjuvant and formulations beyond those currently available.

The foregoing considerations exigently lead to development of new adjuvants which can selectively modulate the immune response to the desired level and type. An adjuvant with potent properties is required for eliciting both – humoral and cell mediated – immune responses against all types of antigens, such as purified, subunit, combination, or synthetic vaccines. In addition, an adjuvant that helps in enhancing plasma cells, leading to generation of antigen-specific B-cell memory response, is a useful objective. This leads to the fact that the magnitude, quality, type, and persistence of antibody response are regulated by innate immunity. The intracellular signaling pathway, and transcription factors that control the innate immunity, are concepts currently under appreciation and investigation.

Incorporating the foregoing essential aspects, many new adjuvants are in the research pipeline. Further, based on preclinical and clinical observations, a number of adjuvants with some of the above properties are expected in the coming years.

References

1. Alving CR (1993) Lipopolysaccharide, lipid A and liposomes containing lipid A as immunologic adjuvants. Immunobiol 187:430–446

2. Alving CR (2002) Design and selection of vaccine adjuvants: animal models and human trials. Vaccine 20:S56–S64
3. Alving CR et al (1993) Novel adjuvant strategies for experimental malaria and AIDS vaccines. Ann N Y Acad Sci 690:265–275
4. Alving CR, Glass M, Detrick B (1992) Summary: adjuvants/clinical trials working group. AIDS Res Hum Retroviruses 8:1427–1430
5. Alving CR, Koulchin V, Glenn GM, Rao M (1995) Liposomes as carriers of peptide antigens: induction of antibodies and cytotoxic T lymphocytes to conjugated and unconjugated peptides. Immunol Rev 145:5–31
6. Aprile MA, Wardlaw AC (1966) Aluminum compounds as adjuvants for vaccines and toxoids in man: a review. Can J Public Health 57(8):343–360
7. Audibert FM, Lise LD (1993) Adjuvants: current status, clinical perspectives and future prospects. Trends Pharmacol Sci 14:174–178
8. Awate S et al (2012) Activation of adjuvant core response genes by the novel adjuvant PCEP. Mol Immunol 51:292–303
9. Awate S, Babiuk A, Mutwiri G (2013) Mechanisms of action of adjuvants. Front Immunol 4(114):1–10. doi:10.3389/fimmu.2013.00114
10. Baylor NW, Egan W, Richman P (2002) Aluminum salts in vaccines- US perspective. Vaccine 20: S18–S23
11. Beveridge T, Li TS, Oomah BD, Smith A (1999) Sea buckthorn products : manufacture and composition. J Agric Food Chem 47(9):3480–3488
12. Bode C, Zhao G, Steinhagen F et al (2011) CpG DNA as a vaccine adjuvant. Expert Rev Vaccines 10(4):499–511
13. Bojang KA et al (2001) Efficacy of RTS, S/AS02 malaria vaccine against Plasmodium falciparum infection in semi-immune adult men in The Gambia: a randomised trial. The Lancet 358:1927–1934
14. Bojang KA et al (2005) Safety and immunogenicity of RTS, S/AS02A candidate malaria vaccine in Gambian children. Vaccine 23:4148–4157
15. Coffman RL, Sher A, Seder RA (2010) Vaccine adjuvants: putting innate immunity to work. Immunity 33:492–503
16. Cooper MA et al (2009) Cytokine induced memory-like natural killer cells. Proc Natl Acad Sci U S A 106:1915–1919
17. Cooper PD (1994) The selective induction of different immune responses by vaccine adjuvants. In: Ada GL (ed) Strategies in vaccine design. Landes, Austin, pp 125–158
18. Cox JC, Coulter AR (1997) Adjuvants-a classification and review of their modes of action. Vaccine 15:248–256
19. Davenport FM (1968) Seventeen years experience with mineral oil adjuvant influenza virus vaccines. Ann Allergy 26:288–292
20. Davies MN et al (2009) Toward the discovery of vaccine adjuvants: coupling *in silico* screening and *in vitro* analysis of antagonist binding to human and mouse CCR4 receptors. PLoS One 4(11):e8084: 1-12. PMCID-PMC2787246
21. Davis ID et al (2004) Recombinant NY-ESO-1protein with ISCOMATRIX adjuvant induces broad integrated antibody and CD4+ and CD8+ T cell responses in humans. Proc Natl Acad Sci U S A 101:10697–10702
22. De Gregorio E, Caproni E, Ulmer JB (2013) Vaccine adjuvant: mode of action. Front Immunol 4:214. doi:10.3389/fimmu.2013.00214. PMCID: PMC3728558
23. Didierlaurent AM et al (2009) AS04, an aluminum salt-and TLR4 agonist- based adjuvant system, induces a transient localized innate immune response leading to enhanced adaptive immunity. J Immunol 183:6186–6197
24. Dubensky TW, Kanne DB Jr, Leong ML (2013) Rational, progress and development of vaccine utilizing STING- activating cyclic dinucleotide adjuvants. Ther Adv Vaccines 1(4):131–143. doi:10.1177/2051013613501988
25. Dupuis M et al (2001) Immunization with the adjuvant MF59 induces macrophage trafficking and apoptosis. Eur J Immunol 31:2910–2918
26. Edelman R (1980) Vaccine adjuvants. Rev Infect Dis 2:370–383
27. Eisenbarth SC et al (2008) Crucial role for the Nalp3 inflammasome in the immunostimulatory properties of aluminum adjuvants. Nature 453:1122–1126
28. Ellebedy AH et al (2011) Inflammasome-independent role of the apoptosis-associated speck-like protein containing CARD (ASC) in the adjuvant effect of MF59. Proc Natl Acad Sci U S A 108:2927–2932
29. EMEA- Guidelines on adjuvants in vaccine for human use, 1/2005 and explanatory note on immunomodulators, 7/2006. Accessed at http://www.emea.europa.eu/pdfs/human/vwp/13471604en.pdf
30. Exley C et al (2009) A role for the body burden of aluminum in vaccine – associated macrophagic myofasciitis and chronic fatigue syndrome. Med Hypotheses 72:135–139
31. Fernandez J et al (2000) Randomized trial of the immunogenicity of fractional dose regimens of PRP-T Haemophilus influenzae type b conjugate vaccine. Am J Trop Med Hyg 62(4):485–490
32. Focetria®, the Novartis pandemic influenza vaccine, receives European union approval, 2007. Accessed 12th Oct 2007, at http://novartisvaccine.com/pressroom/news/20070508focetria.shtml
33. Fraser CK et al (2007) Improving vaccines by incorporating immunological coadjuvants. Expert Rev Vaccines 6:559–578
34. Freund J (1956) The mode of action of immunologic adjuvants. Bibl Tuberc 10:130–148
35. Freytag LC, Clements JD (2005) Mucosal adjuvants. Vaccine 23:1804–1813
36. Galli G et al (2009) Adjuvanted H5N1 vaccine induces early CD4+ T cell response that predicts

long-term persistence of protective antibody levels. Proc Natl Acad Sci U S A 106:3877–3882
37. Galli G, Hancock K, Hoschler K, DeVos J et al (2009) Fast rise of broadly cross-reactive antibodies after boosting long-lived human memory B cells primed by an MF59 adjuvanted prepandemic vaccine. Proc Natl Acad Sci 106:7962–7967
38. Ganju L et al (2005) Anti-inflammatory activity of Seabuckthorn (*Hippophae rhamnoides*) leaves. Int Immunopharmacol 5:1675–1684
39. Gaucher D et al (2008) Yellow fever vaccine induces integrated multilineage and polyfunctional immune responses. J Exp Med 205:3119–3131
40. Gavin AL et al (2006) Adjuvant-enhanced antibody responses in the absence of toll-like receptor signaling. Science 314:1936–1938
41. Germain RN (2010) Vaccines and the future of human immunology. Immunity 33:441–450
42. Ghochikyan A et al (2006) Prototype Alzheimer's disease epitope vaccine induced strong Th2-type anti-Aβ antibody response with Alum to Quil A adjuvant switch. Vaccine 24(13):2275–2282. doi:10.1016/j.vaccine.2005.11.039.PMC 2081151. PMID: 16368167
43. Glenn GM, Taylor DN, Lix FS et al (2000) Transcutaneous immunization a human vaccine delivery strategy using a patch. Nat Med 6:1403–1406
44. Gluck R (1999) Adjuvant activity of immunopotentiating reconstituted influenza virosomes (IRIVs). Vaccine 17(13–14):1782–1787
45. Glück R, Mischler R, Brantschen S et al (1992) Immunopotentiating reconstituted influenza virus virosome vaccine delivery system for immunization against hepatitis A. J Clin Invest 90(6):2491–2495. doi:10.1172/JCI116141
46. Goldenthal K et al (1993) Safety evaluation of vaccine adjuvants. NCVDG meeting working group. AIDS Res Hum Retroviruses 9:S47–S51
47. Golding B (1991) Cytokine regulation of humoral immune responses. In: Sprigs DR, Koff WC (eds) Topics in vaccine adjuvant research. CRC press, Boca Raton, pp 25–37
48. Gonzalez AM, Nguyen TV, Azevedo MSP et al (2004) Antibody response to human rotavirus (HRV) in gnotobiotic pigs following a new prime/vaccine boost strategy using oral attenuated HRV priming and intranasal VP2/6 rotavirus like particle (VLP) boosting with ISCOM. Clin Exp Immunol 135:361–372
49. Guidance for Industry – Clinical Data Needed to Support the Licensure of Pandemic Influenza Vaccines (2007a). Accessed at http://www.fda.gov/cber/gdlns/panfluvac.pdf
50. Guidance for Industry – Clinical Data Needed to Support the Licensure of Trivalent Inactivated Influenza Vaccines (2007b). Accessed at http://www.fda.gov/cber/gdlns/trifluvac.pdf
51. Gupta RK et al (1993) Adjuvants- a balance between toxicity and adjuvanticity. Vaccine 11(3):293–306
52. Gupta RK, Rost BE, Relyveld E, Siber GR (1995) Adjuvant properties of aluminum and calcium compounds. In: Powell MF, Newman MJ (eds) Vaccine design: the subunit and adjuvant approach. Plenum Press, New York, pp 229–248
53. Gupta RK, Siber GR (1995) Adjuvants for human vaccines-current status, problems and future prospects. Vaccine 13:1263–1276
54. Guy B (2007) The perfect mix: recent progress in adjuvant research. Nat Rev Microbiol 5:505–517
55. Hammad H et al (2010) Inflammatory dendritic cells–not basophils–are necessary and sufficient for induction of Th2 immunity to inhaled house dust mite allergen. J Exp Med 207:2097–2111
56. Hardegree MC, Pittman M, Maloney CJ (1972) Influence of mouse strain on the assayed potency (Unitage) of tetanus toxoid. Appl Microbiol 24(1):120–126
57. Hilleman MR (1966) Critical appraisal of emulsified oil adjuvants applied to viral vaccine. Prog Med Virol 8:131–182
58. Hoebe K, Janssen E, Beutler BF (2004) The interface between innate and adaptive immunity. Nat Immunol 5:971–974
59. Hoelscher MA, Garg S, Bangari DS et al (2006) Development of adenoviral-vector-based pandemic influenza vaccine against antigenically distinct human H5N1 strains in mice. Lancet 367 (9509):475–481
60. Hoelscher MA, Jayashankar L, Garg S et al (2007) New pre-pandemic influenza vaccines: an egg- and adjuvant-independent human adenoviral vector strategy induces long-lasting protective immune responses in mice. Clin Pharmacol Ther 82(6):665–671
61. Huleatt JW, Jacobs AR, Tang J et al (2007) Vaccination with recombinant fusion proteins incorporating Toll-like receptor ligands induces rapid cellular and humoral immunity. Vaccine 25:763–775
62. Hutchison S et al (2012) Antigen depot is not required for alum adjuvanticity. The FASEB J 26:1272–1279
63. Jain M et al (2008) Effect of Hippophae rhamnoides leaf extract against Dengue virus infection in human blood–derived macrophages. Phytomedicine 15(10):793–799. doi:10.1016/J.phymed.2008.04.017, Epub 2008 June 30
64. Janeway CA Jr (1989) Approaching the asymptote? Evolution and revolution in immunology. Cold Spring Harb Symp Quant Biol 54:1–13
65. Jordan MB et al (2004) Promotion of B cell immune responses via an alum-induced myeloid cell population. Science 304:1808–1810
66. Kashala O et al (2002) Safety, tolerability and immunogenicity of new formulations of the Plasmodium falciparum malaria peptide vaccine SPf66 combined with the immunological adjuvant QS-21. Vaccine 20(17–18):2263–2277
67. Kasturi SP et al (2011) Programming the magnitude and persistence of antibody responses with innate immunity. Nature 470:543–547

68. Kenney RT, Cross AS (2010) Adjuvants for the future. In: Levine MM, Dougan G, Good MF, Liu MA, Nabel GJ, Nataro JP, Rappuoli R (eds) New generation vaccines. Informa Healthcare USA, Inc., New York, pp 250–262
69. Khurana S, Chearwae W, Castellino F et al (2010) Vaccines with MF59 adjuvant expand the antibody repertoire to target protective sites of pandemic avian H5N1 influenza virus. Sci Transl Med 2:15ra15–15ra15
70. Klinman DM (2003) CpG DNA as a vaccine adjuvant. Expert Rev Vaccines 2(2):305–315
71. Kool M, Soullie T, van Nimwegen M et al (2008) Alum adjuvant boosts adaptive immunity by inducing uric acid and activating inflammatory dendritic cells. J Exp Med 205:869–882
72. Lagos R, Valenzuela MT, Levine OS et al (1998) Economisation of vaccination against Haemophilus influenzae type b: a randomised trial of immunogenicity of fractional-dose and two-dose regimens. The Lancet 351(9114):1472–1476
73. Lavelle EC, Jarnicki A, McNeela E et al (2004) Effects of cholera toxin on innate and adaptive immunity and its application as an immunomodulatory agent. J Leukoc Biol 75:756–763
74. Lavelle EC, McGuipk P, Mills KHG (2004) Molecules of infectious agents as immunomodulatory drugs. Curr Top Med Chem 4:499–508
75. Leibundgut-Landmann S, Groβ O, Robinson MJ et al (2007) Syk-and CARD9-dependent coupling of innate immunity to the induction of T helper cells that produce IL17. Nat Immunol 8:630–638. Pubmed:17450144
76. Leroux-Roles I, Roman F, Forgus S et al (2010) Priming with AS03 A-adjuvanted H5N1 influenza vaccine improves the kinetics, magnitude and durability of the immune response after a heterologous booster vaccination: an open non-randomised extension of a double- blind randomised primary study. Vaccine 28:849–857
77. Li H, Willingham SB, Ting JP, Re F (2008) Cutting edge: inflammasome activation by alum and alum's adjuvant effect are mediated by NLRP3. J Immunol 181:17–21
78. Mackay F, Schneider P (2009) Cracking the BAFF code. Nat Rev Immunol 9:491–502
79. Maraskovsky E, Schnurr M, Wilson NS et al (2009) Development of prophylactic and therapeutic vaccines using the ISCOMATRIX adjuvant. Immunol & Cell Biol 87:371–376
80. Martin S, Daniel SL, Rouse BT (1991) Cytokines and regulation of cellular immune responses to viruses. In: Spriggs DR, Koff WC (eds) Topics in vaccine adjuvant research. CRC Press, Boca Raton, pp 39–50
81. Mata-Haro V, Cekic C, Martin M et al (2007) The vaccine adjuvant Monophosphoryl lipid A as a TRIF- biased agonist of TLR 4. Science 316:1628–1632
82. Matsuhasi T (1991) Influence of mouse strain on the results of potency test of diphtheria and tetanus toxoid components of D, T, DT and DPT vaccines. In: Manclark CR (ed) Proceeding of an informal consultation on the World Health Organization Requirements for diphtheria, tetanus, pertussis and combined vaccine. Department of Health and Human Services, United States Public Health Service, Bethesda. DHHS Publication No. (FDA) 91-1174, pp 55–58
83. Valiante NM et al. (2010) New adjuvants for human vaccines. Curr Opin Immunol 22:411–416
84. McGhe JR, Mestecky J, Dertzbaugh MT et al (1992) The mucosal immune system from fundamental concepts to vaccine development. Vaccine 10:75–88
85. Metzinger P (1994) Tolerance, danger and the extended family. Annu Rev Immunol 12:991–1045
86. Moreira LO, Smith AM, Defreitas AA et al (2008) Modulation of adaptive immunity by different adjuvant- antigen combinations in mice lacking Nod2. Vaccine 26:5808–5813
87. Morel S, Didierlaurent A, Bourguignon P et al (2011) Adjuvant system AS03 containing α-tocopherol modulates innate immune response and leads to improved adaptive immunity. Vaccine 29 (13):2461–2473
88. Mosca F et al (2008) Molecular and cellular signatures of human vaccine adjuvant. Proc Natl Acad Sci U S A 105:10501–10506
89. Neefjes JJ, Schumacher TNM, Ploegh HL (1991) Assembly and intracellular transport of major histocompatibility complex molecules. Curr Opin Cell Biol 3:601–609
90. Nemazee D, Gavin A, Hoebe K, Beutler B (2006) Immunology: toll like receptors and antibody responses. Nature 441:E4. doi:10.1038/nature04875
91. Neuzil KM, Jackson LA, Nelson J et al (2006) Immunogenicity and reactogenicity of 1 versus 2 doses of trivalent inactivated influenza vaccine in vaccine-naive 5-8-year-old children. J Infect Dis 194(8):1032–1039
92. O'Hagan DT, Ott GS, De Gregorio E, Seubert A (2012) The mechanism of action of MF59 – an innately attractive adjuvant formulation. Vaccine 30(29):4341–4348
93. Ohnmacht C et al (2010) Basophils orchestrate chronic allergic dermatitis and protective immunity against helminths. Immunity 33:364–374
94. Okubo HY, Saade F, Petrovsky N (2012) Advax™, a polysaccharide adjuvant derived from delta inulin, provides improved influenza vaccine protection through broad-based enhancement of adaptive immune responses. Vaccine 30(36):5373–5381
95. Padwad Y, Ganju L, Jain M et al (2006) Effect of leaf extract of Seabuckthorn on lipopolysaccharide induced inflammatory response in murine macrophages. Int Immunopharmacol 6(1):46–52

96. Pashine A, Valiante NM, Ulmer JB (2005) Targeting the innate immune response with improved vaccine adjuvant adjuvants. Nat Med 11(4):S63–S68
97. Paul WE (2003) Antigen processing and presentation. In: Paul WE (ed) Fundamental immunology, 5th edn. Lippincott Williams & Wilkins Publishers, Philadelphia
98. Petrovsky N, Aguilar JC (2004) Vaccine adjuvants: current state and future trends. Immunol & Cell Biol 82:488–496
99. Pichyangkul S, Gettayacamin M, Miller RS et al (2004) Pre-clinical evaluation of the malaria vaccine candidate P. falciparum MSP1(42) formulated with novel adjuvants or with alum. Vaccine 22:3831–3840
100. Plotkin SA (2008) Vaccines: correlates of vaccine –induced immunity. Clin Infect Dis 47:401–409
101. Plotkin SA (2010) Correlates of protection induced by vaccination. Clin Vaccine Immunol 17:1055–1065
102. Podda A, Del Giudice G (2003) MF59-adjuvanted vaccines: increased immunogenicity with an optimal safety profile. Expert Rev Vaccines 2:197–203
103. Poovorawan Y, Theamboonlers A, Chumdermpadetsuk S et al (1995) Safety, immunogenicity, and kinetics of the immune response to a single dose of virosome-formulated hepatitis A vaccine in Thais. Vaccine 13(10):891–893
104. Pulendran B, Ahmed R (2011) Immunological mechanisms of vaccination. Nat Immunol 131 (6):509–517
105. Pulendran B, Powell J, Flavell RA (2010) Modulating vaccine responses with innate immunity. In: Levine MM, Dougan G, Good MF, Liu MA, Nabel GJ, Nataro JP, Rappuoli R (eds) New generation vaccines. Informa Healthcare USA, Inc., New York, pp 183–190
106. Querec TD, Akondy RS, Lee EK et al (2009) Systems biology approach predicts immunogenicity of the yellow fever vaccine in humans. Nat Immunol 10:116–125
107. Radošević K, Rodriguez A, Mintardjo R et al (2008) Antibody and T-cell responses to a virosomal adjuvanted H9N2 avian influenza vaccine: impact of distinct additional adjuvants. Vaccine 26:3640–3646
108. Ramon G (1924) Sur la toxine et sur I' anatoxine diphteriques. Ann Inst Pasteur 38:1–10
109. Reed SG, Bertholet S, Coler RN et al (2009) New horizon in adjuvants for vaccine development. Trends Immunol 30(1):23–32
110. Rerks-Ngarm S, Pitisuttithum P, Nitayaphan S et al (2009) Vaccination with ALVAC and AIDSVAX to prevent HIV infection in Thailand. N Engl J Med 361:2209–2220
111. Rimoldi M, Rescigno M (2005) Uptake and presentation of orally administered antigens. Vaccine 23:1793–1796
112. Rock KL (1996) A new foreign policy MHC class I molecules, monitor the outside world. Immunol Today 17:130–137
113. Saenz SA, Noti M, Artis D (2010) Innate immune cell populations function as initiators and effectors in Th2 cytokine responses. Trends Immunol 31:407–413
114. Sesardic D, Dobbelaer R (2004) European Union regulatory developments for new vaccine adjuvants and delivery systems. Vaccine 22(19):2452–2456
115. Seubert A et al (2011) Adjuvanticity of the oil-in-water emulsion MF59 is independent of Nlrp3 inflammasome but requires the adaptor protein MyD88. Proc Natl Acad Sci U S A 108:11169–11174
116. Shenderov K (2010) Inflammasome-dependent IL-1 {beta} production is critical for complete Freund's adjuvant-induced helper T cell polarization. J Immunol 184:136–144
117. Shi Y, Evans IE, Rock KL (2003) Molecular identification of danger signal that alerts the immune system to dying cells. Nature 425:516–521
118. Spriggs DR, Koff WC (1991) Topics in vaccine adjuvant research. CRC Press, pp 119–136
119. Stewart-Tull DES (1989) Recommendations for the assessment of adjuvants (immunopotentiators). In: Gregoriadis G, Allison AC, Poste G (eds) Immunological adjuvants and vaccines. Springer, pp 213–226
120. Stout R, Gutierrez M, Freeland P et al (2007) Needle-free injections using a spring-powered device for subcutaneous, intramuscular & intradermal injections. Drug Deliv Technol 7(2):40–43
121. Stuart Harris CH (1969) Adjuvant influenza vaccines. Bull WHO 41:617–621
122. Su SB, Silver PB, Grajewski RS et al (2005) Essential role of the MyD88 pathway, but nonessential roles of TLRs 2, 4, and 9, in the adjuvant effect promoting Th1-mediated autoimmunity. J Immunol 175:6303–6310
123. Sun JC, Beilke JN, Lanier LL (2009) Adaptive immune features of natural killer cells. Nature 457:557–561
124. Tang H, Cao W, Kasturi SP et al (2010) The T helper type 2 response to cysteine proteases requires dendritic cell-basophil cooperation via ROS-mediated signaling. Nat Immunol 11:608–617
125. Tomljenovic L, Shaw CA (2011) Aluminum vaccine adjuvants: are they safe? Curr Med Chem 18 (17):2630–2637
126. Tomljenovic L, Shaw CA (2012) Mechanisms of aluminum adjuvant toxicity and autoimmunity in pediatric populations. Lupus 21(2):223–230. doi:10.1177/0961203311430221
127. Vandepapeliere P, Horsmans Y, Moris P et al (2008) Vaccine adjuvant systems containing monophosphoryl lipid A and QS21 induce strong and persistent humoral and T cell responses against hepatitis B surface antigen in healthy adult volunteers. Vaccine 26:1375–1386
128. Vogel FR (1995) immunologic adjuvants for modern vaccine formulations. Annals of the New York Academy of Sciences 754:153–160

129. Wack A, Rappuoli R (2005) Vaccinology at the beginning of the 21st century. Curr Opin Immunol 17:411–418
130. Wickelgren I (2006) Mouse studies question importance of toll-like receptors to vaccines. Science 314 (5807):1859–1860. doi:10.1126/Science.314.5807.1859a. PMID 17185572
131. Xie Y, Pan H, Sun H et al (2008) A promising balanced Th1 and Th2 directing immunological adjuvant, saponins from the root of *Platycodon grandiflorum*. Vaccine 26(31):3937–45. doi:10.1016/J.Vaccine.2008.01.061. Epub 2008 May 20
132. Xu M, Sun X, Tong W (1994) Medical research and development of sea buckthorn. Hippophae 7:32–40

Rapid Acclimatization Strategies for High-Altitude Induction

Gopinath Bhaumik, Anju Bansal, Dishari Ghosh, Harish Kumar, Sanjeev Kumar, Deepak Dass, Jitendra Kanujia, Krishan Singh, Utkarsha Kumar, Sarada Suryakumari, Mrinalini Singh, Shweta Saxena, Deepika Saraswat, D. Majundar, and Shashi Bala Singh

Abstract

High altitude (HA) is defined as elevation above 9,000 ft. At this altitude, most people develop acute mountain sickness (AMS). If untreated, this may lead to high-altitude pulmonary oedema (HAPE) or high-altitude cerebral oedema (HACE), both of which are potentially life-threatening. In emergencies/warlike conditions, rapid deployment of military personnel to high altitude frequently occurs without giving the adequate degree of altitude acclimatization, resulting in acute mountain sickness (AMS). Acclimatization to high altitude is the best strategy to prevent AMS, and this can be achieved by hypoxia preconditioning by the use of interventions like hypoxia mimetics. Efficacy of hypoxia mimetics, viz. cobalt chloride ($CoCl_2$), ethyl 3, 4-dihydroxybenzoate (EDHB), sphingosine-1-phosphate (S1P) and other pharmacological agent nanocurmin in facilitating acclimatization to high altitude in animal model, has been discussed. An alternative approach to induce acclimatization and reduce incidence of AMS is the use of intermittent hypoxic exposure (IHE). This study was conducted to evaluate the effect of IHE exposure at sea level on incidence of AMS during acute ascent to 3,500 m altitude in Indian military personnel. The army volunteers were divided into two groups, viz. control and experimental. Experimental group of subjects were

G. Bhaumik (✉) • A. Bansal • D. Ghosh
H. Kumar • S. Kumar • D. Dass • J. Kanujia • K. Singh
U. Kumar • S. Suryakumari • M. Singh • S. Saxena
D. Saraswat • D. Majundar
Scientist 'E', Divisional Head of High Altitude Physiology Group, Defence Institute of Physiology and Allied Sciences (DIPAS), DRDO, Ministry of Defence, Lucknow Road, Timarpur, New Delhi 110 054, India

S.B. Singh
Defence Institute of Physiology and Allied Sciences (DIPAS), New Delhi, India
e-mail: drshashisingh@gmail.com

S.B. Singh et al. (eds.), *Translational Research in Environmental and Occupational Stress*, DOI 10.1007/978-81-322-1928-6_12, © Springer India 2014

exposed to intermittent normobaric hypoxia consisting of 12 % FIO_2 (altitude – air equivalent 4,350 m) for 4 h per day for 4 consecutive days. After giving IHT, the subjects were inducted to 3,500 m altitude (Leh) by air and different physiological parameters like AMS score (LLS), pulse arterial oxygen saturation (SaO_2) and ventilatory parameters (V_E, VO_2, V_T/Ti) were recorded daily. IHE-treated group showed a significant reduction in AMS at HA in comparison to control. IHE may be considered as an alternative approach to induce the altitude acclimatization at low altitude-based soldiers before their deployment to high-altitude operations in emergency-like conditions.

Introduction

Our country is unique in having the most beautiful and loftiest snow clad summits of the majestic Himalayas along the Northern frontiers, which attract numerous tourists all over the world. Due to military and strategic reasons, a large body of troops is also deployed in snowbound areas of high altitude. In modern scenario, military operations frequently require rapid deployment of defence personnel to high altitude with little or no time for physiological acclimatization. Rapid deployment of unacclimatized soldiers to high altitude/mountain environments may cause debilitating effects on operational capabilities (i.e. physical work performance) and force health (i.e. altitude sickness). The 'altitude' is the vertical height above sea level. The term 'high altitude' (HA) has no precise definition. Considering various aspects of mountain ascent, the Indian Army authority has taken an elevation of 2,700 m (9,000 ft) and above as HA (A.O. No. 110/80). They selected this height because above this point, majority of the people develop signs and symptoms associated with low oxygen pressure resulting in decrease in oxygen haemoglobin saturation below 90 %. The most important feature of HA is low barometric pressure, resulting in systemic hypoxia. Other environmental hazards at HA are severe cold, high wind velocity, low humidity, high solar radiation, increased ultraviolet radiation and difficult terrains. Oxygen concentration in air is 20.93 %, which remains the same whether at sea level or on top of the mountains. As one ascends, the barometric pressure decreases with increase in altitude and atmosphere becomes rarefied resulting in reduced number of oxygen molecules entering into lungs per breath creating oxygen deficiency in the body. Reduced oxygen in the blood leads to high-altitude illness like AMS, HAPE and HACE. Besides this, other health hazards are dehydration, dryness, snow blindness, sunburn and cold injuries. People most susceptible to HA illness are (a) low landers, who ascend rapidly to HA and engage in physical activity; (b) both low and high landers who return to HA after spending a period of time at the plains (reinductees) and (c) people who are confident and disinclined to take adequate rest during initial days of their subsequent visit to HA and do not undergo proper acclimatization. The main determinants of AMS are the altitude reached, rate of ascent, physical activity and degree of pre-acclimatization.

High-Altitude Ailments

The most common ailment of high altitude is AMS, and symptoms are headache, nausea, anorexia, insomnia, lassitude, vomiting and dizziness. The magnitude of these symptoms is related to the speed of ascent, the height climbed and the mode of induction. The symptoms of AMS also vary from person to person. The incidence and severity of AMS in unacclimatized soldiers/persons rapidly increase from 20 to 70 % at the altitude above 3,000 m. The symptoms of AMS appear within 4–24 h of exposure and generally resolve after 3–5 days as acclimatization to hypoxia is achieved. Many physiological events associated with the pathophysiology of AMS have been documented including relative hypoventilation, impaired gas exchange (interstitial pulmonary oedema), fluid retention and redistribution and increased sympathetic drive. If AMS is not resolved, it may lead to HAPE or HACE, both of which are potentially life-threatening. HAPE is most dreaded of all mountain maladies and is characterized by headache, shortness of breath, chest pain, mild fever, lethargy, severe cough with blood tinged sputum, crepitation at the time of respiration and blueness in the lips and extremities. HAPE patients must stop all physical activities and put on bed rest. They must be given oxygen and transferred to hospital. If evacuation is not possible, the patient must be treated in recompression chamber or in inflated HAPE bag.

Causative Factors for HA Ailments

The important determinants of acute mountain sickness are the degree of pre-acclimatization, rate of ascent and also the altitude attained. In our earlier study, we have presented the effect of altitude on autonomic control of heart rate [5], cardiovascular system [77], respiratory system [3], chemoreceptor sensitivity [7, 9] and submaximal and maximal exercise responses [6, 8] during initial days of acclimatization at different altitudes. At high altitude, hypoxemia triggers a series of pulmonary and cardiovascular adjustments to maintain an adequate oxygenation of the different organ systems. In the heart, the major adjustments are an increase in heart rate, cardiac contractility and cardiac output. At the vascular level, the main initial adaptive mechanisms to altitude-induced hypoxaemia are pulmonary artery vasoconstriction and peripheral and cerebral artery vasodilatation. The hypoxia-mediated stimulation of cardiovascular system reaches its maximum effects during the initial few days of exposure, and thereafter a new steady state condition is established. Indeed, once these adjustments have reached their optimal effect, any further stimulation may have detrimental effects and induce specific high-altitude-related diseases like HAPE (exaggerated pulmonary hypertension) or HACE (exaggerated cerebral vasodilation). But beyond these factors, there exists a great variation in individual susceptibility to this altitude illness. The changes in autonomic control of cardiovascular system during acclimatization to high altitude have been studied by electrical nerve activity [87], beta-adrenergic blockade [29] and pharmacological interventions [83]. It has also been shown that beta receptors are responsible for most electrocardiographic (ECG) changes at altitude. This is also evident that beta-adrenergic blockade lowers the level of cardiovascular changes without any effect on hypoxia-induced hyperventilation [10]. At high altitude, sympathetic activity and vagal (parasympathetic) withdrawal act synergistically to increase heart rate, blood pressure and cardiac output. Increased sympathoadrenal activity in combination with decreased parasympathetic tone at high altitude accounts for most of the cardiovascular and ECG-related changes [45]. Heart rate variability (HRV), which represents beat-to-beat alterations of the R-R intervals in an ECG, is generally used to assess the sympathovagal balance [74]. The increased sympathoadrenal activity associated with decreased parasympathetic tone during ascent to high altitude is without effect on hypoxic hyperventilation but accounts for most of the cardiovascular and ECG changes [45].

Acute exposure to HA leads to reduction in maximal O_2 consumption and exercise performance. However, altitude does not appear to affect every individual to the same extent. The decrease in SaO_2 with increasing hypoxia accounts for approximately 86 % of the variance in the decrement in VO_{2max} in trained individuals [21]. The degree of O_2 desaturation during exercise at altitude and its relationship with HVR and exercise ventilation have been studied [88]. It has also been reported that high altitude, mainly via chemo reflexes, triggers a series of pulmonary as well as cardiovascular adjustments to maintain adequate oxygen supply to the different organ systems. In the heart, the major adjustments are increase in heart rate, blood pressure and cardiac output. As a direct consequence of these adjustments, myocardial workload and oxygen demand increase. To respond to this increased demand, there is an increase in coronary vasodilatation and enhancement of coronary blood flow.

The susceptibility of the individual's hypoxic tolerance is also determined by hypoxic ventilatory response (HVR). Interindividual differences in HVR at sea level influence ventilatory acclimatization to high altitude [81]. The HAPE susceptible subjects showing low isocapnic HVR have been reported due to low chemo-sensitive response [26, 57]. HVR at sea level accounted for 42 % of the variation in ventilation after acclimatization [81]. Moore et al. [68] demonstrated a clear relationship between a low ventilatory response to hypoxia and AMS. AMS is also associated with higher $AaDO_2$, lower PaO_2 and SaO_2. Rathat et al. [79] reported that 80 % of AMS-susceptible subjects could be predicted by the ventilatory and cardiac responses to hypoxia during exercise. Burtscher et al. [12] showed that hypoventilation seems to be important in the pathophysiology of AMS, and it is directly associated with water and sodium retention that can be attributed to hypoventilation and impaired gas exchange. It has also been reported that the period 20–30 min after exposure to hypoxia corresponding to altitudes between 3,000 and 4,000 m is suitable for detecting subjects which are highly susceptible to AMS. Two key adaptations comprising altitude acclimatization are increased ventilation and decreased total body water, resulting in a reduced plasma volume (i.e. hemoconcentration). Ventilatory acclimatization to altitude is characterized by progressive increase in ventilation, arterial oxygen partial pressure and oxygen saturation (SaO_2) and drop of arterial carbon dioxide partial pressure. Concomitant with the increase in ventilation, the oxygen-carrying capacity of the blood is increased by hemoconcentration resulting from the reduction in plasma volume. The net result of the increased ventilation and hemoconcentration is the increase in arterial oxygen content. Ventilatory acclimatization can be accelerated by the drug acetazolamide [47]. Acetazolamide stimulates the respiration through inhibition of carbonic anhydrase, which builds up carbon dioxide at peripheral and central chemoreceptor level resulting in increase in PAO_2 and SaO_2. Acetazolamide and/or dexamethasone prophylaxis has been tried to rapidly acclimatize the soldiers at 4,578 m altitude and found to be effective for rapid acclimatization. Dexamethasone improves the oxygen saturation of haemoglobin (SaO_2) by reducing the pulmonary oedema and better diffusion of oxygen in the lung [18]. Acclimatization is elevation specific, that is, full acclimatization at one altitude confers only partial acclimatization to a higher altitude. The amount of time required for a person to become fully acclimatized is a function of that individual's physiology and the magnitude of the hypoxic challenge, i.e. altitude attained [81].

Acclimatization to High Altitude

Altitude acclimatization is a series of physiological adjustments that compensates for the reduction in ambient oxygen. Adaptation to high altitude describes changes that have occurred over a number of generations as a result of natural selection in a hypobaric hypoxic environment, and this can be observed in some groups of high-altitude residents. Acclimatization is the set of beneficial processes whereby lowland humans respond to a reduced inspired partial pressure of oxygen. These changes tend to reduce the gradient of oxygen partial pressure from ambient air to the tissues (classical oxygen cascade) and are distinct from the pathological changes that lead to altitude illness. The amount of time required for a person to become acclimatized is a function of that individual's physiology and the magnitude of the hypoxic challenge, as defined by the altitude attained. Individuals with no recent (>1 month) altitude acclimatization require the greatest physiological compensations and thus the longest time to acclimatize. Individuals residing at moderate or high altitudes will achieve acclimatization to a higher altitude more rapidly. In general, 70–80 % of the respiratory component of acclimatization occurs in 4–10 days, and 80–90 % of their overall acclimatization is accomplished by 2 weeks to a month at high altitude.

As millions of visitors' tourists, trekkers, mountaineers or defence personnel travel to high-altitude locations each year, these high-altitude maladies pose a public health problem and have severe economic consequences. Recovery occurs with descent, oxygen inhalation or bed rest, but where descent is not possible and oxygen is not available, deaths continue to occur. Hypoxia is thus a life-threatening stress that has to be dealt with at both cellular and systemic levels. The best way to acclimatize the humans to high-altitude hypoxia is to induce necessary physiological and genetic changes in the body of the humans before they are inducted to high altitude. This can be achieved by hypoxia preconditioning.

Preconditioning

Preconditioning is a process by which a tissue is rendered more tolerant to a subsequent lethal insult such as hypoxia/ischemia. Tolerance can be attained by subjecting tissues to a sublethal stress that results in intracellular adaptation and enhanced endogenous defence mechanism.

Hypoxia preconditioning has been shown to confer tolerance against hypoxic-ischemic brain injury in newborn rats [14, 15, 22, 23]. Gradual ascent, allowing time for acclimatization, is the best strategy for preventing high-altitude illness. Several studies revealed that hypoxia preconditioning protects the brain and heart from several types of injury including ischemia, seizures and oedema. Since, tolerance to hypoxia generally takes several hours to days, it is suggested that the preconditioning stimulus involves adaptive changes in gene and protein expression. Also, because the expression of hypoxia-inducible factor (HIF) is increased throughout the brain after hypoxia preconditioning in newborn rats [4], HIF-1-mediated gene expression may be involved in hypoxia-induced tolerance.

HIF-1: Mediator of Hypoxic Response

Hypoxia elicits a variety of adaptive responses at different levels in the body that enhance cell survival. At the organism level, there is increase in ventilation, increased erythropoiesis and neovascularization, which in combination lead to increased oxygen delivery from the atmosphere to the tissue. At the cellular level, adaptation involves activated glycolysis, increased glucose uptake, thus maintaining ATP despite low oxygen availability and the expression of cell survival- and cell death-related proteins. The regulation of the proteins required for hypoxic adaptation occurs at gene level, and the hypoxic induction of all these diverse genes appears to depend on a common mode of oxygen sensing and signal transduction mechanism mediated by activation of a transcription factor, hypoxia-inducible factor (HIF1) [109]. HIF-1 binds to hypoxia-responsive element in different hypoxia-responsive genes including erythropoietin (Epo), vascular endothelial growth factor (VEGF), inducible nitric oxide synthase (iNOS), heme oxygenase (HO-1), glucose transporter-1 (Glut-1) and several glycolytic enzymes, thus activating their transcription. HIF-1 is a heterodimer composed of two basic helix-loop-helix-PAS (PER-ARNT-SIM) domain (bHLH-PAS) proteins called HIF-1α and HIF-1β (ARNT, aryl hydrocarbon nuclear receptor translocator) [110]. Role of HIFs in adaptation has been recently reviewed by Prabhakar and Semenza [76]. HIF-1α is rapidly degraded under normoxia by hydroxylation of proline residue by prolyl hydroxylases within a highly conserved region in its oxygen-dependent degradation domain (ODDD). HIF-1-prolyl hydroxylases (PHD1-3) hydroxylate the prolyl residues at amino acid 402 and 564 [11]. These enzymes require dioxygen, Fe^{2+}, ascorbate and 2-oxoglutarate for activity [61]. The hydroxylated peptides interact with an E3 ubiquitin-protein ligase complex composed of VHL, elongin B and C and Cullin 2 (CUL2) and then polyubiquitinized, resulting in HIF-1α degradation by the 26S proteasome [41, 59].

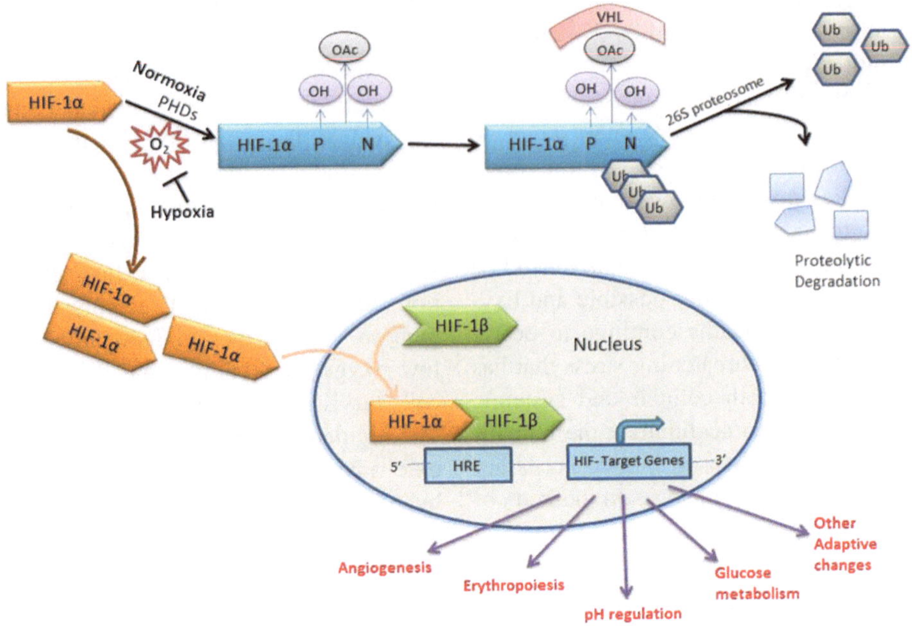

Fig. 1 Regulation of HIF and its target genes

Under hypoxic conditions, HIF-1α is not hydroxylated because of limitation of the major substrate oxygen. The unmodified protein escapes the pVHL-binding, ubiquitination and degradation, dimerizes with HIF-1β and stimulates the transcription of its target genes [90] (Fig. 1).

Hypoxia Mimetics

The hypoxia preconditioning has potential clinical usefulness and can be mimicked by many divalent metals as cobalt, nickel, cadmium and zinc that act as hypoxic mimetics by stabilizing HIF-1α, thus allowing its accumulation, nuclear translocation and binding to HIF-1β to form the transcriptionally active HIF-1 complex even under normoxia. Among all the metals reported, cobalt is one of the classic examples of a hypoxic mimetic.

Cobalt (Co) is a silvery-white, hard metal with an atomic number of 27 and an atomic weight of 58.93. Cobalt is as an essential component of vitamin B12 (cobalamin). Mammals lack the ability to synthesize vitamin B12, and nonruminant animals require a dietary source of vitamin B12. Absorbed cobalt is primarily excreted in urine with small amounts excreted via faecal endogenous routes. A cobalt concentration in tissues is generally low (1 mg/kg DM or less). The amount of cobalt normally stored in the human body is around 1.5 mg.

Cobalt toxicosis in animals is very rare because the concentration of cobalt normally present in animal diets is much lower than that needed to cause toxicosis. Cobalt toxicosis is not likely to occur in ruminants unless environmental contamination of feed or water occurs.

Cobalt Preconditioning

It has been known for a long time that cobalt increases erythropoietin production both in vitro and in vivo in normoxia. Cobalt had also been in use for the treatment of anaemia in infants and women. In the human hepatoma cell lines, production of erythropoietin mRNA was stimulated 6- to 12-fold in response to Co^{2+} in the absence of hypoxia. Chronic oral administration of $CoCl_2$ has been reported to induce polycythaemia without an effect on body or heart weight in animals as well as humans. Administration of $CoCl_2$ (Co^{2+}) in 7 day-old rats was shown to provide protection against ischemia reperfusion injury in the brain [4, 64]. It has been reported that pretreatment with a low dose of cobalt in mice induced cardiac preconditioning, and this protective effect of $CoCl_2$ is achieved through selective activation of HIF-1α signalling [113]. Also, administration of cobalt resulted in a marked protection against ischemic renal injury [58]. Cobalt was also shown to be cytoprotective against tert-Butyl hydroperoxide-induced oxidative stress in HepG2 cells and also resulted in induction of renoprotective genes in rats when $CoCl_2$ was given with drinking water for 13 days. Similarly, it has been reported to improve cardiac contractile function in rats when administered with water containing 0.01 % $CoCl_2$ for 6–7 weeks [19]. Bergeron et al. [4] reported that preconditioning with pharmacological activators of HIF-1 (desferrioxamine or $CoCl_2$) rather than hypoxia confers significant protection in the central nervous system (CNS) by upregulation of hypoxia-inducible factor [42]. It is interesting to consider the finding that the responsiveness of HIF-1 to hypoxia, but not cobalt chloride, wanes with age [13].

Various hypotheses had been proposed to describe the mechanism of action of Co in stabilizing HIF-1α. (1) $CoCl_2$ stabilizes HIF-1α by antagonizing Fe^{+2}, which is an essential cofactor along with oxygen for PHDs that degrade HIF-1α [113]; (2) partial inhibition of PHDs, depletion of ascorbate, which is required to maintain the HIF-PHDs and FIH (factor inhibiting HIF) in an active state and/or (3) direct binding of cobalt to HIF-1α, which may prevent its degradation by VHL-dependent and VHL-independent pathways [32].

Chemical preconditioning has several advantages over physical preconditioning: (i) reduced acclimatization schedule at altitude, leading to decreased loss of man days; (ii) number of people that can be preconditioned is not limited as compared to that of physical preconditioning in simulation chambers; (iii) easy to implement and (iv) is economical. However, most of the studies on hypoxia mimetics, especially on cobalt, are focused on ischemia/reperfusion injury. This led us to investigate the efficacy of cobalt in facilitating acclimatization to high altitude (hypobaric hypoxia) and prevention of HA-induced ailments.

$CoCl_2$ Promotes Hypoxic Tolerance and Facilitates Acclimatization to Hypobaric Hypoxia in Rat Brain/Lung

We determined the hypoxic tolerance by measuring the gasping time (GT) and hypoxic survival time (HST) by exposing rats to simulated hypobaric hypoxia of 10,668 m. $CoCl_2$ at the dose of 12.5 mg cobalt/kg body weight for 7 days significantly improved GT and HST by about three to four times as compared to the control rats. Exposure of Co-supplemented rats to simulated altitude of 7,619 m for 48 h showed increase in expression and DNA-binding activity of HIF-1α and its regulated genes EPO, VEGF, Glut1 and NOS. An increased EPO level in the brain, lung and heart of cobalt-supplemented animals was in line with the observed higher blood haematocrit (53 %) and haemoglobin levels (16.5 mg/dl) which in turn increase the O_2-carrying capacity of the blood [40, 97, 98].

Further, EPO has also been shown to act as a neuroprotective and neurotrophic factor directly in the brain [92]. Higher VEGF levels in the brain and heart enhance capillary density and hence improve oxygen transport to these organs. The most important factor that is reported to be responsible for the occurrence of lung injury in high altitude is VEGF, an angiogenic/permeability factor. We found a marginal decrease in VEGF expression in the lungs of cobalt-pretreated animals both during normoxic and hypoxic conditions [97]. Expression of GLUT-1, mediating the transport of glucose from blood to tissues, was found to increase after cobalt supplementation under hypoxia in the brain, lung and heart, indicating enhanced glucose uptake for continued energy generation in hypoxic environments. Thus, the increased EPO, VEGF, Glut-1 and NOS by cobalt facilitate increased oxygen transport, capillary density, glucose transport and vascular tone, respectively, to cope with the limited oxygen availability during hypoxia. All these changes were responsible for the observed increase in hypoxic tolerance and acclimatization by cobalt supplementation [94, 97, 98].

$CoCl_2$ Attenuates Hypobaric Hypoxia-Induced Oxidative Stress

Hypobaric hypoxia at high altitude is considered as an acute physiological stress often leading to oxidative stress, causing potential damage to proteins, lipids and DNA. Under normal conditions, ROS produced are fully inactivated by an elaborate cellular and extracellular antioxidant defence system [60]. The decrease in available cellular oxygen to be reduced to H_2O by cytochrome oxidase results in accumulation of reducing equivalents in the mitochondrial respiratory chain leading to ROS formation thus causing oxidative stress [56, 67]. The disturbances in oxygen availability and cellular oxidant/antioxidant balance have been implicated in the number of disorders including stroke, head trauma, neoplasia, vascular malformations, neurodegenerative disorders and in high-altitude ailments.

Our studies showed that $CoCl_2$ significantly inhibited ROS levels and oxidation of cellular proteins and lipids in the rats exposed to simulated hypobaric hypoxia (7,619 m, 48 h), via induction of HO-1 and metallothioneins (MT) offering efficient protection to the tissues [95, 97, 98, 104]. HO-1 is known to possess antioxidant and anti-apoptotic activity [72]. Several reports have proposed that HO-1 induction represents an antioxidant defence, operating by decreasing the levels of potential pro-oxidants and increasing the concentration of active bile pigments, such as bilirubin, capable of acting as antioxidants [51]. Moreover, cobalt is known to activate expression of HO-1 [50]. Hence, one of the possible reasons for the observed reduction in oxidative stress might be an increase in HO-1 levels. Metallothioneins constitute a family of metalloproteins involved in cytoprotection during stress situations such as oxidative stress [24, 107]. These proteins are very efficient hydroxyl radical scavengers, and many studies indicate that MT provides protection against oxidative injury in multiple organ systems, strongly implicating its antioxidant function [105]. Regulation of MT gene by cobalt was reported to be mediated by activation of metal response element/metal transcription factor-1 [73], which activates HIF-1 [69]. Thus HO-1 and MT genes are regulated by a single transcriptional factor Hif-1α, and we have reported a significant increase in Hif-1α levels in the brain (tense changed from passive to active), lung and heart of animals supplemented with cobalt under hypoxia. Thus, the observed antioxidant activity of cobalt through HO-1 and MT was found to be mediated via Hif-1α signalling mechanisms.

The findings revealed the potential of cobalt either as drug or nutraceutical for prevention of high-altitude-induced oxidative stress.

$CoCl_2$ Prevents Hypobaric Hypoxia-Induced Cerebral/Pulmonary Oedema in Rats

HAPE that develops two to four days after arrival at high altitude is a form of non-cardiogenic pulmonary oedema with increased pulmonary vascular permeability and pulmonary hypertension due to excessive pulmonary vasoconstriction [44]. This leads to vascular leakage through overperfusion, capillary stress failure or both, resulting in high concentration of vascular proteins and red blood cells in the alveolar fluid [100]. Extracellular vasogenic cerebral oedema had been implicated as the predominant mechanism in HACE [27, 28]. It is characterized by increased permeability of brain capillary endothelial cells (ECs) or alteration in blood-brain barrier (BBB) via generation of ROS or increased VEGF [111]. Inflammation may either be an inciting event or a secondary factor after the initial permeability disruption.

We studied the efficacy of preconditioning with $CoCl_2$ in attenuating pulmonary and cerebral oedema by quantifying hypoxia-induced vascular leakage of sodium fluorescein dye in rat lung and brain which is a measure of vascular permeability [30]. Hypobaric hypoxia resulted in increase in transvascular leakage of sodium fluorescein dye in the lung and brain, lung water content, lavage total protein, albumin, VEGF levels, proinflammatory cytokine levels, tissue expression of cell adhesion molecules and nuclear factor kappa-B (NF-κB) DNA-binding activity. Chemical preconditioning with cobalt significantly attenuated the vascular leakage induced by hypoxia. This was attributed to decrease in VEGF, ROS and NFkB activity and increased oxygen availability via HIF-1 signalling mechanisms. Apart from increased levels of TGF-β and IL-6, cobalt preconditioning resulted in higher levels of anti-inflammatory proteins HO-1 and MT which might also be responsible in attenuating vascular leakage induced by hypoxia. HO-1 has been shown to be protective in a number of pathophysiological states such as ischemia and reperfusion injury, hypertension and pulmonary hypertension [65, 71, 114]. Thus, our data suggest that hypoxic preconditioning with cobalt has protective effect against high-altitude-induced cerebral and pulmonary oedema and other diseases induced by hypoxia [96]. Toxicity studies (biochemical, haematological, histopathological parameters) indicate no adverse effect of $CoCl_2$ supplementation [40].

To conclude, cobalt enhances hypoxic tolerance, prevents transvascular leakage in the brain and lungs, downregulates expression of proinflammatory interleukins, decreases lipid peroxidation and ROS levels, maintains antioxidant status and stimulates the expression of HIF-1α resulting in increased expression of its target genes, thus facilitating acclimatization to HA.

No adverse effect of $CoCl_2$ supplementation was observed.

Cobalt chloride thus has potential to be used in rapid acclimatization to hypobaric hypoxia.

Hypoxia Preconditioning by $CoCl_2$ Enhances Endurance Performance and Protects Skeletal Muscles from Exercise-Induced Oxidative Damage in Rats

The advantages of training performed under hypoxic conditions, e.g. living low-training high or living high-training high, are the subject of much debate and importance to improve the physical performance in athletes, defence personnel or mountaineers [48, 106]. Hypoxic stimulus elicits specific molecular responses in skeletal muscle tissue [33, 108]. It is well established that sustained exposure to severe hypoxia has detrimental effects on muscle structure. Therefore, the potential negative effects of permanent exposure to high altitude could be avoided by exposing the subjects to hypoxia only during exercise sessions. The results of these studies suggest that exercise under hypoxic conditions could induce muscular and systemic adaptation, which either are absent or found to be at a lesser degree after training under normoxic conditions [2, 17]. Exercise training is known to increase lung and cardiac function, muscle build-up and aerobic capacity of skeletal muscle. Terrados et al. [103] noted that training competitive cyclists in hypobaric chamber (2,300 m) for 3–4 weeks improved work capacity at altitude than after sea level training. There was an increase in muscle volume, mitochondrial density, capillary to fibre ratios and muscle fibre cross-sectional area, which were observed only with hypoxia training [17].

Hoppler and Vogt [33] found increased mRNA levels of the regulatory subunit of HIF-1 after training under hypoxic conditions irrespective of training intensity, but not after training in normoxia. As a consequence of this upregulation of HIF-1, the levels of mRNA for myoglobin, VEGF and glycolytic enzymes, such as phosphofructokinase, together with mitochondrial and capillary densities, increased in a hypoxia-dependent manner [31, 33, 34]. These results support the involvement of HIF-1 in the regulation of adaptation processes in skeletal muscle tissue that compensate for the reduced availability of oxygen during training, after training in hypoxia.

Taken together, intermittent hypoxia training has been shown to elicit specific molecular and biochemical responses in skeletal muscle tissue. However, the disadvantage of intermittent hypoxia training is the requirement for technical installations to simulate appropriate altitude conditions either by diluting environmental air with nitrogen or by reducing atmospheric pressure or by individual exposure to hypoxia chamber, which are difficult to attain. Supplementation of hypoxia mimetics is a practical and feasible way of inducing hypoxia.

Therefore, we examined the effect of hypoxic preconditioning by cobalt on skeletal muscle prooxidant-antioxidant balance and physical performance. Preconditioning with cobalt resulted in significant increase in physical performance in rats (33 %, $p < 0.01$)) as determined by swimming time till exhaustion. It also protected against training-induced oxidative damage as observed by increase in GSH/GSSG ratio, reduced lipid peroxidation and enhanced expression of antioxidant proteins HO1 and MT. There was marked reduction in exercise-induced muscle fibre damage as indicated by decreased necrotic muscle fibre, decreased lipofuscin content of muscle and plasma creatine kinase level in rats preconditioned with cobalt [86].

We have also reported augmentation of biological activities of enzymes of TCA cycle, glycolysis and cytochrome c oxidase (COX). Further, there was increase in expression of Glut-1 in muscle showing increased glucose metabolism by aerobic respiration. There was also an increase in mitochondrial biogenesis in skeletal muscle observed by increased mRNA expressions of mitochondrial biogenesis markers which was further confirmed by electron microscopy. Increase in nitric oxide production in skeletal muscle by cobalt seems to be the major reason for peroxisome proliferator-activated receptor-gamma coactivator-1α (PGC-1α) induction and mitochondrial biogenesis. Thus, hypoxia preconditioning by $CoCl_2$ in rats increases mitochondrial biogenesis, glucose uptake and metabolism by aerobic respiration in skeletal muscle, which leads to increased physical performance [85].

Prolyl Hydroxylase Inhibitor as Hypoxia Mimetic

Hydroxylation of HIF-1α subunit by prolyl hydroxylase enzymes was first described in 2001 [11, 20, 35–37, 115]. There are three isoforms of PHDs: PHD1, PHD2 and PHD3 [20], which are also known as HIF-P4Hs: HPH3, HPH2 and HPH1 [11] or EGLN2, EGLN1 and EGLN3, respectively [102]. These proteins differ in size, intracellular localization and tissue distribution [112]. Expression of PHD2 and PHD3 can be regulated by HIF by a negative feedback loop, and these have been characterized as main HIF target genes [25, 66, 101]. PHDs catalyse the hydroxylation of HIF in a two-step reaction, 2-oxoglutarate is first decarboxylated to succinate, and a reactive iron-oxo complex is formed that subsequently hydroxylates the defined amino acid residue of the peptide substrate. 2-oxoglutarate, an intermediate of the tricarboxylic acid (TCA) cycle, is an essential co-substrate for PHDs due to its role in Fe (II) coordination in the catalytic centre. At the lower level of oxygen in the cell, PHD becomes inactivated, causing rapid accumulation of HIF-1αsubunits in the cytoplasm and migrate to nucleus to heterodimerize with HIF-1β to form functional transcriptional factor.

One of the most important class of PHD inhibitors is analogues of 2-oxoglutarate, e.g. ethyl 3, 4-dihydroxybenzoate (EDHB). It is also known as protocatechuic acid and is present in plant foods such as olives, roselle, du-zhong and white grapevine [49]. It has been reported to have antioxidant, cardioprotective [78], neuroprotective [53], antimicrobial [62], anti-inflammatory and myoprotective activity. Recently, it has been reported to possess antiulcer activity also [46]. This plethora of properties of EDHB prompted us to study the efficacy of EDHB in facilitating acclimatization to hypoxia and improving endurance performance in Sprague Dawley rats.

Preconditioning with EDHB (75 mg/kg b.wt, 3 days) was found to be efficacious in increasing hypoxia tolerance as measured by gasping time and survival time of animal under simulated hypoxia conditions. There was appreciable attenuation of hypoxia-induced oedema index and vascular leakage as accessed by wet/dry wt. ratio and sodium fluorescein dye leakage, respectively, in both the lung and brain. Preconditioning with EDHB also resulted in stabilization of HIF-1α, upregulation of EPO expression and antioxidant proteins HO-1, MT-1 along with improved antioxidant and anti-inflammatory status in the brain suggesting the neuroprotective effect of EDHB (our unpublished data).

Further, hypoxia preconditioning with EDHB (50 mg/kg bw for 10 days) boosted the physical performance of rats as assessed by running time till exhaustion by treadmill test which could be due to improved blood oxygen-carrying capacity as indicated by increase in haematocrit and blood haemoglobin level and also by upregulation of VEGF expression in skeletal muscles indicating improved angiogenesis (manuscript communicated). EDHB supplementation could also accord protection to animals against oxidative damage by improvement in antioxidant status. It has been demonstrated that mitochondrial adaptation to endurance training is associated with activation of PGC-1α [16, 82] as well as its downstream transcription factors (NRF-1, mTFA, PRC and PPARγ), which induce coordinated expression of mitochondrial transcripts. We observed increase in the expression of mitochondrial biogenesis markers indicating an increase in mitochondrial density in muscles to sustain from fatigue for a longer period. Also, preconditioning with EDHB resulted in increased expression of HIF and hypoxia-responsive genes (Epo, HO-1, MT-1) and various metabolic and myogenic genes which confirms the activation of oxygen-sensing system in skeletal muscle that leads to hypoxia adaptation (our unpublished data). Although this study is still in progress, the results so far underscore the potential of preconditioning with EDHB in facilitating acclimatization to hypobaric hypoxia and improvement of physical performance in rats.

Sphingosine-1-Phosphate (SIP) as Hypoxia Mimetic

Sphingosine-1-phosphate (S1P), a biologically active pleiotropic lipid, is an emerging signalling molecule involved in several physiological processes especially in the areas of vascular biology and immunology encompassing cell survival, angiogenesis, vascular tone, immune response, etc. by interacting with specific cell surface receptors [54]. S1P is derived from ceramide that is synthesized de novo or as part of the sphingomyelin cycle. The enzyme ceramidase acts upon ceramides to release sphingosine, which is phosphorylated by sphingosine kinases (SphKs) to form S1P [70]. The physiological level of S1P is maintained in low nanomolar to micromolar range, and its plasma pool is maintained in association with high-density lipoproteins (HDL), where it is bound to apoprotein M. It is produced continuously and stored in relatively high concentrations in human platelets and erythrocytes [43]. The signalling effects of S1P are executed via five G-protein-coupled receptors ($S1P_1$–$S1P_5$) known till date – $S1PR_{1-5}$ – and through potential intracellular targets not yet known. The relative intracellular concentration of S1P is an important arm of the 'sphingolipid rheostat' and is a key regulator of cell survival, hypoxia-inducible factor 1α (HIF-1α) kinetics, redox balance and differentiation state of the cell during ischemia-reperfusion injury [55, 75]. Further, recent research studies on S1P and its functional agonists have unfolded the involvement of S1P-SphK-$S1PR_{1-5}$ axis in several pathologies of respiratory, cardiovascular, cerebral and renal system [54]. Interestingly, in most of these disorders underlying hypoxia is a common factor and is either a cause or an outcome.

Most interestingly, it has been observed that exogenous S1P supplementation mimics hypoxia by increasing HIF-1α stability and enhancing its transcriptional activity on hypoxia-regulated genes such as VEGF, GLUT-1 and PAI-1 [63]. These fascinating facts of S1P signalling guided us to explore the potential of S1P as a hypoxic preconditioning agent with a hypothesis that modulation of physiological level of S1P by its exogenous supplementation would boost the acclimatization responses to hypobaric hypoxia since it has been shown to be a hypoxia mimetic in earlier studies.

With this view, protective effect of exogenous S1P was evaluated in murine splenocytes against hypoxia-induced injury [14]. In our study, splenocyte death upon exposure to 0.5 % oxygen for 24 and 48 h duration could be alleviated by preconditioning splenocytes with S1P, in the range of 50–800 nM. The cytoprotective effect of S1P was an outcome of a boost in pro-survival signals – ERK1/2 and AKT activation and cellular ATP level. Further, the S1P pretreatment of splenocytes prior to hypoxia exposure delivered redox homeostasis, reduced oxidative damage and balanced anti-/proinflammatory cytokine profiles and temporal benefit of nitric oxide secretion and intracellular calcium release. We have observed that preconditioning with exogenous S1P has potential to alleviate hypoxia-induced cell death and the associated stresses such as oxidative damage and inflammation in murine splenocytes. Following the in vitro study, evaluation of pharmacological benefit of preconditioning with S1P prior to exposure of Sprague Dawley rats to hypobaric hypoxia (25,000 ft at 28 °C for 6 h) was conducted (paper under review in FEBS Journal). In this study, S1P pretreatment, especially at 1 μg/kg body weight dose, could protect the animals from hypoxia-induced oxidative damage and overt inflammation, and most importantly it enriched the blood oxygen-carrying capacity as indicated by raised haemoglobin, haematocrit and RBC count. This bioavailable oxygen, owing to S1P preconditioning, could restore hepatic aerobic respiration as indicated by a significant boost in the activity of citrate synthase and succinate dehydrogenase and total ATP content. It is well known that during acute hypoxia exposure, diuresis is beneficial as it leads haemoconcentration which further enhances the blood oxygen-carrying potential. Expression of S1P receptor 1 ($S1PR_1$) in renal tissue has been linked with enhanced diuresis, and in our study exogenous supplementation of S1P led to overt expression of $S1PR_1$ in renal tissues of these rats. S1P being a signalling lipid, the underlying mechanism for its preconditioning benefits could be multifactorial in nature; however in the animal studies conducted by us, HIF-1α accumulation and haemoconcentration appear to be the key underlying mechanisms along with antioxidant and anti-inflammatory properties.

Although in-depth studies are still being conducted, our in vitro and in vivo studies on S1P have unfolded preconditioning benefits of exogenously supplemented S1P to improve acclimatization to hypobaric hypoxia. These findings have long-term clinical application in the field of rapid acclimatization to high-altitude conditions.

Other Intervention

Efficacy of Nanocurcumin on Hypobaric Hypoxia-Induced Maladies

Curcumin is a polyphenolic compound isolated from turmeric. It has been shown to possess tremendous pharmacological value due to its anti-inflammatory, antioxidant and anti-proliferative activities [39, 52, 80]. These properties make curcumin effective against various pathological conditions like cancer, cardiovascular diseases, diabetes, arthritis and neurological diseases [1, 99]. Despite these properties, low bioavailability of curcumin downplays the role it plays in today's drug scenario [1]. To overcome this problem of poor bioavailability, many approaches are being acquired such as the use of adjuvant like piperine, association with liposome and making the nano-sized curcumin particles, to name a few [84, 91, 93].

Curcumin nanoparticles are nano-sized particles of curcumin that enhance the bioavailability of the curcumin so as to improve its pharmacological effects. Our approach involved enhanced and rapid acclimatization of the individual in high-altitude conditions by using curcumin nanoparticles. Hypobaric hypoxia also increases the oxidative stress along with accounting for various pathological conditions [26, 38, 56, 89]. Preliminary studies in our laboratory using in vitro and in vivo models have shown nanocurcumin to be effective in improving mental and physical performance and cardiovascular function under hypoxic conditions with marked decrease in the oxidative stress. These effects are more profound than the effects registered by using curcumin (our unpublished data).

These preclinical studies have shown immense potential of these interventions in improving tolerance to high-altitude hypoxia and thus are promising agents for clinical trials.

Current Practices for HA Acclimatization in Humans

The present acclimatization schedule framed by the Indian Army is of three successive stages at varying heights to avoid altitude sickness (AMS). The first stage of acclimatization is for 6 days and the altitude above 2,700–3,600 m. The second stage of acclimatization is for 4 days and the altitude above 3,600–4,500 m. The third stage of acclimatization is for 4 days and the altitude above 4,500 m. Staging and slow ascent protocols effectively induce acclimatization and reduce the incidence and severity of AMS. Altitude acclimatization is the best strategy for the prevention of acute mountain sickness (AMS) and allows people to achieve the maximal physical and cognitive work performance. Slow and staged ascent has been found to be most effective to induce altitude acclimatization and reduces the chances of the development of AMS. Due to emergencies/warlike conditions, military personnel may not get adequate time for staged acclimatization and to be inducted to high altitude within a short period of time. Acetazolamide and/or dexamethasone prophylaxis is commonly prescribed for rapid acclimatization to high altitude. In modern military operations, rapid deployment of military personnel to high altitude frequently occurs with little or no time for physiological acclimatization. Another recent approach to induce altitude acclimatization is the use of daily intermittent hypoxic exposure (IHE) in lieu of continuous stay at high altitude. IHE helps to preconditioning of an individual at sea level to reduce incidence of high-altitude maladies as well as to improve physical work performance at high altitude.

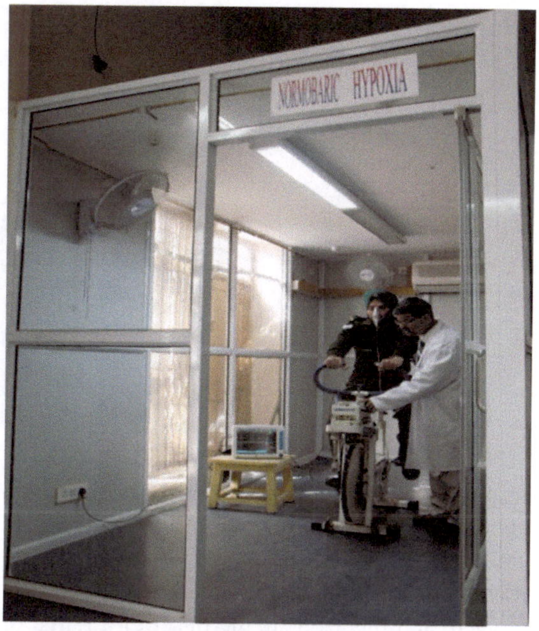

Fig. 2 Normobaric hypoxia room

Intermittent Hypoxic Exposure (IHE)

IHE involves breathing low oxygen air periodically, living in or exercising in reduced air for the purpose of pre-acclimatization/preconditioning to high altitude so as to reduce the incidence of AMS. IHE can be administered using either normobaric hypoxia (Fig. 2) or hypobaric hypoxia (Fig. 3). Hypobaric hypoxia is induced by decreasing the barometric pressure, whereas normobaric hypoxia is induced by lowering the percentage of oxygen in inspired air (FIO_2). Compared with hypobaric chambers, normobaric hypoxia rooms/tents are relatively inexpensive most importantly due to their light weight and setup to operate anywhere. IHE treatment consists of three basic elements, i.e. IHE simulated altitude, IHE session duration and the total number of the IHE sessions over the treatment period. The short duration of IHE session is usually paired with a greater number of IHE sessions per trial period, and the long duration of IHE session (i.e. overnight sleep) is paired with lower simulated altitudes. In IHE treatment, our body develops a series of physiological

Fig. 3 Hypobaric hypoxia chamber

responses due to temporary decrease in oxygen content in the blood. In response to low oxygen content, the physiological system tries to get required amount of energy with less available oxygen, and this triggers the onset of a range of physiological adaptation process of the cardiovascular, respiratory and oxygen utilization systems. The major IHE-induced altitude acclimatization is to increase arterial oxygen saturation through ventilatory acclimatization.

The IHE study was conducted on Indian Army subjects with the aim (a) to reduce HA illness (AMS, HAPE and HACE) on induction to HA, (b) to reduce the present acclimatization schedule in terms of days, (c) to develop a screening method for high-altitude illness susceptibility of an individual and (d) to formulate methods of pre-acclimatization during emergency situation/warlike conditions. The study was carried out on large numbers of army volunteers, and they were divided into control and experimental groups. The sea level, base line study was carried out in Delhi (barometric pressure 740 mmHg) where the laboratory temperature was maintained between 20 and 24 °C with a relative humidity range of 40–50 %. After recording the base line study at sea level for 2 days, subjects were exposed to normobaric hypoxia chamber and allowed to breathe 13.5 % FIO_2 (altitude – air equivalent 3,500 m, final SaO_2 in the blood was around 90 %) for 1 h. During hypoxic air breathing, pulse oxygen saturation (SaO_2) level in the blood and heart rate was continuously monitored for 1 h (pre-hypoxic challenge). On the next four consecutive days, the subjects were exposed to normobaric intermittent hypoxia exposure at 12 % FIO_2 (altitude – air equivalent 4,350 m, final SaO_2 in the blood was around 87–88 %) for 4 h per day for four consecutive days. On the fifth day, subjects were again exposed to 13.5 % FIO_2 (altitude – air equivalent 3,500 m, final SaO_2 in the blood was 92 %) for 1 h (post-hypoxic challenge). On the following day (sixth day), the subjects were inducted in the early morning (within 24 h) to an altitude of 3,500 m at Leh (barometric pressure 483 mmHg) in the Western Himalayas, India, in 55–60 min by pressurized aircraft. The ambient temperature at this altitude (3,500 m) varied between 10 and 20 °C during the period and relative humidity 40 %. At 3,500 m altitude, all the tests were conducted in the morning for six consecutive days where the temperature was maintained between 20 and 25 °C in the recording room. The first recording of the responses was made early next morning (within 24 h of arrival at HA). A finger pulse oximeter probe was set on the right index finger to measure resting oxygen saturation (SaO_2) level (Model MU 300, China). Ventilatory parameters like VE, VO2 and ventilatory drive (V_T/Ti, where Ti is the inspiratory time) were recorded in sitting position at both the locations using breath-by-breath, open-circuit metabolic measurement system (model $K4b^2$ mobile breath-by-breath metabolic system, Cosmed, Italy) calibrated with certified gases and volume standard.

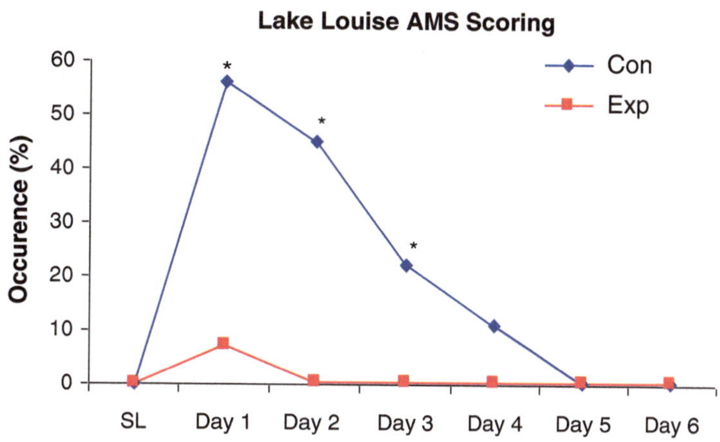

Fig. 4 Symptoms of AMS score in different days at HA (3,500 m), *Con* control, *Exp* experimental, (*) significant difference between groups, $P < 0.05$

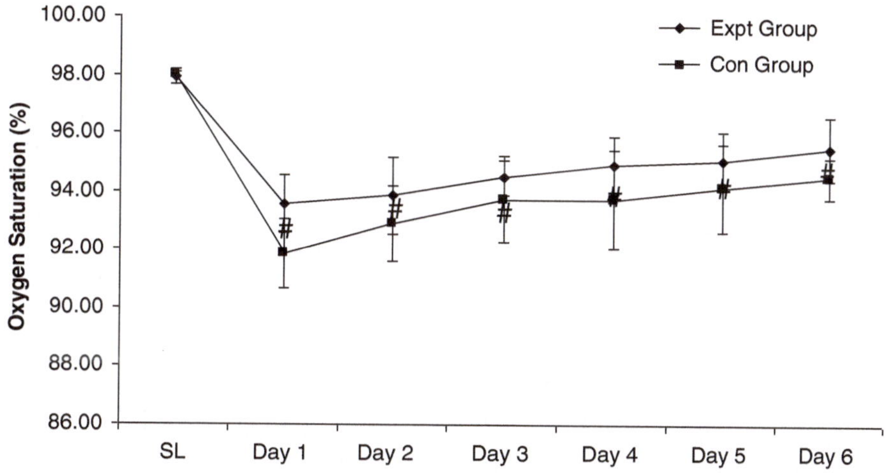

Fig. 5 SaO_2 changes at sea level and different days at HA (3,470 m), *Con* control, *Exp* experimental, (#) significant difference between groups value, $P < 0.05$

The results indicate that IHE at sea level significantly reduces the incidence of AMS. The incidence and severity of the symptoms of AMS was significantly less in IHE-treated group of subjects in comparison to control on initial two days of exposure at altitude. On day 1 at HA, 60 % of control group suffered from AMS, whereas the IHE-treated group showed only 5 %. On day 2 at HA, the incidence of AMS in control group was further decreased to 40 %, whereas the experimental group did not show any symptoms of AMS (Fig. 4). IHE-treated group also showed the significant increase in resting SaO_2, VE, VO_2 and ventilatory drive (Fig. 5). The present study also showed that IHE-treated group at sea level may reduce the acclimatization tenure at 3,500 m altitude from 6 to 4 days and prevent occurrence of high-altitude maladies.

To conclude, IHE treatment at sea level may be used as an alternative approach to pre-acclimatize the military personnel before their deployment to high mountain regions in emergency/warlike conditions.

References

1. Anand P, Kunnumakkara AB, Newman RA, Aggarwal BB (2007) Bioavailability of curcumin: problems and promises. Mol Pharm 4:807–818
2. Bailey DM, Davies B (1997) Physiological implications of altitude training for endurance performance at sea level: a review. Br J Sports Med 31:183–190
3. Basu CK, Selvamurthy W, Bhaumik G et al (1996) Respiratory changes during initial days of acclimatization to increasing altitudes. Aviat Space Environ Med 71:685–691
4. Bergeron M, Gidday JM, Yu AY, Semenza GL, Ferriero DM, Sharp FR (2000) Role of hypoxia inducible factor-1 in hypoxia induced ischemia tolerance in neonatal rat brain. Ann Neurol 48:285–296
5. Bhaumik G, Dass D, Bhattacharyya D, Sharma YK, Singh SB (2013) Heart rate variability changes during first week of acclimatization to 3500 m altitude in Indian military personnel. Indian J Physiol Pharmacol 57:16–22
6. Bhaumik G, Dass DK, Lama H et al (2008) Maximum exercise responses of men and women mountaineering trainees on induction to high altitude by trekking. Wilderness Environ Med 19:151–156
7. Bhaumik G, Purkayastha SS, Sharma RP et al (2005) Chemoreceptor sensitivity in women mountaineering trainees of different altitudes inducted by trekking to 4350 m. Def Sci J 55:427–435
8. Bhaumik G, Purkayastha SS, Selvamrthy W et al (2003) Oxygen saturation responses to exercise VO2 at 2100 m and 4350 m in women mountaineering trainees. Indian J Physiol Pharmacol 47:43–51
9. Bhaumik G, Sharma RP, Dass DK et al (2003) Hypoxic ventilatory response changes of men and women 6 to 7 days after climbing from 2100 m to 4350 m altitude and after descent. High Alt Med Biol 4:341–348
10. Boutellier U, Koller EA (1981) Propranolol and the respiratory, circulatory and ECG responses to high altitude. Eur J Appl Physiol 46:105–119
11. Bruick RK, McKnight SL (2001) A conserved family of prolyl-4-hydroxylases that modify HIF. Science 294:1337–1340
12. Burtscher M, Flatz M, Faulhaber M (2004) Prediction of susceptibility to acute mountain sickness by SaO_2 values during short-term exposure to hypoxia. High Alt Med Biol 5:335–340
13. Chavez JC, LaManna JC (2003) Hypoxia inducible factor-1α accumulation in rat brain in response to hypoxia and ischemia is attenuated during aging. Adv Exp Med Biol 510:337–341
14. Chawla S, Sahni C, Tulswani R, Singh M, Saraswat D, Bansal A, Saxena S (2013) Exogenous sphingosine 1-phosphate protects murine splenocytes against hypoxia-induced injury. Lipids. doi:10.1007/s11745-013-3860-9
15. Chen J, Simon R (1996) Ischemic tolerance in the brain. Neurology 48:306–311
16. David CW, Dong-Ho H, Pablo M, Garcia R, Paige CG, Terry EJ, John OH (2007) Exercise-induced mitochondrial biogenesis begins before the increase in muscle PGC-1α expression. J Biol Chem 282 (1):194–199
17. Desplanches D, Hoppeler H (1993) Effects of training in normoxia and hypoxia on human muscles ultrastructure. Pflugers Arch 425:263–267
18. Elleswerth AJ, Mayer EF (1991) Acetazolamide or Dexamethasone use versus placebo to prevent acute mountain sickness on Mount Rainier. West J Med 154:289–293
19. Endoh H, Kaneko T, Nakamura H, Doi K, Takahashi E (2000) Improved cardiac contractile functions in hypoxia-reoxygenation in rats treated with low concentration Co^{2+}. Am J Physiol Heart Circ Physiol 279:H2713–H2719
20. Epstein AC, Gleadle JM, McNeill LA, Hewitson KS, O'Rourke J, Mole DR, Mukherji M, Metzen E et al (2001) C. elegans EGL-9 and mammalian homologs define a family of dioxygenases that regulate HIF by prolyl hydroxylation. Cell 107(1):43–54
21. Ferretti G, Mioa C, Thomet JM, Kayser B (1997) The decrease of maximal oxygen consumption during hypoxia in man: a mirror image of the oxygen equilibrium curve. J Physiol 498:231–237
22. Gidday JM, Fitzgibbon JC, Shah AR, Park TS (1994) Neuroprotection from ischemic brain injury by hypoxic preconditioning in the neonatal rat. Neurosci Lett 168:221–224
23. Gidday JM, Shah AR, Maceren RG, Wang Q, Pelligrino DA, Holtzman DM, Park TS (1999) Nitric oxide mediates cerebral ischemic tolerance in a neonatal rat model of hypoxic preconditioning. J Cereb Blood Flow Metab 19:331–340
24. Giedroc DP, Chen X, Apuy JL (2001) Metal response element (MRE)-binding transcription factor-1 (MTF-1): structure, function and regulation. Antioxid Redox Signal 3:577–596
25. Ginouves A, Ilc K, Macias N, Pouyssegur J, Berra E (2008) PHDs overactivation during chronic hypoxia "desensitizes" HIFalpha and protects cells from necrosis. Proc Natl Acad Sci U S A 105 (12):4745–4750
26. Hackett PH, Roach RC (2001) High-altitude illness. N Engl J Med 345:107–114
27. Hackett PH, Roach RC et al (1988) Abnormal control of ventilation in high altitude pulmonary oedema. J Appl Physiol 64:1268–1272
28. Hackett PH, Roach RC, Wood RA et al (1988) Dexamethasone for prevention and treatment of acute mountain sickness. Aviat Space Environ Med 59:950–954
29. Halliwill JR, Minson CT (2002) Effect of hypoxia on arterial baroreflex control of heart rate and muscle sympathetic nerve activity in humans. J Appl Physiol 93:857–864

30. Heike JS, Silvia F, Hugo HM (2002) Hypoxia-induced vascular endothelial growth factor expression causes vascular leakage in the brain. Brain 125:2549–2557
31. Helene A, Gustafsson T, Sundberg CJ, Okamoto K, Jansson E, Poellinger L, Makino Y (2005) Physiological activation of HIF-1 in human skeletal muscle. FASEB J 19:1009–1011
32. Hirsilä M, Koivunen P, Xu L, Seeley T, Kivirikko KI, Myllyharju J (2005) Effect of desferrioxamine and metals on the hydroxylases in the oxygen sensing pathway. FASEB J 19:1308–1310
33. Hoppler H, Vogt M (2001) Muscle tissue adaptation to hypoxia. J Exp Biol 204:3133–3139
34. Hoppler H (2003) Response of skeletal muscle mitochondria to hypoxia. Exp Physiol 88:109–119
35. Ivan M, Haberberger T, Gervasi DC, Michelson KS, Gunzler V, Kondo K, Yang H, Sorokina I, Conaway RC, Conaway JW, Kaelin WG Jr (2002) Biochemical purification and pharmacological inhibition of a mammalian prolyl hydroxylase acting on hypoxia-inducible factor. Proc Natl Acad Sci U S A 99 (21):13459–13464
36. Ivan M, Kondo K, Yang H, Kim W, Valiando J, Ohh M, Salic A, Asara JM, Lane WS, Kaelin WG Jr (2001) HIFα targeted for VHL-mediated destruction by proline hydroxylation: implications for O2 sensing. Science 292(5516):464–468
37. Jaakkola P, Mole DR, Tian YM, Wilson MI, Gielbert J, Gaskell SJ, Kriegsheim A, Hebestreit HF, Mukherji M, Schofield CJ, Maxwell PH, Pugh CW, Ratcliffe PJ (2001) Targeting of HIF-α to the von Hippel-Lindau ubiquitylation complex by O2-regulated prolyl hydroxylation. Science 292 (5516):468–472
38. Joanny P, Steinberg J, Robach P, Richalet J, Gortan C, Gardette B et al (2001) Operation Everest III (Comex'97): the effect of simulated severe hypobaric hypoxia on lipid peroxidation and antioxidant defence systems in human blood at rest and after maximal exercise. Resuscitation 49:307–314
39. Joe B, Lokesh B (1994) Role of capsaicin, curcumin and dietary *n*-3 fatty acids in lowering the generation of reactive oxygen species in rat peritoneal macrophages. Biochim Biophys Acta 1224:255–263
40. Kalpana S, Bansal A, Singh B, Sairam M, Ilavazhagan G (2010) Sub-chronic oral toxicity study in Sprague Dawley rats with hypoxia mimetic cobalt chloride, towards the development of promising neutraceutical for oxygen deprivation. Exp Toxicol Pathol 62(5):489–496
41. Kamura T, Sato S, Iwai K, Czyzyk-Krzeska M, Conaway RC, Conaway JW (2000) Activation of HIF1alpha ubiquitination by a reconstituted von Hippel-Lindau (VHL) tumor suppressor complex. Proc Natl Acad Sci U S A 97:10430–10435
42. Kerendi F, Kirshbom PM, Halkos ME, Wang NP, Kin H, Jiang R, Zhao Z, Kanter KR, Guyton RA, Vinten-Johansen J (2006) Cobalt chloride pretreatment attenuates myocardial apoptosis after hypothermic circulatory arrest. Ann Thorac Surg 81:2055–2062
43. Kobayashi N, Kobayashi N, Yamaguchi A, Nishi T (2009) Characterization of the ATP-dependent sphingosine 1-phosphate transporter in rat erythrocytes. J Biol Chem 284:21192–21200
44. Kobayashi T, Koyama S, Kubo K, Fukushima M, Kusama S (1987) Clinical features of patients with high-altitude pulmonary edema in Japan. Chest 92:814–821
45. Koller EA, Bischoff M, Buhrer A (1991) Respiratory, circulatory and neuropsychological responses to acute hypoxia in acclimatized and non-acclimatized subjects. Eur J Appl Physiol 62:67–72
46. Kore KJ, Shete RV, Kale BN, Borade AS (2011) Protective role of hydroalcoholic extract of Ficus carica in gentamicin induced nephrotoxicity in rats. Int J Pharm Life Sci 2:978–982
47. Kronenberg RS et al (1968) Hastening respiratory acclimatization to altitude with benzolamide. Aerosp Med 39:296–300
48. Levin BD, Stray-Gundersen J (1997) "Living high-training low": effect of moderate-altitude acclimatization with low-altitude training on performance. J Appl Physiol 83:102–112
49. Lin CY, Huang CS, Huang CY, Yin MC (2009) Anticoagulatory, antiinflammatory, and antioxidative effects of protocatechuic acid in diabetic mice. J Agric Food Chem 57:6661–6667
50. Lin JH, Villalon P, Martasek P, Abraham NG (1990) Regulation of heme oxygenase gene expression by cobalt in rat liver and kidney. Eur J Biochem 192:577–582
51. Llesuy SF, Tomaro ML (1994) Heme oxygenase and oxidative stress: evidence of involvement of bilirubin as physiological protector against oxidative damage. Biochim Biophys Acta 1223:9–14
52. Lokesh B (1994) Studies on anti-inflammatory activity of spice principles and dietary n-3 polyunsaturated fatty acids on carrageenan-induced inflammation in rats. Ann Nutr Metab 38:349–358
53. Lomb DJ, Desouza LA, Franklin JL, Freeman RS (2009) Prolyl hydroxylase inhibitors depend on extracellular glucose and hypoxia-inducible factor (HIF)-2alpha to inhibit cell death caused by nerve growth factor (NGF) deprivation: evidence that HIF-2alpha has a role in NGF-promoted survival of sympathetic neurons. Mol Pharmacol 75:1198–1209
54. Maceyka M, Harikumar KB, Milstien S, Spiegel S (2012) Sphingosine-1-phosphate signaling and its role in disease. Trends Cell Biol 22(1):50–60
55. Maceyka M, Milstien S, Spiegel S (2007) Shooting the messenger: oxidative stress regulates sphingosine-1-phosphate. Circ Res 100:7–9
56. Maiti P, Singh SB, Sharma AK, Muthuraju S, Banerjee PK, Ilavazhagan G (2006) Hypobaric

hypoxia induces oxidative stress in rat brain. Neurochem Int 49:709–716

57. Mathew L, Gopinathan PM et al (1983) Chemoreceptor sensitivity and mal adaptation to high altitude in men. Eur J Appl Physiol 51:137–144
58. Matsumoto M, Makino Y, Tanaka T, Tanaka H, Ishizaka N, Noiri E, Fujita T, Nangaku M (2003) Induction of renoprotective gene expression by cobalt ameliorates ischemic injury of the kidney in rats. J Am Soc Nephrol 14:1825–1832
59. Maxwell PH, Wiesener MS, Chang GW, Clifford SC, Vaux EC, Cockman ME, Wykoff CC, Pugh CW, Maher ER, Ratcliffe PJ (1999) The tumour suppressor protein VHL targets hypoxia-inducible factors for oxygen-dependent proteolysis. Nature 399:271–275
60. Maxwell SR (1995) Prospects for the use of antioxidant therapies. Drugs 49:345–361
61. McNeill LA, Hewitson KS, Gleadle JM, Horsfall LE, Oldham NJ, Maxwell PH, Pugh CW, Ratcliffe PJ, Schofield CJ (2002) The use of dioxygen by HIF prolyl hydroxylase (PHD1). Bioorg Med Chem Lett 12:1547–1550
62. Merkl R, Hrádková I, Filip V, Šmidrkal J (2010) Antimicrobial and antioxidant properties of phenolic acids alkyl esters. Czech J Food Sci 28:275–279
63. Michaud MD, Robitaille GA, Gratton JP, Richard DE (2009) Sphingosine-1-phosphate: a novel nonhypoxic activator of hypoxia-inducible factor-1 in vascular cells. Arterioscler Thromb Vasc Biol 29:902–908
64. Miller BA, Perez RS, Shah AR, Gonzales ER, Park TS, Gidday JM (2001) Cerebral protection by hypoxic preconditioning in a murine model of focal ischemia-reperfusion. Neuroreport 12:1663–1669
65. Minamino T, Christou H, Hsieh CM, Liu Y, Dhawan V, Abraham NG, Perrella MA, Mitsialis SA, Kourembanas S (2001) Targeted expression of heme-oxygenase-1prevents the pulmonary inflammatory and vascular responses to hypoxia. Proc Natl Acad Sci U S A 98:8798–8803
66. Minamishima YA, Moslehi J, Padera RF, Bronson RT, Liao R, Kaelin WG Jr (2009) A feedback loop involving the Phd3 prolyl hydroxylase tunes the mammalian hypoxic response in vivo. Mol Cell Biol 29(21):5729–5741
67. Mohanraj P, Merola AJ, Wright VP, Clanton TL (1998) Antioxidants protect rat diaphragmatic muscle function under hypoxic conditions. J Appl Physiol 84:1960–1966
68. Moore LG, Harrison GL et al (1986) Low acute hypoxic ventilatory response and hypoxic depression in acute altitude sickness. J Appl Physiol 60:1407–1412
69. Murphy BJ, Sato BG, Dalton TP, Laderoute KR (2005) The metal-responsive transcription factor 1 contributes to HIF-1 activation during hypoxic stress. Biochem Biophys Res Commun 337:860–867
70. Ogretmen B, Hannun YA (2004) Biologically active Sphingolipids in cancer pathogenesis and treatment. Nat Rev Cancer 4(8):604–616
71. Otterbein LE, Kolls JK, Mantell L, Cook JL, Alam J, Choi AMK (1999) Exogenous administration of heme oxygenase-1 by gene transfer provides protection against hyperoxia-induced lung injury. J Clin Invest 103:1047–1054
72. Otterbein LE, Soares MP, Yamashita K, Bach FH (2003) Heme Oxygenase-1: unleashing the protective properties of heme. Trends Immunol 24:449–455
73. Palmiter RD (1994) Regulation of metallothionein genes by heavy metals appears to be mediated by zinc-sensitive inhibitor that interacts with a constitutively active transcription factor, MTF-1. Proc Natl Acad Sci U S A 91:1219–1223
74. Parati G, Saul JP, De Rienzo M, Mancia G (1995) Spectral analysis of blood pressure and heart rate variability in evaluating cardiovascular regulation. Hypertension 25:1276–1286
75. Pchejetski D, Kunduzova O, Dayon A, Calise D, Seguelas MH, Leducq N, Seif I, Parini A, Cuvillier O (2007) Oxidative stress-dependent sphingosine kinase-1 inhibition mediates monoamine oxidase A-associated cardiac cell apoptosis. Circ Res 100:41–49
76. Prabhakar NR, Semenza GL (2012) Adaptive and non-adaptive cardiorespiratory responses to continuous and intermittent hypoxia mediated by hypoxia inducible factors 1 and 2. Physiol Rev 92:967–1003
77. Purkayastha SS, Bhaumik G, Sharma RP et al (2000) Effects of mountaineering training at high altitude (4350 m) on physical work performance of women. Aviat Space Environ Med 71:685–691
78. Raphael J, Rivo J, Beeri R, Abedat S, Gozal Y (2004) Mechanisms of myocardial ischemic preconditioning: a potential protective role for HIF-1 in a rabbit model of regional myocardial ischemia. Anesthesiology 101:A717
79. Rathat C, Richalet JP, Herry JP, Larmignat P (1992) Detection of high risk subjects for high altitude disease. Int J Sports Med 13:76–79
80. Reddy ACP, Lokesh B (1992) Studies on spice principles as antioxidants in the inhibition of lipid peroxidation of rat liver microsomes. Mol Cell Biochem 111:117–124
81. Reeves JT, McCullogh RE et al (1993) Sea level PCO2 relates to ventilatory acclimatization at 4350 m. J Appl Physiol 75:1117–1122
82. Renée VC, Anne G, Vladimir V (2008) Transcriptional control of mitochondrial biogenesis: the central role of PGC-1α. Cardiovasc Res 79:208–217
83. Richalet JP, Kacimi R, Antezana AM (1992) The control of cardiac chronotropic functions in hypobaric hypoxia. Int J Sports Med 13:S22–S24
84. Sasaki H, Sunagawa Y, Takahashi K, Imaizumi A, Fukuda H, Hashimoto T et al (2011) Innovative

preparation of curcumin for improved oral bioavailability. Biol Pharm Bull 34:660–665
85. Saxena S, Shukla D, Bansal A (2012) Augmentation of aerobic respiration and mitochondrial biogenesis in skeletal muscle by hypoxia preconditioning with cobalt chloride. Toxicol Appl Pharmacol 264:324–334
86. Saxena S, Shukla D, Saxena S, Khan YA, Singh M, Bansal A, Sairam M, Jain SK (2010) Hypoxia preconditioning by cobalt chloride enhances endurance performance and protects skeletal muscles from exercise induced oxidative damage in rats. Acta Physiol (Oxf) 200:249–263
87. Schneider M, Bernasch D et al (2002) Acute mountain sickness: influence of susceptibility, preexposure and ascent rate. Med Sci Sports Exerc 34:1886–1891
88. Schoene RB, Lahiri S, Hackett PH, Jr P, Milledge JS, Pizzo CJ, Sarnquist FH, Boyer SJ, Graber DJ, Maret KH, West JB (1984) Relationship of hypoxic ventilatory response to exercise performance on Mt Everest. J Appl Physiol 56:1478–1483
89. Schoonman GG, Sándor PS, Nirkko AC, Lange T, Jaermann T, Dydak U et al (2007) Hypoxia-induced acute mountain sickness is associated with intracellular cerebral edema: a 3 T magnetic resonance imaging study. J Cereb Blood Flow Metab 28:198–206
90. Semenza GL (1999) Regulation of mammalian O2 homeostasis by hypoxia-inducible factor 1. Annu Rev Cell Dev Biol 15:551–578
91. Shaikh J, Ankola D, Beniwal V, Singh D, Kumar M (2009) Nanoparticle encapsulation improves oral bioavailability of curcumin by at least 9-fold when compared to curcumin administered with piperine as absorption enhancer. Eur J Pharm Sci 37:223–230
92. Shingo T, Sorokan ST, Shimazaki T, Weiss S (2001) Erythropoietin regulates the *in vitro* and *in vivo* production of neuronal progenitors by mammalian forebrain neural stem cells. J Neurosci 21:9733–9743
93. Shoba G, Joy D, Joseph T, Majeed M, Rajendran R, Srinivas P (1998) Influence of piperine on the pharmacokinetics of curcumin in animals and human volunteers. Planta Med 64:353–356
94. Shrivastava K, SaiRam M, Bansal A, Singh SS, Ilavazhagan G (2008) Cobalt supplementation promotes tolerance and facilitates acclimatization to hypobaric hypoxia in rats. High Alt Med Biol 9 (1):63–75
95. Shrivastava K, Dhananjay S, Bansal A, SaiRam M, Ilavazhagan G (2008) Neuroprotective effect of cobalt-chloride on hypobaric hypoxia induced oxidative stress. Neurochem Int 52:368–375
96. Shukla D, Saxena S, Purushothaman J, Shrivastava K, Singh M, Shukla S, Malhotra VK, Mustoori S, Bansal A (2011) Hypoxic preconditioning with cobalt ameliorates hypobaric hypoxia induced pulmonary edema in rat. Eur J Pharmacol 656:101–109
97. Shukla D, Shrivastava K, Saxena S, Jayamurthy P, Sairam M, Jain SK, Bansal A, Ilavazaghan G (2009) Hypoxic preconditioning with cobalt attenuates hypobaric hypoxia induced oxidative damage in rat lungs. High Alt Med Biol 10(1):57–69
98. Singh M, Shukla D, Thomas P, Saxena S, Bansal A (2010) Hypoxic preconditioning facilitates acclimatization to hypobaric hypoxia in rat heart. J Pharm Pharmacol 62:1729–1739
99. Srinivasan K (2005) Role of spices beyond food flavoring: nutraceuticals with multiple health effects. Food Rev Int 21:167–188
100. Stelzner TJ, O'Brien RF, Sato K, Weil JV (1988) Hypoxia-induced increases in pulmonary transvascular protein escape in rats. Modulation by glucocorticoids. J Clin Invest 82:1840–1847
101. Stiehl DP, Wirthner R, Köditz J, Spielmann P, Camenisch G, Wenger RH (2006) Increased prolyl 4-hydroxylase domain proteins compensate for decreased oxygen levels. Evidence for an autoregulatory oxygen-sensing system. J Biol Chem 281(33):23482–23491
102. Taylor MS (2001) Characterization and comparative analysis of the EGLN gene family. Gene 275:125–132
103. Terrados N, Melichna J, Sylven C, Jansson E, Kaijser I (1988) Effects of training at simulated altitude on performance and muscle metabolic capacity in competitive road cyclists. Eur J Appl Physiol 57:203–209
104. Thomas P, Bansal A, Singh M, Shukla D, Saxena S (2011) Preconditioning effect of cobalt chloride supplementation on hypoxia induced oxidative stress in male albino rats. Biomed Prev Nutr 1:84–90
105. Thornally PJ, Vasak M (1985) Possible role for metallothioneins in protection against radiation induced oxidative stress. Kinetics and mechanism of its reaction with superoxide and hydroxyl radicals. Biochim Biophys Acta 827:36–44
106. Truijens MJ, Toussaint HM, Dow J, Levine BD (2003) Effect of high-intensity hypoxic training on sea –level swimming performances. J Appl Physiol 94:733–743
107. Vallee BL (1995) The function of metallothionein. Neurochem Int 27:23–33
108. Vogt M, Puntschart A, Geiser J, Zuleger C, Billete R, Hoppeler H (2001) Molecular adaptations in human skeletal muscle to endurance training under hypoxic conditions. J Appl Physiol 91:173–182
109. Wang GL, Semenza GL (1995) Purification and characterization of hypoxia-inducible factor 1. J Biol Chem 270:1230–1237
110. Wang GL, Jiang BH, Rue EA, Semenza GL (1995) Hypoxia inducible factor 1 is basic-helix-loop-helix-PAS heterodimer regulated by cellular O2 tension. Proc Natl Acad Sci U S A 92:5510–5514

111. Wang W, Dentler WL, Borchardt RT (2001) VEGF increases BMEC monolayer permeability by affecting occluding expression and tight junction assembly. Am J Physiol Heart Circ Physiol 280: H434–H440
112. Wenger RH, Camenisch G, Stiehl DP, Katschinski DM (2009) HIF prolyl-4- hydroxylase interacting proteins: consequences for drug targeting. Curr Pharm Des 15(33):3886–3894
113. Xi L, Tahler M, Yin C et al (2004) Cobalt chloride induces delayed cardiac preconditioning in mice through selective activation of Hif-1 and AP-1 and iNOS signaling. Am J Physiol Heart Circ Physiol 287:2369–2375
114. Yet SF, Tian R, Layne MD, Wang ZY, Maemura K, Solovyeva M, Ith B, Melo LG, Zhang L, Ingwall JS, Dzau VJ, Lee ME, Perrella MA (2001) Cardiac-specific expression of heme oxygenase-1 protects against ischemia and reperfusion injury in transgenic mice. Circ Res 89:168–173
115. Yu F, White SB, Zhao Q, Lee FS (2001) HIF-1alpha binding to VHL is regulated by stimulus-sensitive proline hydroxylation. Proc Natl Acad Sci U S A 98 (17):9630–9635

Noise, the Silent Killer

Neeru Kapoor

Abstract

Noise, the 'unwanted sound', is now realised as one of the most important environmental hazard in the present day industrial world. Exposure to noise is imminent, spearheaded by man's casual attitude and ignorance of its effects. Thus, the indiscriminate hooting of horns and shrieking of loudspeakers continue unabated and disturb man's peace and tranquility. Noise leads to impaired hearing and causes general stress resulting in rise of blood pressure, enhanced sweating, muscle cramps, changes in cardiac functions and biorhythm, and increase in blood cortisol and cholesterol. Defence Institute of Physiology & Allied Sciences (DIPAS), DRDO, has been working towards combating noise pollution for over the last four decades for both the Defence Forces and for the civilians. In the Armed Forces, personnel are exposed to noise resulting from arms and ammunitions, gun and artillery firings, movement of heavy armoured vehicles, flying of aircraft and helicopters, engine rooms of naval ships and submarines etc. The high level of noise experienced by the personnel in their working environment is responsible for impairment in hearing functions and liable to trigger adverse reactions in extra-auditory systems. The auditory effects of noise directly influence the peripheral auditory system and the hearing function. Noise encountered by man in his occupational environment could be either loud or impulsive, of short intermittent duration or continuous noise of mixed intensities over a prolonged period. It acts as a biological stressor which results in various non-auditory system responses, leading to physiological and psychological stress. Researches on physiological and psychological effects and on performance and comfort have provided an extensive technology pool that serves as the basis of exposure guidelines, criteria and standards.

N. Kapoor (✉)
Occupational Health Division, Defence Institute of Physiology and Allied Sciences (DIPAS), Lucknow Road, Timarpur, New Delhi 110 054, India
e-mail: neeru0260@yahoo.co.uk

S.B. Singh et al. (eds.), *Translational Research in Environmental and Occupational Stress*,
DOI 10.1007/978-81-322-1928-6_13,

Controlling noise and safeguarding hearing are important environmental issues that need to be addressed with care. The ill effects of noise on the auditory system may be alleviated by adequate hearing protection with the use of appropriate ear defenders, electronic noise reducing head sets, inhalation of carbogen (gas mixture of 5% CO_2 and 95% O_2), supplementation with antioxidants and restricted working time in noise environment as per the laid down environmental safety guidelines. These and other management techniques including use of musical sounds may also be useful in coping with the extra auditory effects of noise.

Introduction

Noise is a wrong sound in a wrong place at a wrong time. It is defined as irritating, not understandable, and nonrhythmic sound. It has become an unavoidable by-product of modern technological developments which favor more mechanization, faster traffic, and closer packing of population. The peaceful environment of every city, town, and even the country side is being shattered from morning till night by the roar of jet planes, honking of cars and pressure horns of trucks, wailing of emergency vehicle sirens, whistling of railway engines, the rattle of mechanized farm implements, and the thunder of construction machinery. All this has given rise to the term noise pollution.

Hazardous sound levels are an accompaniment of even many enjoyable pastimes like music. Availability of high powered electronic amplifiers has led to much greater sound levels than were earlier encountered. Fireworks, a common feature in many countries during religious or national festivals, produce unpleasant noise. Soldiers are not spared from this insidious enemy either. Armed forces personnel are exposed to high intensities of continuous and impulsive noise from fighting vehicles, guns, small arms, heavy machines in workshops, engines and other machinery of naval ships and aircraft, etc. The rapport from a firing gun, for example, produces very high impulsive noise levels. Noise also interferes with communication.

Despite the cacophony all around us, noise pollution is a newly recognized occupational hazard. This is in part due to the fact that the problem till recently has not been studied systematically and also due to the general ignorance about the injurious effects of noise on human biology and behavior [1]. Noise has been now recognized as one of the important stress-producing factors in modern life.

Effects of Noise Pollution

Noise interferes with human activities in three ways:

1. Auditory apparatus: There is definite evidence that a person exposed to high intensities may suffer from some degree of hearing impairment [2, 3], generally assessed through audiometric tests. In this regard, impulsive noise is more damaging than continuous noise [4], which can cause damage to the eardrum [5]. Continuous noise of about 85 dB may cause some damage to the hearing performance. This may be of transient nature and may last for a few hours only. This is Temporary Threshold Shift (TTS). The hearing gets shifted to higher dB values. Given a certain amount of rest in quiet environs, this gets restored. For example, working in 85–90 dBA noisy places for 8 h results in appreciable hearing impairment. However, rest and sleep in environ of 65 dBA for subsequent 16 h or so restores hearing. If this rest in quiet environs is not available, the TTS may get converted into Permanent Threshold Shift (PTS) [6].

   The magnitude of hearing loss depends on the level of noise, its frequency content, the duration of one's exposure, and individual susceptibility. The loss in hearing sensitivity caused by noise exposure is insidious in its development, initially beginning at the higher frequencies (around 4–8 kHz). With

continuing exposures this loss slowly progresses towards the speech frequencies (300 Hz–4 kHz) and becomes evident to the individual.
2. Physiological: Noise interferes in the normal physiological functioning of the body. It may cause an increase in blood pressure, sweating, and muscular contractions [7–10].
3. Behavioral: Noise affects the sociological behavior of the individuals as it produces irritation, interferes with sleep, and impairs physical and mental performance [11–13].

Of the many health hazards related to noise, hearing loss is the most clearly observable and measurable by health professionals. The other hazards are harder to pin down. For many of us, there may be a risk that exposure to the stress of noise increases susceptibility to disease and infection. The more susceptible among us may experience noise as a complicating factor in hearing problems and other diseases. Noise that causes annoyance and irritability in healthy persons may have serious consequences for those already ill in mind and body. Other pollutants present in the environment such as air and water may enhance the ill effects of noise.

Noise affects us throughout our lives. For example, there are indications of effects on unborn child when mothers are exposed to industrial and environmental noise [2, 14]. During infancy and childhood, youngsters exposed to high noise levels may experience learning difficulties and generally suffer poor health [15–19]. Later in life, the elderly may have trouble falling asleep and obtaining necessary amounts of rest.

Why, then, is there not greater alarm about these dangers? Perhaps it is because the link between noise and many disabilities has not yet been conclusively demonstrated. Perhaps it is because we tend to dismiss annoyance as a price to pay for living in the modern world. And maybe we still think noise as an occupational hazard only.

The effects of noise on health are often misunderstood or unrecognized. Well-documented studies to clarify the role of noise as a public health hazard are still required, but we at least know from existing evidence that the danger is real.

Except for the serious problem of hearing loss, there is no human illness known to be directly caused by noise. But in many studies, noise has been clearly identified as an important factor of physical and psychological stress, and stress has been directly linked with many of our most common health problems. Thus, noise can be associated with many of these disabilities and diseases, which include heart disease, high blood pressure, headaches, fatigue, and irritability.

Noise Pollution in Military Service

The sources of noise in the military are as varied as the activities carried out by the personnel of the Army, Navy, Air Force, Marine Corps, and Coast Guard. Obvious sources of potentially hazardous noise are weapon systems and jet engines. However, ground vehicles, other aircraft, watercraft, communication systems, and industrial-type activities also serve as sources of potentially damaging noise.

The Defence Institute of Physiological and Allied Sciences (DIPAS), one of the laboratories of DRDO belonging to the Life Sciences group, has conducted several studies on armed forces personnel and civilians exposed to steady-state noise as well as impulsive noise and gained considerable expertise in understanding how noise affects the physiology of human beings.

Studies conducted in the Indian Naval work environment revealed that exposures to continuous engine room turbine noise in the range of 105–116 dB "A" during sailing of the ships led to a generalized loss of hearing sensitivity in 76 % of the engine room crew, with the most affected frequencies being at 4, 6 and 8 kHz [20]. The incidence of hearing impairment in naval personnel engaged in duties outside the engine room with SPL ranging from 73.4 to 91.0 dB "A" was not of concernable degree, being 27 %. Investigations conducted on a group of young audiometrically normal men exposed to noise of moderate intensity (100 dB "A") for short duration of 20 min developed TTS_2 (TTS, 2 min after the cessation of noise) of 22.63 ± 1.06 dB at 4 kHz. TTS vanished after resting for 2 h in quiet [21]. In another study on 100 volunteers, it was revealed that the TTS_2 increased with exposure time (15–90 min.) and intensity (80–105 dB "A") [22]. The higher the value of TTS, the longer is the time taken for complete recovery. The upper safe limit and safe exposure period was also worked out by giving due weight to the development of TTS and its recovery in different steady-state noise conditions. Studies conducted on a large number of audiometrically normal human volunteers showed that a linear correlation exists between the noise level and log of exposure time. The safe exposure level for 8 h was found to be 89 dB "A" and permitted the increase of noise level by 3 dB "A" for every halving of duration, in concert with studies conducted by Burns and Robinson [23].

In the Naval Dockyards, noise produced due to the operation of different types of woodcutting machines and running of air compressor ranged from 91.0 to 114.2 dB "A." The maximum sound energy was in the speech range with sharp tonal components that are injurious to hearing as well as a source of irritability [24]. In the industrial work environment of the Army Base Workshops, the noise levels ranging from 90.0 to 113.5 dB during the maintenance and operations of different engines and lathe machines, testing of tanks, and the use of air compression facilities exceeded the upper safe limit of exposure [25]. Studies conducted on the noise generated by heavy vehicles like tanks and armored personnel carriers, infantry, and combat vehicles have revealed that they produce high levels of continuous noise in the range of 117.5–123.5 dB [20].

The problem of noise hazard is equally serious in military aviation environment. Sources of noise due to the operation of combat aircrafts and helicopters at any Air Force base are ground running/operations of aircraft for routine check and maintenance (120–146 dB "A" at 5–10 m), during takeoff (110–139 dB "A" at 30 m), taxying (106–112 dB), aircraft flying over head (84–120–82 dB "A" at 100 m), and during landing (100 dB "A" and below at 30 m) [26]. The major noise sources in an aircraft are power sources, transmission systems, propellers, and jet efflux, the interaction between the aircraft and the medium through which it flows (flow of air over aircraft surfaces). Other noise sources are the subsidiary noises arising from cabin conditioning and pressurizing systems, hydraulic systems and communication equipment, sonic booms, and armament discharge. Helicopter noise, during ground operations and hovering, poses problems of different nature because of the characteristic differences in the nature of their operations. The helicopter maintenance crew is frequently subjected to noise levels between 100 and 120 dB "A" while attending to check and maintain jobs. Noise levels at Air Force Radar Stations too exceeded the acceptable limits ranging from 82.9 to 131.2 dB. Intense noise exposures to air and ground crews constitute a great hazard, posing difficulties with aircrew communications, affecting performance, and even contributing to accidents [27]. Examination of the incidences of noise-induced hearing loss among fliers, ground duty personnel, families, and children living in the close vicinity of the runway has indicated widespread impairment among the ground crew and their families. In a total of 451 persons investigated, pilots indicated 46 % hearing loss out of 52 tested, and ground crew and maintenance staff showed 71 % hearing loss out of 278 tested with a characteristic notch at 6 kHz. Even the families and children residing in close proximity to the runway were not spared, as out of the 62 family members tested for their hearing status, 90 % indicated hearing disability, while of the 59 children examined, 36 % presented hearing impairments to different degrees.

The effect of noise on the hearing of workers in military aviation has been studied by various researchers. Malhotra et al. [28] examined the hearing of 92 flight deck personnel of an air craft carrier and observed that about 70 % of them had elevated hearing level. Murthy et al. [29] found that incidence of noise-induced hearing loss (NIHL) among 931 jet aircraft maintenance personnel was 45.6, 49.1 and 68.1 % in the age group of 30 years and below, between 30 and 40 years, and above 40 years, respectively. In another study, Soodan and Rao [30] found that incidence of NIHL was 25.7 % in fighter aircrew and 74.3 % in transporter/helicopter group. Studies conducted by Deshmukh et al. [31] revealed that out of the 400 IAF aircrew investigated, only 28.3 % were unaffected by NIHL, while 56.4 % had mild, 11.8 % moderate, and 3.5 % had severe involvement. Extensive noise survey has been conducted by Rao et al. [32] at different Air Force stations. In most of the places, noise level was more than the upper safe limit of 90 dBA and in some places it reached 135.0 dBA, which is the threshold of pain to the ears. Audiometric study conducted on 3,391 airmen of various technical trades revealed that 11.2 % had hearing loss in low frequencies (0.5–2 kHz) and 10.7 % in the high frequencies (3–8 kHz). As compared to this, exposures to industrial noise (95 dBA) in workers with 15 years work experience did not affect the hearing at 1 and 2 kHz, while it produced an average hearing loss of 45 dB at 4 kHz [33].

Studies on the influence of traffic noise exposures on the control personnel and civilians at prominent traffic circles/junctions in Delhi during peak hours showed that the continuous equivalent noise levels on busy roads and at some major road crossings ranged from 78 to 92 dB "A." The effect of this noise on the hearing problems was evident in the traffic police personnel who suffered high-frequency hearing loss of up to 80 % that increased with the length of service [34]. Similar investigations undertaken on traffic policemen in the city of Hyderabad revealed that 76 % had NIHL [35].

Reid [36] studied the effect of impulsive noise on the hearing of gunners of the Australian armed forces. He found acoustic trauma in 50 % personnel. Later, Salmivalli [37] reported incidence of acoustic trauma in 43.2–68.7 % in different branches of Finnish army. Studies conducted on 117 Indian gunners revealed that 50.9 % of the tested population had hearing loss of various degrees [38]. The authors surveyed a bigger sample of 1,917 personnel from artillery and armored corps and found that 55.5–62.1 % of them had hearing loss [39]. Incidence of hearing loss was attributed to the prevalence of high-intensity steady-state noise of fighting vehicles (109–116 dB) and impulse noise of guns (174–182 dB). It was also observed that hearing loss that was initially confined to high-frequency region (4–6 kHz) had passed to speech frequencies after 10 years of exposure. This was confirmed in case of gunners from navy as well [20]. A decade later Singh et al. [40, 41] observed that noise level of fighting vehicles (117.5–123.5 dB) and guns (162.8–189.2 dB) was higher than reported by us earlier. Noise, even of small arms has been found to be injurious to the hearing system. The noise level of small arms ranged between 153 and 166 dB that increased and covered more frequencies with the length of service. The highest value of 51 dB was observed at 8 kHz in subjects having more than 15 years of service [42]. A study conducted on the workers of an ammunition testing establishment revealed that those who are frequently exposed to high-intensity impulsive noise (over 160 dB) had suffered hearing loss of 34–41 dB in high-frequency region (4–6 kHz) [43]. It has also been observed in men that those who could voluntarily control their aural reflex developed 10–20 dB less TTS during impulse noise exposure [44].

Our studies on the igniting of nine varieties of firecrackers revealed that the impulse level ranged from 117 to 131 dB AI and peak pressure level from 137 to 154 dB at a distance 4 m in different dwelling conditions of our populace. The data generated by this study was instrumental in the formulation of noise standards by Bureau of Indian Standards (BIS). The impact of 30 min of these exposures on the hearing was examined on adults and children. It was observed that 60 % of the developed TTS was recovered in the first hour following the exposures and there was complete recovery in children in 2 h [45].

In some working environments, noise is a combination of continuous (steady-state) and impulse noise. Several workers have conducted trials on animals and men utilizing various combinations of time, intensity, and characteristics of noise to decipher effect of mixed noise on hearing system.

Studies conducted on chinchillas [46] and men [47] revealed that the combination of continuous and impulsive noise did not produce greater damage to hearing as compared to the one produced separately by these noises.

In animal experiment, exposure duration was 8 weeks (5 days/week, 8 h/day) and levels of continuous and impulsive noises were 75–85 dB and 103 dB, respectively. The respective values in case of men were 20 min, 78–96 dB and 96–132 dB. On the other hand, combined exposure of octave band noise (95 dB, center frequency 3 kHz) and impulse noise (137–158 dB) even for 1 h produced greater damage than either of them separately in case of chinchillas [48, 49]. Similarly in men, the combined exposure of Broad Band Noise (BBN alone, 104–120 dB) for 40–60 min resulted in larger TTS as compared to BBN alone [50, 51]. It appears that the three basic parameters that influence the degree of hearing loss are: bandwidth of noise, intensity of noise, and exposure duration. It has been reported that thinner bandwidths, longer exposure duration, and higher intensities produced larger TTS in man [21, 22].

TTS produced in man after exposure to octave band (0.75–1.2 kHz) noise of 90–110 dB, impulse noise of 124–127 dB, and a combination of two classes of noise was measured [52]. It was found that impulse noise alone produced greatest TTS, while addition of 90 or 100 dB background noise to impulse noise actually reduced the degree of TTS. Authors reasoned that acoustic reflex provided reduction in TTS with combination exposure. Another study also found that the level of TTS produced by impulses (up to 132 dB) was reduced when background noise of 78–96 dBA was added [47]. We are also of the opinion that the presentation of moderate level of continuous noise prior to exposure to impulse may reduce the degree of TTS by activating the aural reflex.

In addition to the above experiments, surveys on hearing acuity were carried out by different investigators in places where noise of both types was encountered. In one such survey [53], it was found that Permanent Threshold Shift (PTS) in hearing of workers of a forging plant exposed to impulse noise of 127–134 dB with a background noise of 110–114 dB occurred within 1–2 years. The PTS was of the order of 50 dB in the frequency range of 4–6 kHz. Surveys conducted on weavers of a textile mill [54] and pharmaceutical firm [55] with sound pressure levels of 102–104 dBA and 100–105 dBA, respectively, revealed the occurrence of significant NIHL that increased with the length of service. NIHL at 4 kHz was 30 dB in the age group 25–29 years, 40 dB in the age group 30–34 years, and 45 dB in the age group 35–39 years [56]. The prevalence of NIHL in miners was 12.8 % with 10.2 % having moderate and 2.6 % severe hearing impairment [57]. Another group of workers [58] measured hearing acuity of 215 soldiers of armored corps having 10 years of active duty and found that 63 % had sustained hearing losses. Out of this, 41 % had losses in excess of 50 dB at 4–6 kHz. Comparable hearing losses in other branches of army were 23 % in infantry, 28 % in artillery, and 16.3 % in aviation. High incidence of hearing loss in armored branch was attributed to the combined effect of high-intensity continuous and impulsive noise. In another study on 83 personnel (42 civil and 41 Army) from an ammunition testing laboratory [59], exposure to both impulse noise from guns and continuous noise due to running of armored vehicles and machinery in workshop indicated hearing impairment of 73.8 % and 58.6 % in the civil and defense staff, respectively. The higher percentage of civil staff being affected was attributed to their longer duration of exposures.

Attempts have also been made to find out which type of noise is more injurious to hearing. Results of animal experiments [60, 61] revealed that impulsive noise produced greater threshold shift and cochlear hair cell damage than continuous noise having equivalent energy. In another study, the hearing data of noise-exposed workers was thoroughly examined, and it was found that impulse noise below 100 dB caused about 10 dB higher hearing loss than continuous noise at an equivalent level [62]. Similar findings have been reported on industrial workers exposed to impulse and continuous noise [63–67].

Noise Pollution Control and Hearing Conservation Techniques

Noise in a military environment constitutes a multifactorial problem and needs to be addressed with maximum care. A good hearing

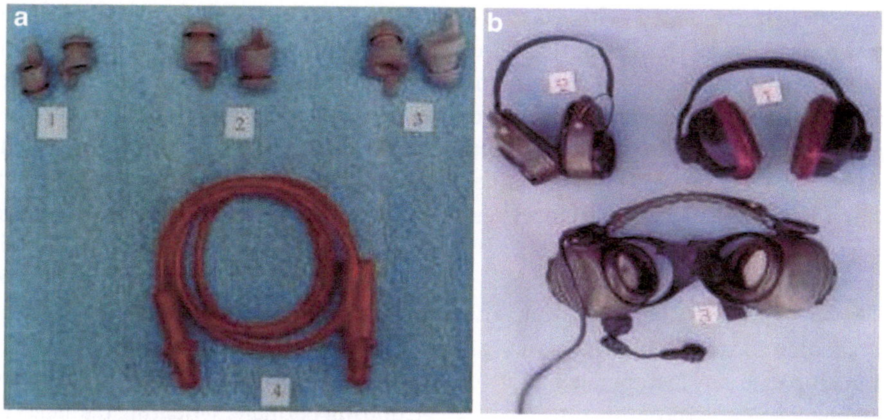

Fig. 1 (**a**) Earplugs (*1*) Sonex (imported), (*2*) Gunfender (imported), (*3*) Gunfender (indigenous), (*4*) Ear Seal (indigenous). (**b**): Earmuffs. (*1*) Earmuff (indigenous), (*2*) slimvalve earmuff (imported), (*3*) sonovalve earmuff (imported)

for the armed forces personnel can never be compromised because of vital communication needs and other activities requiring high auditory sensitivity. The modern-day technological advances in the weapon and military systems development to increase lethality have made the noise exposure inescapable, strenuous, and widespread. It attracts the attention of the technologists and noise control specialists to prevent the exposure of the troops to hazardous noise.

Developed countries have their own noise control measures and legislative laws. Taking into consideration the intensity of noise and time duration of exposure to steady-state and impulsive noises, certain safety measures have been formulated by different countries and different groups of workers. These are known as Damage Risk Criteria (DRC). The DRC permits the determination of allowable permissible duration of exposure in the noisy environment when the noise level is known. The safe limit of exposure to continuous noise of 85 dB is 8 h, and if the noise level increases by 3 dB, the safe time duration reduces to half, that is, 4 h. It is more difficult to formulate the norms for impulsive noise. However, in a broad sense, 140 dB impulsive noise has been found to be the upper safe limit for 100 impulses per day.

The interventions for controlling noise pollution include the design and development of various devices and technologies for the reduction of noise from automobiles, aircrafts, machines, etc.; use of sound absorbers, deflectors, and buffers; and various personal protective devices such as earplugs, earmuffs headphones, and helmets with protective cups for curtailing the entry of noise. These interventions help conserve the hearing of personnel working in noisy environs.

Extensive studies to evolve specifications of hearing protective devices for use in different noisy environments have been undertaken by DIPAS. The earmuffs and earplugs available from indigenous and foreign sources have been evaluated for their attenuation characteristics and acceptability by personnel [68–70].

Some of the commonly used earplugs and earmuffs are shown in Fig. 1. The earplugs are all premolded. Generally soft nonsticky materials such as molded plastic, plastic foam, vinyl, and silicone rubber are used for making earplugs. In addition to these, some disposable materials like cotton, cotton impregnated with wax, and glass wool are also used. The earmuffs are of basically hemispherical shape in design and circumscribe the pinna. All around the edges, special foam padding is provided to obtain proper seal with skin. Both right and left ear cups are joined together with a metallic headband of proper tension. Slimvalve has a special mechanical valve system to provide high degree of noise attenuation. Sonovalve is an improved version of slimvalve earmuff. In addition to mechanical valve, an earphone and microphone are incorporated in it for two-way communication purpose.

The attenuation performance of the ear defenders was evaluated in the laboratory using Monaural Pure Tone Threshold Shift (MPTS), Temporary Threshold Shift Reduction (TTSR), and Artificial Ear Coupler methods. The attenuation values at different frequencies by MPTS method ranged between 1.8 and 29.6 dB for earplugs and 9.5–46.5 dB for earmuffs. In case of TTSR method, the Temporary Threshold Shift (TTS_2) at different frequencies developed in ears protected with the ear defenders ranged from 5.9 to 13.4 dB and was significantly lower (ranging from 20.9 to 23.8 dB) than when no ear defender was worn by the individuals, i.e., in unprotected ears. Further, the magnitude of noise attenuation of earplugs tested by Artificial Ear Coupler ranged from 12.0 to 29.5 dB [69].

Though the attenuation offered by these hearing protective devices is adequate, they have met with limited compliance by the users, partly due to the lack of enforcement as well as restricted utility and due to some limitations like insufficient attenuation, inability to decipher commands/spoken words, and causing irritation and discomfort. This has led to the search for cost-effective, simple-to-use alternatives to conserve hearing that may be easily administered and complied with.

Carbogen Breathing: A Chemical Protective Method

In our continuing efforts towards evolving remedial measures for conserving the hearing and effectively counteracting the adverse interactions of noise on the efficiency and performance of people working in noisy environment, the mechanism of hearing impairment following exposure to different types of noise has been studied. It is a well-known fact that Permanent Threshold Shift (PTS) occurs when a previous Temporary Threshold Shift does not recover completely and the ear is further and continuously exposed to noise in the working place. In cases where the magnitude of TTS is not high and full recovery takes place during the rest periods, the risk of acquiring permanent hearing damages is substantially reduced. Intense acoustic stimulation has also been reported to produce discernible change in blood supply and oxygen tension of hair cells [70]. Thus, it has been postulated that TTS produced by noise exposure may be due to increased oxygen consumption by hair cells coupled with depletion of blood supply caused due to vasoconstriction. Hence any system that could counteract these effects may be able to mitigate the degree of hearing loss. Carbogen, a gas mixture of 95 % oxygen and 5 % carbon dioxide, is a well-known powerful vasodilator of the cerebral capillary beds, and this potential was utilized to counteract the vasoconstrictive effect of noise. Carbon dioxide acts with oxygen synergistically in the oxygenation of the cochlear tissue. At DIPAS it has been observed that the administration of gas even for short duration of 5 min. before and after exposure to occupational noise provided dual benefit by reducing the magnitude of TTS development and accelerating the recovery process [71]. These findings were utilized in the design and development of a safe, simple-to-operate, portable, lightweight, and aesthetically appealing carbogen delivery system in order to make the system available to the masses (Fig. 2).

The present system was designed to deliver carbogen at regulated flow rate of 10 l/min catering to the resting minute ventilation and for a fixed time period of 5 min. The basic design of the system incorporates the source of carbogen which is a cylinder of 10 l water capacity containing 1,320 l of carbogen at 1 atm, coupled with flow valve and time regulating electronic circuit.

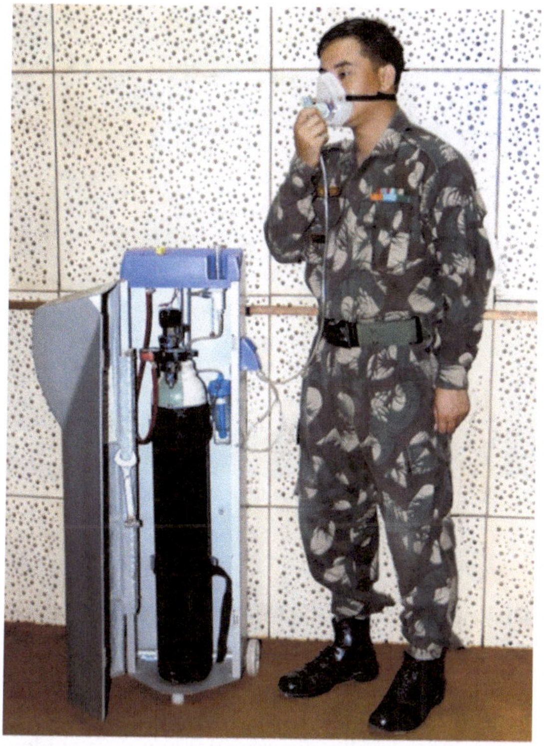

Fig. 2 Portable carbogen breathing system (Patent No. 236936)

The main structure is based on the minimal design concept to reduce weight and cost. A triangular-shaped outer body was conceptualized to reduce floor space area. Mounted on wheels, the system is portable and easy to maneuver. The control panel on the top has been fitted with red and green LED indicators to indicate the start and continuation of the carbogen supply. A buzzer warns the user on the completion of the 5-min installment of carbogen breathing. Housed in the body of the equipment, between the carbogen cylinder and a double valve breathing mask, a nebulizer humidifies the gas before inspiration and thus reduces the dryness of the air passage and complications such as coughing arising thereof. The chassis of the system is developed from lightweight and noncorrosive material. It is based on the minimal design concept for reducing weight, cost, and space. A handgrip enables the user to push the trolley with ease. The cylinder is firmly secured in place in the body with the help of straps, thus facilitating its easy replacement.

Efficacy of carbogen and the functional aspects of the prototypes have been successfully evaluated in the laboratory setup as well as in the industrial environment of the Army, Navy, Air Force, and engine rooms of ships for protection against high levels of continuous noise and against impulse noise of guns at proof and firing ranges [72–75]. Carbogen showed its suitability in providing protection against noise-induced hearing loss while maintaining intelligibility of speech [76]. Further, the administration of carbogen has both protective and therapeutic action [71, 77, 78]. Carbogen breathing will be helpful as a general means of prevention of NIHL in all kinds of noisy environments where it is not possible to control noise otherwise both in the armed forces and the civil sector, such as industries, airports, ordinance factories, railway workshops, boiler rooms, engine rooms, and welding workshops. Such a system has been developed for the first time in the country. The "Integrated Carbogen Delivery System" has been granted Indian Patent (No. 236936). A multiuser Kiosk workstation system was also developed to facilitate the administration of carbogen to ten individuals at a time (Fig. 3).

Breathing of carbogen for longer durations has medical applications and may be used as therapeutic in the management of sudden sensorineural hearing loss, as adjuvant to radiation therapy in cancer, in alcohol withdrawal, and in cigarette smokers.

The stress imposed by noise may lead to complex metabolic and physiological changes causing a "metabolic overload" by way of lipid peroxidation and free radical formation, which may result in membrane damage to hair cells and lead to loss of hearing sensitivity [79]. Supplementation with antioxidants like vitamin "E" that serve as scavengers of free radicals may prove effective in controlling the noise-induced deafness [25]. The investigations highlighted the benefit of the supplementation with vitamin E and/or carbogen alone in combating the oxidative stress imposed due to exposure to noise [25]. Musical sounds can be used as a management technique for conditioning the body and mind to cope with the stress imposed by noise and

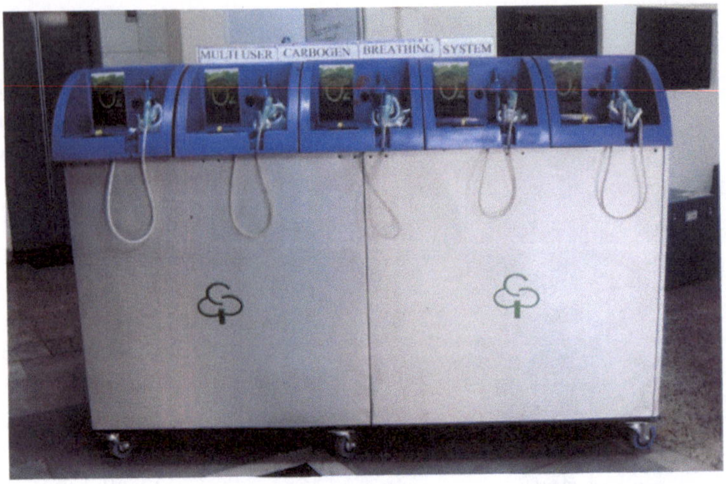

Fig. 3 Multiuser system

promote well-being and performance. Studies have shown that exposures to pleasant sounds are indicative of parasympathetic arousal with music, resulting in the lowering of the heart rate, blood pressure, and peripheral vascular responses coupled with increased galvanic skin response [80].

Active Noise Cancellation/Electronic Noise Reduction (ANC)

Proper communication with the traffic control tower is vital for pilots. They do so by radio communication through microphones and receivers fitted inside their helmets. The conventional passive hearing protective devices such as earmuffs and earplugs are not effective against low-frequency noise. Hence an electronic method to reduce/cancel the noise at both ends—transmission and receiving—is needed. The principle of ANC is destructive interference between the undesired noise and a deliberately produced antinoise of the same amplitude but opposite in phase. This technique works best where the noise source is repetitive and the noise is propagated in a closed loop. Recent advances elsewhere include noise cancellation headsets; mufflers for use in automotive, industrial applications; and a variety of consumer appliances [81, 82].

Policies and Strategies for Noise Abatement

The steady growing concern for and adoption of means to control environmental noise are evident everywhere. The proliferation of highway noise barrier walls along the nations' intestates and major thoroughfares is but one visible manifestation of the success of this landmark environmental legislation drawn decades ago. Hundreds of residential communities near these major transportation routes are significantly quieter because of these noise abatement measures. Highway engineers and architects are even developing noise barrier walls and landscaped beams that are aesthetically pleasing to both motorists travelling the highways and the residents on the other side of these barriers.

Addressing and solving any environmental noise problem involves two initial steps:

1. Quantifying the problem using noise measurements or analytical means
2. Determining the applicable criteria, goals, or noise limits

The first step—quantifying sound—is usually straightforward; the second step, finding an applicable limit, is also simple if the community affected has in place a well-written and workable environmental noise ordinance or guideline.

With "global" environmental noise sources, such as highways, railroads, and aircraft, the primary responsibility lies with federal authorities under the supervision of the appropriate federal agencies. The knowledge of how to measure and control environmental noise is a professional expertise that is readily available throughout the country. Many practicing acoustical consultants, architects, and engineers, and those working at universities and federally supported research centers in the country, agree that we are well prepared to make the twenty-first century a "quiet" one. Yes, the invisible pollutant of environmental noise can be tamed. In India a beginning has been made in this direction. The noise pollution (regulation and control) rules 2000 prescribes the standards of day and night noise levels in industrial, commercial, residential, and silence zones [83].

Conclusions

Environmental noise pollution is a problem of growing extent and concern. Concerted actions on a technological and political level are required to improve the present situation and to reverse existing trends. The elements required to set up a successful noise control strategy are known and, as far as technology is concerned, readily available, but greater efforts are still needed to set up and enforce noise legislation and to support this by education and information programs and by promoting research and development.

References

1. Kryter KD (1985) The effects of noise on man, IIth edn. Academic Press, New York
2. EPA (1978) Noise: a health problem. United States Environmental Protection Agency – Office of Noise Abatement and Control, Washington, DC
3. Berglund B, Lindvall T, Schwela DH (1999) Guidelines for community noise. WHO, Geneva
4. Jeanne Mager S (1998) Encyclopedia of occupational health & safety, vol II, 4th edn. ILO, Geneva, pp 47.2–47.5
5. Clifford RE, Rogers RA (2009) Impulse noise: theoretical solutions to the quandary of cochlear protection. Ann Otol Rhinol Laryngol 118:417–427
6. Kazi Rehan A (2002) Noise induced hearing loss-an occupational hazard. Indian J Occup Environ Med 6:207–212
7. Spreng M (2000) Possible health effects of noise induced cortisol increase. Noise & Health 2:59–64
8. Stephen SA, Matheson MP (2003) Noise pollution: non-auditory effects on health. Br Med Bull 68:243–257
9. Ising H, Braun C (2000) Acute and chronic endocrine effects of noise: review of the research conducted at the Institute for Water, Soil and Air Hygiene. Noise & Health 2:7–24
10. Goyal S, Gupta V, Walia L (2010) Effect of noise stress on autonomic function tests. Noise & Health 12:182–186
11. Gitanjali B, Ananth R (2003) Effect of acute exposure to loud occupational noise during day time on the nocturnal sleep architecture, heart rate and cortisol secretion in healthy volunteers. J Occup Health 45:146–152
12. Taylor SE (2006) Health psychology, 6th edn. Tata McGraw Hill, New York
13. Carlier IV, Voerman AE, Gersons BP (2000) The influence of occupational debriefing on post-traumatic stress symptomatology in traumatized police officers. Br J Med Psychol 73:87–98
14. Evans GW, Lercher P, Meis M, Ising H, Kofler WW (2001) Community noise exposure and stress in children. J Acoust Soc Am 109:1023–1027
15. Cunningham J, Nicol T, Zecker SG, Bradlow A, Kraus N (2001) Neurobiologic responses to speech in noise in children with learning problems: deficits and strategies for improvement. Clin Neurophysiol 112:758–767
16. Matsui T, Stansfeld S, Haines M, Head J (2004) Children's cognition and aircraft noise exposure at home-the West London Schools Study. Noise & Health 7:49–57
17. Hiramatsu K, Tokuyama T, Matsui T, Miyakita T, Osada Y, Yamamoto T (2003) The Okinawa study: effect of chronic aircraft noise exposure on memory of school children. In: Proceedings of the 8th international conference on noise as a public health problem. Foundation ICBEN, Schiedam, pp 179–180
18. Hygge S, Evans GW, Bullinger M (2002) A prospective study of some effects of aircraft noise on cognitive performance in schoolchildren. Psychol Sci 13:469–474
19. Stansfeld SA, Berglund B, Clark C et al (2005) Aircraft and road traffic noise and children's cognition and health: a cross-national study. Lancet 365 (9475):1942–1949
20. Rai RM, Chaturvedi RC, Malhotra MS (1976) Audiometric study of Indian Naval gunnery crew. Indian J Otolaryngol 26:143–151
21. Chaturvedi RC, Rai RM, Sharma RK (1991) Influence of noise spectra on auditory frequencies and degree of temporary threshold shift. Indian J Med Res 94:107–114
22. Chaturvedi RC, Rai M, Sharma RK (1982) Safety criteria of noise exposure. Indian J Med Res 76:758–765

23. Burns W, Robinson DW (1970) Hearing noise in industry. HMSO, London, 241 pp
24. Kapoor N et al. (2004) Occupational noise induced oxidative stress and remedial measures. Tech Report No. DIPAS/13/2004, Defence Institute of Physiology and Allied Sciences, New Delhi, India
25. Kapoor N, Mani KV, Shyam R, Sharma RK, Singh AP, Selvamurthy W (2011) Effect of Vitamin E supplementation on carbogen-induced amelioration of noise induced hearing loss in man. Noise & Health 13:452–458
26. Kapoor N, Singh AP, Bhatia MR, Sharma RK, Banerjee PK (2005) Noise in military aviation. In: Proceedings in challenges of occupational safety and health, section-IV: occupational safety, health and environment. Concept Publishing Company, New Delhi, India, pp 280–289
27. Rood GM, James SH (1999) Noise and communication. In: Ernsting J, Nicholson AN, Rainford DJ (eds) Aviation medicine, 3rd edn. Butterworth-Heinemann, Oxford, pp 489–521
28. Malhotra MS, Bal Krishna, Sampath Kumar R, Chaturvedi RC (1968) Effect of noise on naval personnel working on Board INS Vikrant. Technical report no. DIPAS/7/68
29. Murty VSN, Rao PLN, Keshava SL (1975) Acoustic trauma amongst jet aircraft maintenance personnel. Aviat Med 19:16–22
30. Soodan KS, Rao Pranesh BN (1980) A preliminary study of noise induced hearing loss in serving air crew of IAF. Institute of Aviation Medicine, Bangalore, Tech Report no. 106/80
31. Deshmukh SP, Rao AN, Soodan KS, Rao Pranesh BN (1983) Study of noise induced hearing loss amongst serving aircrew of IAF. Institute of Aviation Medicine, Bangalore, Tech Report no. 1288/81
32. Rao AN, Rao Pranesh BN, Soodan KS, Kapur R (1984) Study of noise environment and audiometric survey of technical airman at various fighter bases in IAF. Institute of Aviation Medicine, Bangalore, Report no. 1467/84
33. Henderson D (1985) Effects of noise on hearing. In: Fieldman AS, Grimes CT (eds) Hearing conservation in industry. Williams & Wilkins, Baltimore, Chapter 2
34. Sharma RK, Kapoor N, Mani KV, Kain TC, Banerjee PK (2007) Assessment of risk of noise induced hearing loss and annoyance due to traffic noise. Tech Report DIPAS/01/2007. Defence Institute of Physiology and Allied Sciences, New Delhi, India
35. Society to aid the hearing impaired (SAHI). http://www.sahiercare.org/trafficpolice.html.
36. Reid G (1948) Permanent deafness due to gunfire. J Laryngol Otol 62:76–87
37. Salmivalli A (1967) Acoustic trauma in regular army personnel. Acta Otolaryngol Suppl (Stockh) 222:1–85
38. Kessar SRV (1974) Audiometric survey of artillery personnel. MJAFI 30:150–154
39. Rai RM, Chaturvedi RC, Singh AP, Malhotra MS (1978) To survey the hearing acuity of the artillery and armoured corps personnel exposed to excessive noise. Tech Report DIPAS/5/78. Defence Institute of Physiology and Allied Sciences, New Delhi, India
40. Singh AP, Bhatia MR, Rai RM (1985) Measurement and analysis of field of gun noise for the purpose of suggesting a suitable ear defender for impulse noise. Tech Report DIPAS/1/85. Defence Institute of Physiology and Allied Sciences, New Delhi, India
41. Singh AP, Chaturvedi RC, Bhatia MR, Rai RM (1985) Noise and vibration problems in vehicles used in armed forces. Tech Report DIPAS/4/85. Defence Institute of Physiology and Allied Sciences, New Delhi, India
42. Chaturvedi RC, Srivastava KK, Sharma RK, Arora KB (1996) Evaluation of nature and magnitude of noise problem in infantry soldiers using small arms with a view to recommend suitable protective measures. Tech Report DIPAS/28/96. Defence Institute of Physiology and Allied Sciences, New Delhi, India
43. Rai RM, Chaturvedi RC, Bhatia MR, Malhotra MS (1975) Occupational hearing loss in workers frequently exposed to gun noise. MJAFI 34:260–265
44. Fleer R (1963) Protection afforded against impulsive noise by voluntary contraction of middle ear muscles. US Army Medical Research Laboratories, Fort Knox, Report No. 576
45. Kapoor N, Singh AP, Bhatia MR, Sharma RK, Banerjee PK (2004) Fire cracker noise and its' auditory implications. Paper published in the proceedings of the 11th international congress on sound and vibration held in St. Petersburg, Russia
46. Hamernik RP, Henderson D, Salvi R (1981) Potential for interactions of low level impulses and continuous noise. Air Force Aerospace Medical Research Laboratories Wright Patterson Air Force Base, Dayton, Report No. AFAMRL 80–68
47. Walker JG (1972) Temporary threshold shift caused by combined steady state & impulse noises. J Sound Vib 24:493–502
48. Hamernik RP, Henderson D, Grossley JJ, Salvi R (1974) Interaction of continuous and impulse noise. Audiometric & histological effects. J Acoust Soc Am 55:117–121
49. Hunt WJ, Hamernik RP, Henderson D (1976) The effects of impulse level on the interaction between impulse and continuous noise. Trans Am Acad Opthalmol Otolaryngol 82:305–308
50. Okada A, Fukuda K, Yamamura K (1972) Growth and recovery of temporary threshold shift at 4 kHz due to a steady state noise and impulse noise. Int Z Angew Physiol 30:105–111
51. Yamamura K, Takashima H, Miyabe H (1974) Effect of combined impact and steady state noise on temporary threshold shift. Med Lav 65:215–223
52. Cohen A, Bengt K, Lenenz RJ (1966) Temporary threshold shift from exposure to combined impact/steady noise conditions. J Acoust Soc Am 40:1371–1380
53. Ceypek T, Kuzniarz J, Lipowczan A (1975) The development of permanent hearing loss due to industrial percussion noise. Med Pr 2:53–59
54. Bhattacharya SK, Saiyed HN, Roy A, Chatterjee SK (1981) Hearing acuity in weavers of a textile mill. Indian J Med Res 74:779–785

55. Bhattacharya SK, Tripathi SR, Kashyap S (1990) Heat and noise problems in a drug and pharmaceutical firm in India. Ind Health 28:203–207
56. World Health Organization (2008) Prevention of blindness and deafness. http://www.who.int/pdp/deafness/heairngimpairment grades/en/indes.html
57. Walden BE, Worthington DW, McCurdy HW (1971) The extent of hearing loss in army – a survey report. Walter Reed Army Medical Centre Report no 21
58. Kapoor N, Sharma RK, Mani KV, Banerjee PK (2007) Assessment of auditory threshold status of employees at PXE Balasore. Tech. Report DIPAS/03/2007. Defence Institute of Physiology and Allied Sciences, New Delhi, India
59. Nilsson R, Dunn DE, Erlandsson B, Brenner J, Hakanson H, Ivarsson A, Wersall J (1983) Conclusions from animal experiments on the effects of steady state and impulse noise. In: Rossi G (ed) Noise as a public health problem: proceedings of fourth international congress. Centro Ricerche E Studi Amplifon, Milano, pp 159–170
60. Dunn DE, Davis RR, Herry CJ, Franks JR (1996) Hearing loss in Chinchilla from impact and continuous noise exposure. J Acoust Soc Am 90:1979–1985
61. Passchier-Vermeer W (1983) Measurement and rating of impulse noise in relation to noise induced hearing loss. In: Rossi G (ed) Proceedings of fourth international congress. Centro Ricerchee Studi Amplifon, Milano, pp 143–157
62. Thiery L, Birch CM (1988) Hearing loss due to partly impulsive industrial noise exposure at levels between 87–90 dBA. J Acoust Soc Am 84:651–659
63. Taylor W, Lempert B, Pelmear P, Hemstock I, Kershaw J (1984) Noise levels and hearing threshold in drop-forging industry. J Acoust Soc Am 76:807–819
64. Sulkowshi WJ, Lipowszan A (1982) Impulse noise induced hearing loss in drop forging operators and energy concept. Noise Control Eng 18:224–229
65. Passchier-Vermeer W (1974) Hearing loss due to continuous exposure to steady state broadband noise. J Acoust Soc Am 56:1585–1593
66. Voight P, Glodenhelm B, Ostlund E (1980) Impulse noise measurement and assessment of risk of noise induced hearing loss. Scand Audiol Suppl 12:1319–1325
67. Chaturvedi RC, Sharma RK, Arora KB, Srivastava KK (1994) To identify suitable ear defenders for the personnel of artillery and armoured corps with special reference to auditory perception. Tech Rept DIPAS/13/94. Defence Institute of Physiology and Allied Sciences, New Delhi, India
68. Chaturvedi RC, Srivastava KK, Arora KB, Sharma RK (1996) Evaluation of nature and magnitude of noise problem in infantry soldiers using small arms with a view to recommend suitable protective measures. Tech Rept DIPAS/28/96. Defence Institute of Physiology and Allied Sciences, New Delhi, India
69. Kapoor N, Chaturvedi RC, Singh AP, Sharma RK, Srivastava KK, Selvamurthy W (2001) A critical analysis of the noise attenuation and speech intelligibility performance of hearing protective devices. J Acoust Soc Ind 29:364–374
70. Vertes D, Axxelson A, Miller J, Liden G (1980) Some vascular effect of noise exposure in the chinchilla cochlea. Acta Otolaryngol 88:47
71. Chaturvedi RC, Sharma RK, Lakhera SC, Tiwary RS, Rai RM (1984) Role of carbogen in protection against noise induced hearing loss in man. Indian J Med Res 80:583–589
72. Singh AP, Kapoor N, Sharma RK, Banerjee PK, Selvamurthy W (2004) Report on user trial of carbogen breathing system for protection against noise induced hearing loss in Indian Navy. Tech Report DIPAS/2004. Defence Institute of Physiology and Allied Sciences, New Delhi, India
73. Kapoor N, Mani KV, Sharma RK (2006) An audiological assessment of crew members of AN-32 Aircraft and effect of carbogen gas breathing on temporary threshold shift amongst ground crew exposed to continuous high ambient noise. Tech. Report No. DIPAS/2006 on AFMRC Project 3294/2004. Defence Institute of Physiology and Allied Sciences, New Delhi, India
74. Kapoor N, Mani KV, Sharma RK, Shyam R, Banerjee PK (2007) Effect of carbogen breathing on TTS of hearing of armed forces personnel engaged in firing of guns. Tech. Report No. DIPAS/04/2007. Defence Institute of Physiology and Allied Sciences, New Delhi, India
75. Kapoor N, Sharma RK, Mani KV, Singh SB (2013) Field evaluation of carbogen breathing system for protection against noise induced hearing loss in the artillery. Tech. Report No. DIPAS/01/2013. Defence Institute of Physiology and Allied Sciences, New Delhi, India
76. Kapoor N, Sharma RK, Mani KV (2004) Effect of breathing carbogen on speech intelligibility in noise. J Acoust Soc Ind 32:409–415
77. Chaturvedi RC, Rai RM, Sharma RK (1988) Prophylactic action of carbogen against noise induced hearing loss. Indian J Med Res 88:60–63
78. Chaturvedi RC, Rai RM, Sharma RK (1990) Therapeutic role of carbogen in impaired hearing. Indian J Med Res 92:420–42371
79. Prasher D (1998) New strategies for prevention and treatment of noise-induced hearing loss. Lancet 352:1240–1242
80. Kapoor N, Singh AP, Sharma RK, Das G, Srivastava KK, Selvamurthy W (2000) Physiological effects of pleasant and unpleasant sound. J Acoust Soc Ind 28:127–134
81. Kuo SM (2006) Active noise control system for headphone applications control systems technology. IEEE Trans 14
82. Crabtree RB (1998) Real world performance of headset active noise reduction systems. J Acoust Soc Am 104:1806
83. CPCB Publication: The Noise Pollution (Regulation & Control) Rules 2000. http://cpcb.nic.in/divisionsofheadoffice/pci2/noise_rules_2000.pdf

Yoga for Preventive, Curative, and Promotive Health and Performance

M. Saha, K. Halder, O.S. Tomar, A. Pathak, and R. Pal

Abstract

Yoga is an ancient Indian system of philosophy, culture, tradition, and way of maintaining better life, established in India thousands of years ago. The Sanskrit word Yoga means union of body and mind through breath control methods, asanas and meditation. The ashtang yogic practices, very popularly known today, are derived from Patanjali's Yoga Sutra. Asana (postural exercises), pranayama (breathing maneuver), and dhyana (meditation) are mostly practiced in different combinations for physical and mental well being. It gradually develops the spiritual harmony of the individual through the control of mind and body. The practice of yoga uses eight methods, known as "limbs," thus being known as "Ashtanga Yoga": yama (restraint), niyama (observance), asana (posture), pranayama (breath control), pratyahara (sensory deprivation), dharana (contemplation), dhyana (fixing the attention), and samadhi (absolute concentration). Regular yogic practices endow perfect physical and mental health to its practitioner. It improves aerobic capacity, anaerobic capacity, joint flexibility, and muscle strength. Evidence shows that the regular execution of these practices provides the practitioner with more physical flexibility, muscle strengthening, increased vitality, delineated psychological stress, and reduced cardiovascular risks. Yogic techniques are known to improve one's overall performance and work capacity. During yoga session, the postural maneuvers are executed without repetition and are connected to each other by passages that establish links between the exercises in a sequence. Yoga is not only a discipline to be practiced by saints or spiritual aspirants but also has relevance to the spirit of military activities.

M. Saha (✉) • K. Halder • O.S. Tomar • A. Pathak
Exercise Physiology & Yoga Laboratory, DIPAS
(DRDO), Timarpur, Delhi-110054, India
e-mail: msaha0365@gmail.com

R. Pal
Centre for Advanced Research and Training in Yoga,
DIPAS, Timarpur, Delhi-110054, India

S.B. Singh et al. (eds.), *Translational Research in Environmental and Occupational Stress*,
DOI 10.1007/978-81-322-1928-6_14, © Springer India 2014

Introduction

Development of Customized Yoga Packages for Armed Forces

During last 30 years, Defence Institute of Physiology and Allied Sciences (DIPAS) has conducted a series of research projects to explore the prophylactic, promotive, and curative potentials of yoga with particular reference to its application for the armed forces. These studies were conducted on the soldiers who practiced a battery of yogic practices including asanas, pranayama, and meditation and on people who have been practicing it for years. The research showed that yogic practices help to achieve a better autonomic balance, orthostatic tolerance, relative hypometabolic state, and thermoregulatory efficiency to cold stress. Yoga has been shown to improve almost all aspects of physiological performance and efficiency across different ages. These studies have helped to a great extent to formulate yoga packages for the armed forces.

DIPAS, DRDO has developed customized yoga packages to keep the health sound and morale high of the armed forces after extensive scientific studies as it is the only beneficial process for improving soldier's mental and physical condition under different adverse climatic conditions. Different yoga training packages have been developed for the soldiers suitable for Siachen area serving at high-altitude, low-intensity conflict zones and Jammu and Kashmir. Similarly, different yoga packages are also formulated for Navy and Air Force personnel. Regular yogic training is being imparting to thousands of soldiers as per their field unit locations.

It is well established that yogic practice helps in the upliftment of various functions of body and mind through its effects on cardiovascular, respiratory, metabolic, hormonal, and neural systems [3, 24, 44, 51]. Regular yogic practice provides the practitioner with more physical flexibility [35], muscle endurance [2, 7], maximal work output and oxygen consumption [19, 36], increased vitality, alleviated psychological stress, reduced cardiovascular risks [38, 44], carbon dioxide elimination, minute ventilation, etc. Studies on the effects of various yogic postures on oxygen consumption, increased vitality, alleviated psychological stress, and reduced cardiovascular risks have revealed that yogic asana besides manifesting broadly similar trends appears to have some degree of specificity in terms of magnitude of influence. Improvement in cardiac recovery index was reported [23] after 10-week yoga training. Yoga helps in the management of stress, and our research has proved that it is an all-round exercise for toning up cardiovascular [40], respiratory [51], endocrine [15], and nervous system [44] to maintain optimum physical fitness and mental health [37].

It was reported [44, 15] that after yogic training, positive developments were found in cardiorespiratory, biochemical, and psychological functions. Yogic practice helps to achieve a stable autonomic balance with better orthostatic tolerance, to develop a relative hypometabolic state and better thermoregulatory efficiency to cold stress [43] thereby improving performance in cold environment. An improvement was observed in body flexibility, and endurance time to complete a standard physical task with reduction in EMG buildup by yogic exercise was reported in middle aged men [35, 36]. Recent studies [28, 37–40] indicate if yogic exercises are supplemented with games and other kind of conventional physical exercises, it helps to improve aerobic capacity as well as to reduce the sense of exertion to physical work in a better way. The studies [43] in our laboratory indicate that the practice of yogic exercises facilitates acclimatization to high altitude in troop and physical performance in subjects with high-level endurance training from sea level itself. Studies [38] reported that at high altitude after 4 days of continuous strenuous military operation, subjects who were practicing yoga for 6 months had better recovery heart rate after exercise and improved VO_{2max} as compared to those who followed usual physical training program. Studies [37] on a group of young fellowship course trainees (DRDO new recruits) showed that along with the improvement in physical performance,

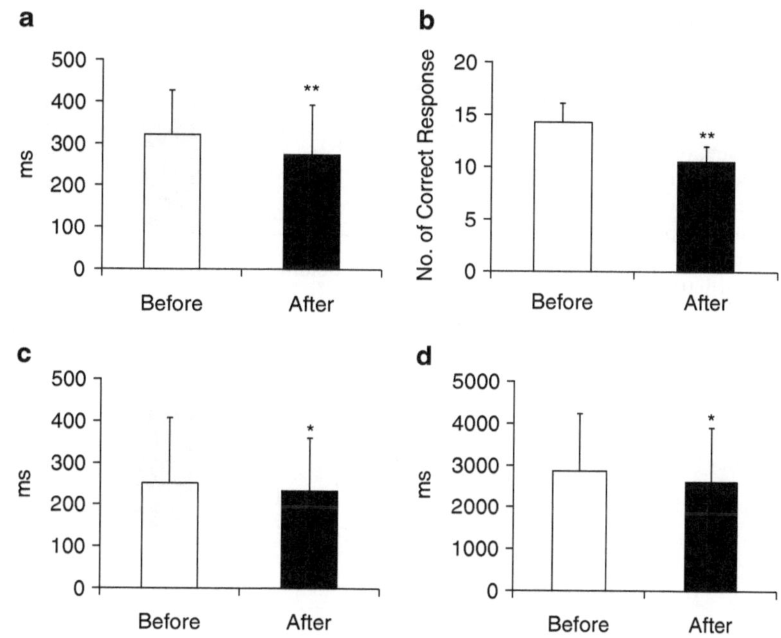

Fig. 1 Effect of 3 months yogic practice on cognitive function. (**a**) Choice reaction time, (**b**) pattern recognition memory, (**c**) simple reaction time, (**d**) paired associates learning. Values are presented as Mean ± SD. *$P < 0.05$, **$P < 0.01$

there were improvements in psychological parameters like cognitive functions (memory, learning efficiency) and faster reaction time, psychomotor performance, and reduction in anxiety and depression among the trainees practicing yogic exercises. These improvements would be helpful for the trainees to cope with different stresses. By practicing yoga, they can achieve the cardiorespiratory conditioning with the help of very moderate level of oxygen consumption without much exertion, i.e., at a rate of 15–33 % of an individual's maximal aerobic capacity. At high altitude as there is already oxygen want, yogic practices may be appropriate.

Impact of Yogic Practice On

Brain Function, Nervous System, and Cognitive Performance

Voluntary regulation in yoga breathing has distinct effects on the autonomic nervous system and higher brain functions. The effects of meditation on nervous system and mental state are even clearer [54]. Hatha yoga practice may be associated with the promotion of neuroplastic changes in executive brain systems such as increase gray matter volume [9].

Yoga asanas session would be able to increase gamma-aminobutyric acid level of the practitioner [53]. Agnihotra pranayama was shown to influence power of alpha band and improve relaxation [45]. Yogic practice is helpful to improve cognition and cognitive speed during emotion [9]. Auditory information transmission at the level of the medial geniculate and primary auditory cortex was delayed following meditation [55]. Long-term Vihangam Yoga meditators have better cognitive abilities than non-meditators in the old age group [30].

Cardiovascular Physiology

Several studies have reported improvement in physiological, physical, and psychological functions [37]. Yogic practice decreases skin temperature, oral temperature, and respiratory rate and provides some degree of resistance against physical and cold climate environmental stress in both male and female practitioners [38, 43, 44]. Yogic practice has a beneficial effect on cardiovascular system. Studies reported that long-term as well as short-term yogic practice decreased heart rate and blood pressure. It is also reported [55] that right nostril breathing (Suryavedi and anuloma-viloma pranayama) decreases heart rate and blood pressure [55] (Fig. 2).

Fig. 2 Effect of 3 months yogic practice on cardiovascular system. (**a**) Systolic blood pressure, (**b**) diastolic blood pressure, (**c**) resting respiratory rate, (**d**) breath-holding time. Values are presented as Mean ± SD. * $P < 0.05$, ** $P < 0.01$

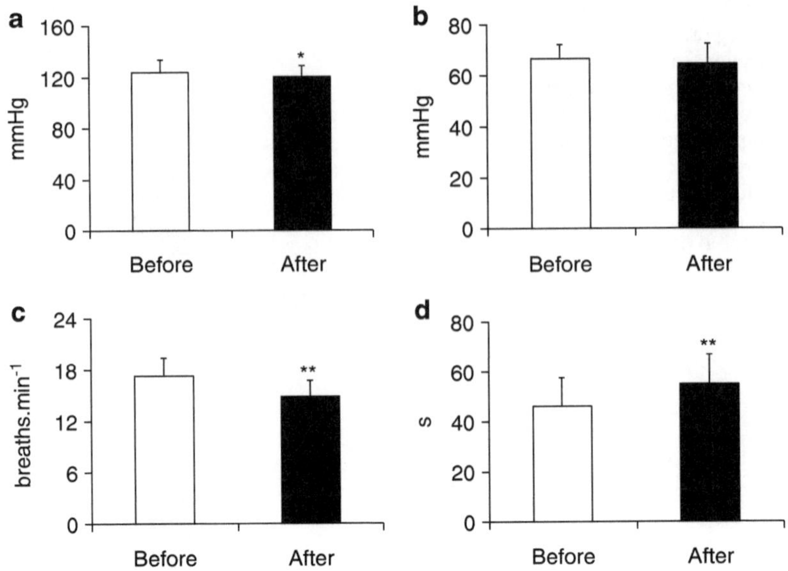

Improvement of cardiovascular parameters after 3 weeks of slow breathing was reported, whereas 3 weeks of fast breathing increased heart rate and blood pressure and shifted the practitioner towards parasympathetic state [20]. Heart rate variability (HRV) is a noninvasive tool to study cardiac autonomic activity. Iyengar yoga is associated with significant increase of cardiac vagal modulation and cardiac parasympathetic modulation among healthy yoga practitioners [18]. Iyengar yoga practitioners showed a lower heart rate and blood pressure and low-frequency power of HRV [48]. The beneficial and relaxing effects of OM meditation on heart rate, respiratory rate, skin resistance, and finger plethysmography amplitude and oxygen consumption have been demonstrated [54]. Savasana, a relaxing technique, has the ability to decrease heart rate, blood pressure, and rate pressure product, whereas no significant alteration was recorded in low-frequency power, high-frequency power, and total spectral power [20]. Long duration yogic intervention was associated with improve mood, and decreased anxiety, and thalamic gamma aminobutyric acid levels [53].

Lung Function and Respiratory Profile

Yogic practices improve lung function parameters. Reports are available on the effects of Hatha yoga practice to improve force vital capacity (FVC), force expiratory volume (FEV), maximum voluntary ventilation (MVV), and peak expiratory flow rate (PEFR) [30]. Yogic practices also decrease respiratory rate or breathing frequency and increase breath-holding time [16] (Fig. 2).

Reports of single asana practice demonstrated that virasana and siddhasana increases the rate of respiration or breathing frequency, minute ventilation, and tidal volume [32, 33]. Short-term yoga practice quickly improved respiratory functions in relatively elder people (age 41–50 years) by means of improving respiratory parameters [2]. In a randomized control trial conducted at Brazil, the elderly healthy population had improved respiratory function and sympathovagal balance following yoga respiratory exercises [42]. Earlier studies [11] showed significant improvement in MVV and a marginal improvement in FVC and FEV in 1st second following yogic practice on border security force.

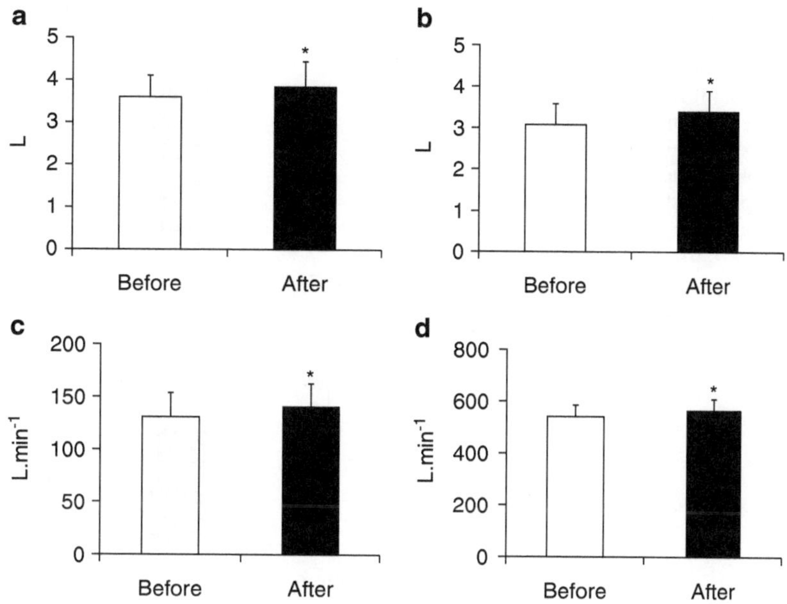

Fig. 3 Effect of 3 months yogic practice on lung function. (**a**) Forced vital capacity, (**b**) forced expiratory volume in 1st second, (**c**) maximum voluntary ventilation, (**d**) peak expiratory flow rate. Values are presented as Mean ± SD. $*P < 0.05$

Body Composition, Physical Fitness, and Performance

Regular yogic practice provides the practitioner with more physical flexibility [35, 37], muscle endurance [19, 36] maximal work output, and oxygen consumption [34, 38]. Yogic practice also improves physical performance in terms of aerobic performance, anaerobic performance, body composition, and cardiovascular endurance [4, 5, 38]. Reports [37] are also available on the effects of Hatha yoga training in shifting the lactate threshold towards higher workload of exercise and improvement in work capacity. Studies showed that yoga has a beneficial role in energy cost or energy expenditure and perceived exertion [38] (Fig. 4).

Virasana and siddhasana increases oxygen consumption, carbon dioxide elimination, respiratory exchange rate, and energy expenditure during the practice session [32, 33]. Changes in cardiorespiratory and energy expenditure upon practicing different postures (asanas), breathing maneuvers (pranayama), and meditation have been reported [4, 51]. It has been stated that yoga improves cardiorespiratory functions, aerobic capacity, and body composition by decreasing fat and increasing lean body mass and affects anaerobic power and anaerobic threshold [4, 5, 37, 38]. Effects of various yogic postures in terms of oxygen consumption, carbon dioxide elimination, minute ventilation, etc. have been reported [40]. Yogic practices reduced exercise-induced cardiovascular responses [20] (Fig. 5).

Improvement in handgrip strength of both hands after practicing various pranayama was observed [31]. Studies showed that yogic practices improve handgrip strength and handgrip endurance [19, 21, 31]. Flexibility of hip, trunk, neck, knee, and shoulder improved progressively after 3 months and 6 months of yogic practices in males and females [37]. Improvement in body composition, cardiovascular endurance, and anaerobic power was described [5]. The enhancement of physical fitness and improvement of body composition in the older adults following yogic practice has been reported [7]. Bikram yoga associated with increased head lift strength, substantially increased lower back/hamstring flexibility, increased shoulder flexibility, and modestly decreased body fat, while no changes reported in handgrip strength, cardiovascular measures, or maximal aerobic fitness [57]. A recent review summarized that yoga may engender improvements in some components of fitness in older adults also suggesting that more

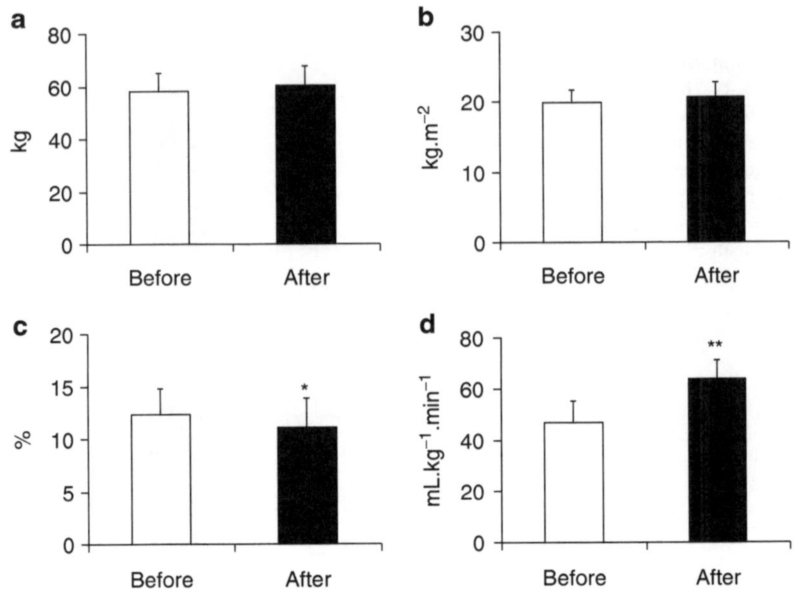

Fig. 4 Effect of 3 months yogic practice on body composition and physical fitness. (**a**) Body weight (**b**) body mass index, (**c**) body fat percentage, (**d**) maximal aerobic capacity. Values are presented as Mean ± SD. $*P < 0.05$; $**P < 0.01$

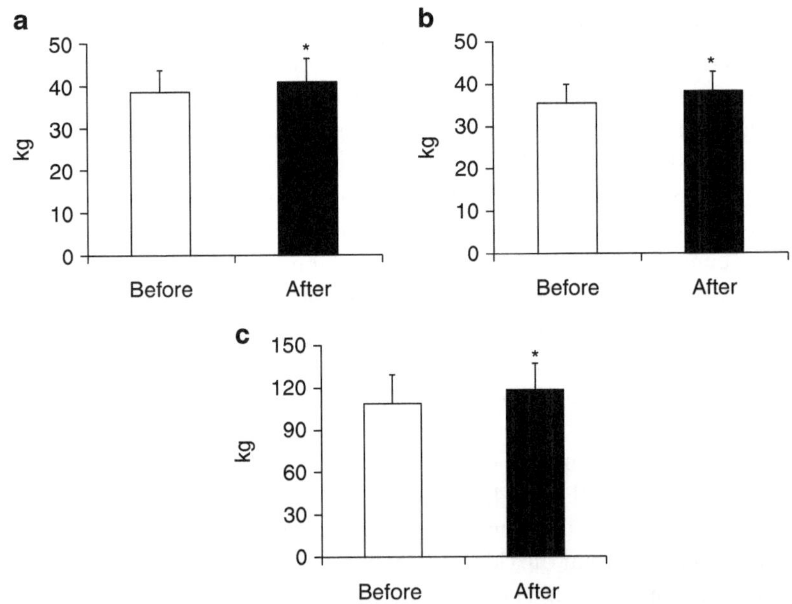

Fig. 5 Effect of 3 months yogic practice on muscular strength. (**a**): Right-hand grip strength, (**b**) left-hand grip strength, (**c**) back leg strength. Values are presented as Mean ± SD. $*P < 0.05$

evidence is required to determine its effectiveness as an alternative exercise to promote fitness in older adults [41]

Oxidative Stress and Other Age-Related Biochemical Variables

Studies [6, 52, 58] showed that yoga can decrease oxidative stress and improve antioxidant and redox status. A review [22] summarized the beneficial effect of short-term and long-term meditation such as Zen meditation and transcendental meditation to reduce oxidative stress in terms of glutathione level and activity of antioxidant enzymes (catalase, superoxide dismutase, glutathione peroxidase, and glutathione reductase). This review also highlighted the effect of

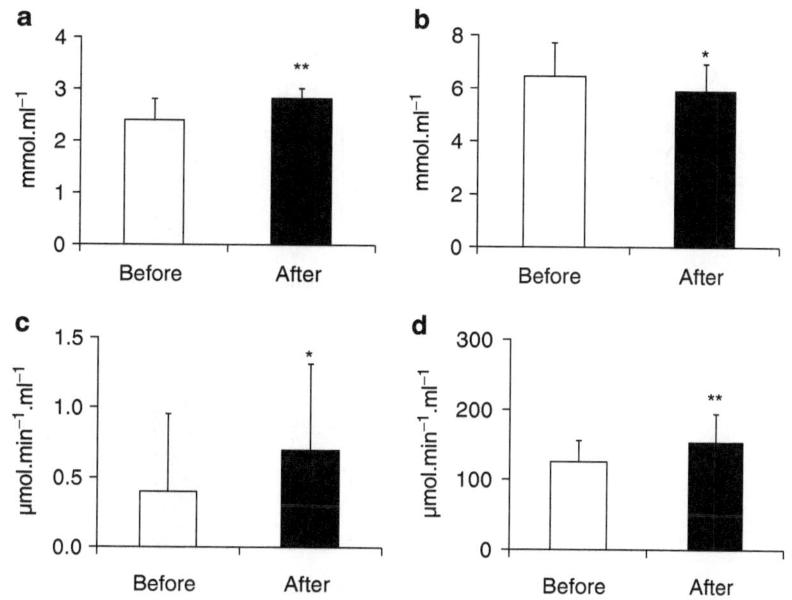

Fig. 6 Effect of 3 months yogic practice on antioxidant status. (**a**) Total antioxidant status, (**b**) malondialdehyde, (**c**) glutathione reductase, (**d**) catalase. Values are presented as Mean ± SD. $*P < 0.05$; $**P < 0.01$

Sudarshan Kriya yoga, Tai Chi training, and diaphragmatic breathing to reduce oxidative stress. It has been reported that 6-month yogic practice improves glutathione as well as the total antioxidant status [52]. Yogic practice shifted the practitioner towards reduced state of antioxidant and redox status. The ratio of reduced glutathione (GSH) to oxidized glutathione (GSSG) decreased after yogic practice. Exercise-induced stress and oxidative stress are also reduced by yogic practice [52] (Fig. 6).

It has been reported [58] that 10-day yoga-based lifestyle modification program (YLMP) decreased the serum concentration of thiobarbituric acid-reactive substances (Fig. 7).

Comprehensive lifestyle changes including yogic intervention increased telomerase activity and consequently telomere maintenance capacity in human immune-system cells in low-risk prostate cancer patients [27]. An increment in glutathione and vitamin C, decrement in malondialdehyde was observed following yogic practice in diabetic volunteers [12]. Another research [28] showed the beneficial effect of yoga to prevent age-related changes in oxygen metabolism, oxidative stress, and antioxidant redox status of physically active Indian Air Force personnel.

Therapeutic Implications

Yoga practice have a beneficial role in prevention and enhancing survival from many lifestyle-related psychosomatic disorders such as obesity, hypertension, cardiovascular diseases, coronary heart disease, and diabetes. Neurological disorders like stress, anxiety, depression, sleep disorders, and insomnia may be reduced or cured following yogic practice [56]. Yogic practices help to manage asthma, chronic obstructive pulmonary disorders, and other respiratory disorders [13]. It helps to manage hypertension and prevent heart attack [1, 24, 46]. It is also beneficial to decrease the discomfortness of irritable bowel syndrome [8]. Life-threatening diseases like breast cancer, lung cancer, and prostate cancer may enhance survival through yogic practice [25, 49]. Yoga can reduce risk profiles induced by stress in geriatric patients with type 2 diabetes and may aid prevention or delay in diabetic complications [17, 29]. Yogic practice reduces risk factors associated with pregnancy such as decrease platelet count, increase uric acid, hypertension, etc. [14]. Hatha yoga exercise has therapeutic, preventative, and protective effects in end-stage renal disease subjects by decreasing oxidative stress in terms of modulating the oxidative stress

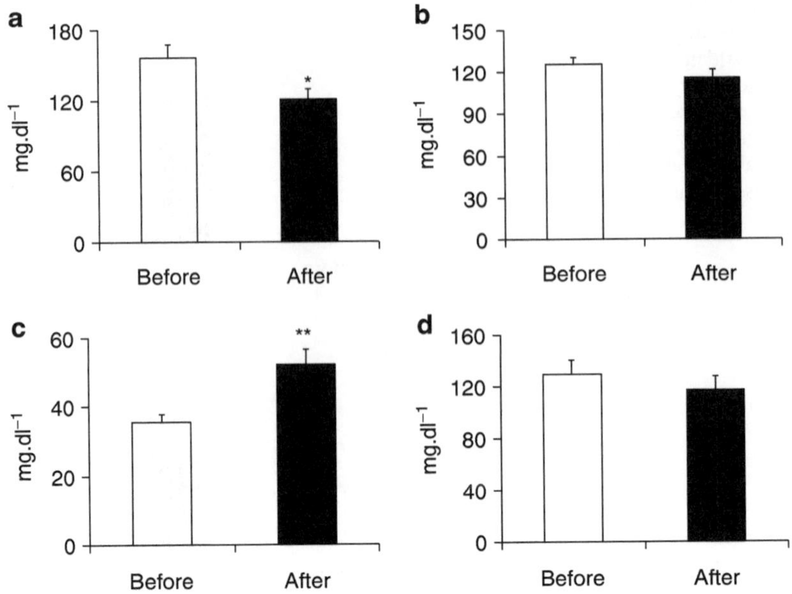

Fig. 7 Effect of 3 months yogic practice on lipid profile. (**a**) Triglyceride, (**b**) total cholesterol, (**c**) HDL cholesterol, (**d**) LDL cholesterol. Values are presented as Mean ± SD. $*P < 0.05$; $**P < 0.01$

indicators malondialdehyde (MDA), protein oxidation, phospholipase A2 (PLA2), superoxide dismutase (SOD), and catalase in blood [10]. A holistic yoga program for 12 weeks was better than physical exercise in reducing antimüllerian hormone, luteinizing hormone, and testosterone. It also improved menstrual frequency in adolescent and polycystic ovarian syndrome [26]. An 8-week yoga training improved body composition and total cholesterol levels in obese adolescent boys [47]. Yoga is valuable in helping the hypothyroid patients to manage their disease-related symptoms of hyperthyroidism [50].

Contribution in Yoga Research by Defence Institute of Physiology and Allied Sciences

It has been reported [35–38, 40] that physical and cognitive performance improves through yogic practice in terms of oxygen consumption, flexibility, anaerobic threshold, and strength. Physiological responses were recorded at 10 °C on the yoga participants [45]. In addition, DIPAS has engaged in the preparation of yogic modules for Army, Navy, and Air Force for optimization of performance at different climacteric conditions such as high-altitude and deep-sea diving. Another field of yoga research in DIPAS is the operational stress management in Army, Navy, and Air Force. Molecular level studies have been conducted to understand how a physical and cognitive performance improvement occurs through yogic practice. Yogic practice has helped to prevent age-related degenerative changes in the oxygen metabolism and antioxidants redox status in Indian Air Force personnel [28]. Reports [59] based on the study of Indian Air Force personnel showed that yoga enhances physical and cognitive performance in terms of cardiovascular, lung functions, body compositions, muscular strength, flexibility, aerobic capacity, cardiorespiratory performance, reaction time, memory leaning, and biochemical responses (Figs. 1, 2, 3, 4, 5, 6, 7).

Development of Yoga Packages

Studies conducted by DIPAS with modern technologies have facilitated formulation of yoga packages for the armed forces (Fig. 8). Each yoga posture were carefully observed and evaluated during yogic practice and validated by taking online recording of various parameters

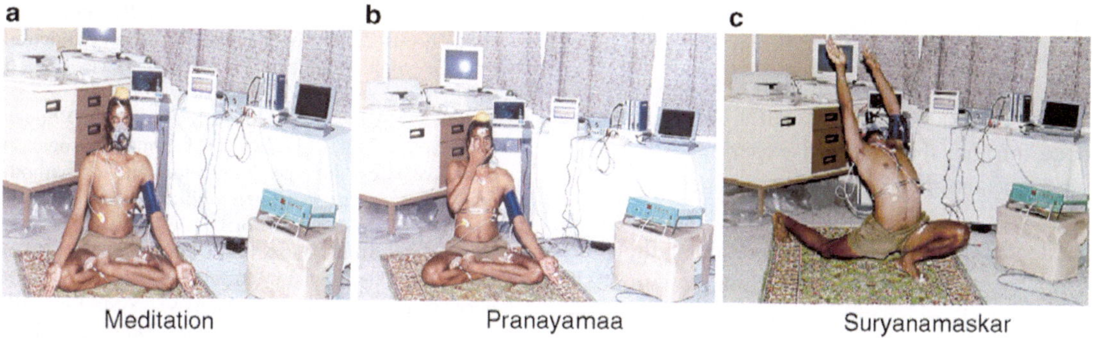

Fig. 8 (**a**) Meditation, (**b**) pranayama, (**c**) surya namaskar

like heart rate, blood pressure, oxygen saturation, aerobic capacity, oxygen consumption, carbon dioxide output, breathing frequency, pulmonary ventilation, respiratory quotient, and also by cycle ergometry for the assessment of physical work capacity. After validation and on the basis of trade in various job-specific tasks in Air Force and Navy and also for the Army at high altitude, different yoga packages have been formulated. The yoga packages for the Navy and Army at high altitude as validated by DIPAS were found to be effective in the trials conducted by the concerned authorities for its introduction in their respective setup. Army has consented for imparting yoga training to soldiers at field locations and accepted the beneficial role of yoga in soldier system.

Energy cost and intensity of exercise in terms of the individual's maximal aerobic capacity (% VO_{2max}) were calculated for all the yogic practices. During various practices if any of the parameter was found to be above normal range in any of the yogic practice, it was excluded from the package. Due consideration was given for the environment and also the place in which one would practice it. For example, in high-altitude and cold conditions, one is already stressed with sympathetic hyperactivity and prone to be hypertensive. So, if any yogic practice elevates blood pressure abnormally, it should be avoided. At high altitude, for a normal healthy individual, chemoreceptor sensitivity should be such that any change in oxygen tension in the blood due to hypoxic environment should be detected, and natural physiological reaction like hyperventilation should take place. The optimization of this ventilatory response is a must. In yogic breathing maneuvers (pranayama), if breath hold (kumbhaka) is practiced, it will help to blunt the sensitivity of oxygen receptors and may put extra stress to the system. So, kumbhaka (breath hold) was not incorporated in the yoga package designed for high altitude. On the contrary, for the yoga packages of Navy, it was included with the advice for careful practice with adequate training for sea diving, and the blunted chemoreceptor sensitivity would help to prolong the operation. For comparatively aged individuals, complicated yoga postures were not included to avoid additional stress over cardiovascular, respiratory, and musculoskeletal system. Among the pranayamas, simple ones like Anuloma-viloma, and Bhramari were included in all the packages as those had better effects on bringing down heart rate, blood pressure, and various respiratory parameters. Bhramari pranayama and Omkar meditation showed many effects on the sensory neural processing during and immediately after its practice. It indicated positive developments in cardiorespiratory, biochemical, and psychological functions. It has been observed that yogic practice helps to achieve a stable autonomic balance with better orthostatic tolerance, to develop a relative hypometabolic state and better thermoregulatory efficiency to cold stress.

Modern warfare with its sophisticated technologies requires a balanced and alert mind with quick decisions at right moment along with a high level of endurance, muscle power, and optimal level of aggression during combat situations. Yogic training is helpful for the decision makers involved in intellectual activities as well. The practice of yoga although has developed through the practical experiences of great yoga masters through ages, still when considering it for practice by the army, special attention is needed to be given for its validation with scientific approach. Thus, the amalgamation of ancient Indian wisdom of yoga with modern technologies helped to explore its potentials scientifically and also to use it effectively for the Armed Forces.

References

1. Agte VV, Tarwadi K (2004) Sudarshan kriya yoga for treating type 2 diabetes: a preliminary study. Altern Complement Ther 10:220–222
2. Ahmed QR, Sau SK, Kar SK (2010) An evaluation of pulmonary parameters in two groups of subjects during Yoga practice. Nepal Med Coll J 12(3):180–182
3. Anand BK, Chinna GS, Singh B (1961) Studies on Sri Ramanand Yogi during his stay in an air tight box. Indian J Med Res 49:82–89
4. Balasubramaniam M, Telles S, Doraiswamy PM (2012) Yoga on our minds: a systematic review of yoga for neuropsychiatric disorders. Front Psychiatry 3:117. doi:10.3389/fpsyt.2012.00117
5. Bera TK, Rajapurkar MV (1993) Body composition, cardiovascular endurance and anaerobic power of yogic practitioner. Indian J Physiol Pharmacol 37:225–228
6. Bhattacharya S, Pandey US, Verma NS (2002) Improvement in oxidative status with yogic breathing in young healthy males. Indian J Physiol Pharmacol 46:349–354
7. Chen KM, Chen MH, Hong SM, Chao HC, Lin HS, Li CH (2008) Physical fitness of older adults in senior activity centres after 24-week silver yoga exercises. J Clin Nurs 17(19):2634–2646
8. Evans S, Cousins L, Tsao JC, Sternlieb B, Zeltzer LK (2011) Protocol for a randomized controlled study of Iyengar yoga for youth with irritable bowel syndrome. Trials 12:15. doi:10.1186/1745-6215-12-15
9. Froeliger BE, Garland EL, Modlin LA, McClernon FJ (2012) Neurocognitive correlates of the effects of yoga meditation practice on emotion and cognition: a pilot study. Front Integr Neurosci 6:48. doi:10.3389/fnint.2012.00048
10. Gordon L, McGrowder DA, Pena YT, Cabrera E, Lawrence-Wright MB (2013) Effect of yoga exercise therapy on oxidative stress indicators with end-stage renal disease on hemodialysis. Int J Yoga 6(1):31–38. doi:10.4103/0973-6131.105944
11. Halder K, Chatterjee A, Kain TC, Pal R, Tomer OS, Saha M (2012) Improvement in ventilatory function through yogic practices. Al Ameen J Med Sci 5 (2):197–202
12. Hegde SV, Adhikari P, Kotian S, Pinto VJ, D'Souza S, D'Souza V (2011) Effect of 3-month yoga on oxidative stress in type 2 diabetes with or without complications: a controlled clinical trial. Diabetes Care 34(10):2208–2210. doi:10.2337/dc10-2430
13. Holland AE, Hill CJ, Jones AY, McDonald CF (2012) Breathing exercises for chronic obstructive pulmonary disease. Cochrane Database Syst Rev 10, CD008250. doi:10.1002/14651858.CD008250.pub2
14. Jayashree R, Malini A, Rakhshani A, Nagendra HR, Gunasheela S, Nagarathna R (2013) Effect of the integrated approach of yoga therapy on platelet count and uric acid in pregnancy: a multicenter stratified randomized single-blind study. Int J Yoga 6(1):39–46. doi:10.4103/0973-6131.105945
15. Joseph S, Sridharan K, Patil SKB, Kumaria ML, Selvamurthy W, Joseph NT, Nayar HS (1981) Study of some physiological and biochemical parameters in subjects undergoing yogic training. Ind J Med Res 74:120
16. Joshi LN, Joshi VD, Gokhale LV (1992) Effect of short term 'Pranayam' practice on breathing rate and ventilatory functions of lung. Indian J Physiol Pharmacol 36(2):105–108
17. Jyotsna VP, Joshi A, Ambekar S, Kumar N, Dhawan A, Sreenivas V (2012) Comprehensive yogic breathing program improves quality of life in patients with diabetes. Indian J Endocrinol Metab 16(3):423–428. doi:10.4103/2230-8210.95692
18. Khattab K, Ortak J, Richardt G, Bonnemeier H (2007) Iyengar yoga increases cardiac parasympathetic nervous modulation among healthy yoga practitioners. Evid Based Complement Alternat Med 4 (4):511–517. doi:10.1093/ecam/nem087
19. Madanmohan, Thombre DP, Balakumar B, Nambinarayanan TK, Thakur S, Krishnamurthy N, Chandrabose A (1992) A effect of yoga training on reaction time, respiratory endurance and muscle strength. Indian J Physiol Pharmacol 36(4):229–233
20. Madanmohan UK, Bhavanani AB, Shatapathy CC, Sahai A (2004) Modulation of cardiovascular response to exercise by yoga training. Indian J Physiol Pharmacol 48(4):461–465
21. Madanmohan MSK, Balakrishnan S, Gopalakrishnan M, Prakash ES (2008) Effect of six weeks yoga training on weight loss following step test, respiratory pressures, handgrip strength and handgrip endurance in young healthy subjects. Indian J Physiol Pharmacol 52(2):164–170
22. Mahagita C (2010) Roles of meditation on alleviation of oxidative stress and improvement of antioxidant system. J Med Assoc Thai 93(Suppl 6):S242–S254
23. Muralidhara DV, Ranganathan KV (1982) Effect of yoga practice on cardiac recovery index. Indian J Physiol Pharmacol 26:279–283

24. Murugesan R, Govindarajulu N, Bera TK (2000) Effect of selected yogic practices on the management of hypertension. Indian J Physiol Pharmacol 44:207–210
25. Nakau M, Imanishi J, Imanishi J, Watanabe S, Imanishi A, Baba T, Hirai K, Ito T, Chiba W, Morimoto Y (2013) Spiritual care of cancer patients by integrated medicine in urban green space: a pilot study. Explore (NY) 9(2):87–90. doi:10.1016/j.explore.2012.12.002
26. Nidhi R, Padmalatha V, Nagarathna R, Amritanshu R (2013) Effects of a holistic yoga program on endocrine parameters in adolescents with polycystic ovarian syndrome: a randomized controlled trial. J Altern Complement Med 19(2):153–160. doi:10.1089/acm.2011.0868
27. Ornish D, Lin J, Daubenmier J, Weidner G, Epel E, Kemp C, Magbanua MJ, Marlin R, Yglecias L, Carroll PR, Blackburn EH (2008) Increased telomerase activity and comprehensive lifestyle changes: a pilot study. Lancet Oncol 9(11):1048–1057. doi:10.1016/S1470-2045(08)70234-1
28. Pal R, Singh SN, Halder K, Tomer OS, Mishra AB, Saha M (2013) Effect of yogic practices on age related changes in oxygen metabolism and antioxidant-redox status. J Exp Integr Med 3(4):305–312. doi:10.5455/jeim.060613.or.079
29. Pandey A, Tripathi P, Pandey R, Srivatava R, Goswami S (2011) Alternative therapies useful in the management of diabetes: a systematic review. J Pharm Bioallied Sci 3(4):504–512. doi:10.4103/0975-7406.90103
30. Prakash R, Rastogi P, Dubey I, Abhishek P, Chaudhury S, Small BJ (2012) Long-term concentrative meditation and cognitive performance among older adults. Neuropsychol Dev Cogn B Aging Neuropsychol Cogn 19(4):479–494
31. Raghuraj P, Nagarathna R, Nagendra HR, Telles S (1997) Pranayama increases grip strength without lateralized effects. Indian J Physiol Pharmacol 41 (2):129–133
32. Rai L, Ram K (1993) Energy expenditure and ventilatory responses during Virasan – a yogic standing posture. Indian J Physiol Pharmacol 37(1):45–50
33. Rai L, Ram K, Kant U, Madan SK, Sharma SK (1994) Energy expenditure and ventilatory responses during Siddhasana – a yogic seated posture. Indian J Physiol Pharmacol 38(1):29–33
34. Raju PS, Prasad KV, Venkata RY, Murthy KJ, Reddy MV (1997) Influence of intensive yoga training on physiological changes in 6 adult women: a case report. J Altern Complement Med 3:291–295
35. Ray US, Hegde KS, Selvamurthy W (1983) Effects of yogic asanas and physical exercises on body flexibility in middle-aged men. Yoga Rev 3:75–79
36. Ray US, Hegde KS, Selvamurthy W (1986) Improvement in muscular efficiency as related to a standard task after yogic exercises in middle-aged men. Indian J Med Res 83:343–348
37. Ray US, Mukhopadhyaya S, Purkayastha SS, Asnani V, Tomer OS, Prashad R, Thakur L, Selvamurthy W (2001) Effect of yogic exercises on physical and mental health of young fellowship course trainees. Indian J Physiol Pharmacol 45:37–53
38. Ray US, Sinha B, Tomer OS, Pathak A, Dasgupta T, Selvamurthy W (2001) Aerobic capacity & perceived exertion after practice of Hatha yogic exercises. Indian J Med Res 114:215–221
39. Ray US (2010) Development of yoga module for common man. National Yoga Week, p 126
40. Ray US, Pathak A, Tomer OS (2011) Hatha yoga practices: energy expenditure, respiratory changes and intensity of exercise. Evid Based Complement Alternat Med. 2011:241–294. eCAM advanced access published June 21. doi: 10.1093/ecam/neq046.1-12
41. Roland KP, Jakobi JM, Jones GR (2011) Does yoga engender fitness in older adults? A critical review. J Aging Phys Act 19(1):62–79
42. Santaella DF, Devesa CR, Rojo MR, Amato MB, Drager LF, Casali KR, Montano N, Lorenzi-Filho G (2011) Yoga respiratory training improves respiratory function and cardiac sympathovagal balance in elderly subjects: a randomised controlled trial. BMJ Open 1(1):e000085. doi:10.1136/bmjopen-2011-000085
43. Selvamurthy W, Ray US, Hegde KS, Sharma RP (1980) Physiological responses to cold (10°C) in men after six months practice of yogic practice. Int J Biometrol 12:188
44. Selvamurthy W, Nayer HS, Joseph TN, Joseph S (1983) Physiological effects of yogic practice. Nimhans J 1:71–80
45. Selvamurthy W, Ray US, Hedge KS, Sharma RP (1988) Physiological responses to cold (10 °C) in men after six months' practice of yoga exercises. Int J Biometeorol 32:188–193
46. Selvamurthy W, Sridharan K, Ray US, Tiwary RS, Hegde KS, Radhakrishan U, Sinha KC (1998) A new physiological approach to control essential hypertension. Indian J Physiol Pharmacol 42:205–213
47. Seo DY, Lee S, Figueroa A, Kim HK, Baek YH, Kwak YS, Kim N, Choi TH, Rhee BD, Ko KS, Park BJ, Park SY, Han J (2012) Yoga training improves metabolic parameters in obese boys. Korean J Physiol Pharmacol 16(3):175–180
48. Shapiro D, Cook IA, Davydov DM, Ottaviani C, Leuchter AF, Abrams M (2007) Yoga as a complementary treatment of depression: effects of traits and moods on treatment outcome. Evid Based Complement Alternat Med 4(4):493–502
49. Sharma M, Haider T, Knowlden AP (2013) Yoga as an alternative and complementary treatment for cancer: a systematic review. J Altern Complement Med 19 (11):870–875, PMID: 23480018 [Epub ahead of print]
50. Singh P, Singh B, Dave R, Udainiya R (2011) The impact of yoga upon female patients suffering from hypothyroidism. Complement Ther Clin Pract 17(3):132–134
51. Sinha B, Ray US, Pathak A, Selvamurthy W (2004) Energy cost and cardiorespiratory changes during the practice of suryanamaskar. Indian J Physiol Pharmacol 48:184–190

52. Sinha S, Singh S, Monga YP, Saha M, Ray US (2007) Improvement of glutathione and total antioxidant status with yoga. J Alternat Complement Med 13 (10):1085–1090. doi:10.1089/acm.2007.0567
53. Streeter CC, Whitfield TH, Owen L, Rein T, Karri SK, Yakhkind A, Perlmutter R, Prescot A, Renshaw PF, Ciraulo DA, Jensen JE (2010) Effects of yoga versus walking on mood, anxiety, and brain GABA levels: a randomized controlled MRS study. J Altern Complement Med 16(11):1145–1152
54. Telles S, Nagarathna R, Nagendra HR (1995) Autonomic changes during "OM" meditation. Indian J Physiol Pharmacol 39(4):418–420
55. Telles S, Nagarathna R, Nagendra HR (1996) Physiological measures of right nostril breathing. J Altern Complement Med 2(4):479–484
56. Telles S, Singh N, Joshi M, Balkrishna A (2010) Post traumatic stress symptoms and heart rate variability in Bihar flood survivors following yoga: a randomized controlled study. BMC Psychiatry 13(18):1–10
57. Tracy BL, Hart CE (2013) Bikram yoga training and physical fitness in healthy young adults. J Strength Cond Res 27(3):822–830
58. Yadav RK, Ray RB, Vempati R, Bijlani R (2005) Effect of a comprehensive yoga based lifestyle modification programme on lipid peroxidation. Indian J Physiol Pharmacol 49(3):358–362
59. Saha M, Tomer OS, Halder KH, Chatterjee A, Pal R (2013) Effects of yoga on performance of Air Force personnel and correlation with psycho-physiological parameters. Report no. DRDO-DIPAS-DLS-TR-09-2013. DIPAS, DRDO, Min of Defence, Delhi

Technology Translation from Heat Physiology Research

Thiruthara Pappu Baburaj, Bharadwaj Abhishek, Amir Chand Bajaj, Gulab Singh, Pooja Chaudhary, Laxmi Prabha Singh, Medha Kapoor, Usha Panjawani, and Shashi Bala Singh

Abstract

A major effect of high-ambient-temperature exposure on human body is rise in body temperature and sweating, leading to dehydration with loss in electrolytes. Keeping these effects in focus, research was conducted to address the following issues: (a) body dehydration by developing a fluid replenishment drink and (b) auxiliary cooling system for the human body. The outcome of these studies is formulation of replenishment drink DIP-SIP and development of a man-mounted air conditioning system (MMACS).

Replenishment Fluid

Working in high ambient temperature, human body tries to regulate body temperature. The main thermoregulatory cooling mechanism is sweating. In this process, the body compromises with its fluid content. Along with fluids, electrolytes are also lost with sweat. Depletion of body fluids and electrolytes at length leads to various heat-related ailments. Keeping this aspect in focus, the need to develop replenishment fluid was envisaged.

Introduction

Civilians, soldiers, and athletes are predisposed to heat-related illnesses due to prolonged exercise in hot and dry environment. This not only results in loss of body water and thermoregulation compromise [21] but also results in mild conditions like heat cramps and syncope to potentially fatal conditions such as heat exhaustion and heat stroke [35]. Failing to replenish the lost fluid leads to impairment of performance/task. However, recommendations concerning replacement of body fluids are often complicated due to paucity of knowledge [19]. Reports pertaining to long-term exercise and intense operation in warm environment showed increased susceptibility to heat stress maladies due to suboptimum levels of electrolytes and physio-hematological essentials [15]. Thus, high incidences of illness/causalities are found emerging apart from conventional heat sickness [24, 38]. An earlier publication from Adolph also

T.P. Baburaj (✉) • B. Abhishek • A.C. Bajaj • G. Singh
P. Chaudhary • L.P. Singh • M. Kapoor • U. Panjawani
S.B. Singh
Defence Institute of Physiology and Allied Sciences (DIPAS), New Delhi, India
e-mail: drtpbaburaj@yahoo.com; drshashisingh@gmail.com

S.B. Singh et al. (eds.), *Translational Research in Environmental and Occupational Stress*,
DOI 10.1007/978-81-322-1928-6_15,

mentioned that prolonged exercise under heat leads to significant loss in fluid and electrolyte, resulting in dehydration, which impairs endurance capabilities [1].

There had been a considerable focus on thermal stress and body fluid loss and replenishment by researchers for the past several decades [23, 26, 29]. Thermal sweating causes decrease in body fluid with a concomitant increase in osmolality. Such loss in body fluid can become a threat to human life in warm climate, particularly during work [16, 31]. Though endurance and coordination with respect to body water deficit is not well reported in literature, psychophysiological performance decrement is noticed for mission-oriented operations.

International Scenario

Though there are several studies available on various replenishment fluids and hydration strategies for people working in warm climate, yet recommendations/products offering physiological benefits are not developed. Burgess attempted to study the effect of carbohydrate substrate on perceived exertion during prolonged exposure under heat stress [16]. Utter and his co-workers have reported a study on effect of carbohydrate ingestion and hormonal response during heat stress. The findings suggested that physiological response of carbohydrate as replenishment fluid under heat stress require further investigation [39].

Lyons et al. [28] reported that glycerol and water hydration may have some effect on improving thermoregulation during exercise under heat stress. They found that the rectal temperature rise was attenuated and sweating rate was elevated above control levels. However, these thermoregulatory benefits during exercise under heat stress have not been confirmed. There is paucity of literature available on the beneficiary effect of glycerol and water mixture.

McNaughten et al. [30] attempted to assess the ergogenic potential of sodium bicarbonate in high-intensity, competitive cycle ergometry of 1-hour bout. Coyle et al. [21] had shown that ingesting carbohydrate drink during prolonged exercise can delay fatigue. Aim of his study was to test the hypothesis that carbohydrate fluid intake maintained at regular intervals during exercise under heat delays fatigue (as measured by perceived exertion) and decreased need for rehydration after exercise. Unfortunately, the study lacks insight into ergogenic aspect of the fluid [18, 20, 21].

Cade et al. [17] carried out a study to determine the need for fluid replacement during and after exercise in heat. The study showed a large difference in rating of perceived exertion at 2 and 3 h of exercise between the groups exercising with no fluid intake and gap-up fluid intake. These substantial differences re-emphasize the importance of ingesting fluids at regular intervals when engaged in strenuous work under heat [14, 28].

Though the need for an effective replenishment fluid has been identified, identification/development of such a fluid is yet to be achieved to combat heat stress maladies. Sporadic studies in this area showed that during compensable exercise heat stress, thermoregulatory responses were identical regardless of whether volunteers were euhydrated, water hyperhydrated, or glycerol hyperhydrated, and pre-hyperhydration delayed development of body water deficits, if fluids were not replaced during exercise under heat stress [16, 22, 32, 33]. These studies have shown that hyperhydration does not provide advantage for maintenance of hydration during exercise under heat [8, 10, 21, 28, 30, 39].

National Scenario

Most of the studies in heat physiology division were conducted on volunteers from temperate zone. Observations and conclusion from such studies are difficult to match with tropical population, where individuals are always exposed to high ambient temperature. A global standardization is not feasible in these conditions due to diversity. Under such circumstances, heat-physiological study in India with respect to hypohydration is imperative. Innately tropical population has the benefit of enhanced tolerance to heat stress. Referring to Indian subcontinent, a wide range of seasonal variations from extreme heat in

Rajasthan to extreme cold in Himalayas is a topographical characteristic. Variation of hot environment from hot humid to hot dry exposes a large population to high ambient temperature. Though the tropical population has enhanced heat tolerance capacity, heat dissipation process is not satisfactory during extreme heat. Topographically, nearly half of Indian subcontinent is arid or semiarid. "Thar desert" is one of the most populated deserts in world. In hot summer, high ambient temperature of above 50 °C increases aridity of the region. This terrain and hot climate is consequential for present study and deployment of defense personnel at such areas [8, 10].

Defence Institute of Physiology and Allied Sciences (DIPAS) had conducted several studies to combat heat stress under high ambient temperatures. Apart from habitability survey and plotting heat stress problems, this institute also investigated physiological response of hypohydration under heat stress. One of the earlier studies was on changes in body fluid compartment and electrolyte balance during exposure to heat under different levels of dehydration. During forced dehydration to 1 and 2 %, more than 80 % of total body loss was observed from extracellular compartment. With an increase in level of dehydration, fluid loss from intracellular compartments is also increased. A maximum of 14 % plasma volume loss was observed at 3 % and 4 % levels of dehydration. At 3 %, plasma volume (PV) loss reached the upper limit and no further loss at higher dehydration levels was observed. Significant changes in blood and plasma viscosity were observed at 3 and 4 % levels of dehydration. The changes were not significant at 1 and 2 % levels. The mean exercise heart rate recorded at 3 and 4 % resulted in significant increase as compared to euhydration state indicating decrease in work capacity. On partial rehydration of 3 % dehydrated subjects, PV was restored and even slight hyperhydration was observed, but in the 4 % dehydrated subjects, PV was not restored indicating that water is going to intracellular compartment from plasma compartment and more water than required should be replenished for restoration of PV. Partial rehydration significantly reduced the circulatory strain under 4 % dehydration levels. Significant decreases were observed in sweat rate under 3 and 4 % dehydration levels during the work-heat test. A significant increase in sweat rate was also observed during standard work test with respect to partial replenishment. Loss in sweat at 3 % was maximum, while loss in K^+ was maximum at 4 % level of dehydration indicating that 3 % level water loss is from extracellular compartment.

Keeping pace with international research, DIPAS continued studies in the replenishment fluid area with a view to produce a tangible product as replenishment fluid. One of the earlier studies was aimed at determining the efficacy of ergogenic herbal replenishment fluids for improving thermoregulation during endurance in heat. Nannari, Brahmi, Roohafza, and water were the four fluids selected for this study. Statistically, Latin square design was adapted to evaluate the potential of herbal fluid against water as well as electoral [8, 10, 34]. This lab study was conducted in human climatic chamber (HCC) simulated for 34 °C WBGT. The study was conducted on batch of volunteers who underwent heat acclimatization for 8 days. The acclimatization schedule was 50 min submaximal exercise and 10 min rest followed by 50 min submaximal exercise in the simulated environment of HCC. After acclimatization 1 day rest was offered to subjects, and thereafter actual experiment on replenishment fluid was carried out. In this study, 2 % pre-hypohydration was achieved from simulated HCC through moderate exercise for a period of approximately 2 h. Subsequently, volunteers shifted to thermoneutral room for a period of 60 min to stabilize the physiological variables. During this period one of the above replenishment fluids (randomized) was administered equivalent to 1 % body weight and left 1 % residual hypohydration before standard work-heat test (SWHT). Standard work-heat test was conducted in HCC simulated for heat stress for duration of 60 min; it involves initial 20 min rest followed by 40 min submaximal exercise on a bicycle ergometer set for 60 W. In the above study, physiological variables like heart rate, oral temperature, mean skin temperature, and sweat rate were recorded during SWHT. Brahmi preparation intake resulted in lowering the values on core temperature and sweat rate in comparison with other fluids, but these findings were statistically insignificant [10]. Therefore, further studies were continued using different kinds of replenishment fluid, which are being used by people staying in desert and warm coastal areas.

Technology Translation

Subsequent studies at DIPAS looked into the efficiency of several replenishment fluids under heat stress in simulated laboratory environment and actual field conditions. Data from studies carried out in the institute helped us understand several aspects of fluid replenishment. From compiling all the findings, an effort has been made to formulate a replenishment drink. Based on earlier studies and sweat analysis, it was hypothesized that replenishment of only fluid may not be enough to tackle hypohydration conditions. It is also necessary to compensate the electrolytes lost while sweating.

The magnitude of sweat loss incurred during work in a warm environment is dependent primarily on work intensity and duration. Sweating results in loss of electrolytes, mainly sodium (Na^+) and chloride (Cl^-), but also to lesser degree magnesium (Mg^{++}), zinc (Zn^{++}), phosphorous (P^+), copper (Cu^+), iron (Fe^{++}), calcium (Ca^{++}), etc. If sweat loss is replaced only by water, this may lead to a situation where the body contains fewer electrolytes than in the normal state (hypo-osmotic). Commercially prepared sports drinks and energy drinks have varying concentrations of glucose and sodium, ranging from hypertonic to hypotonic with respect to plasma. Apart from sodium and carbohydrate, drinks fail to replenish other electrolytes. Some drinks have certain chemical entities such as caffeine added to make them more acceptable to individuals. These components, especially caffeine, have antagonistic effect on the user. It is seen that consumption of caffeine can lead to further water loss from the body. This added water loss is not desirable under heat stress conditions. If plain water is used to replace sweat loss, dilution of plasma may occur making the person hyponatremic and hypokalemic. In such circumstances, DIPAS realized the need to develop a "replenishment drink" to cater to the armed forces. Depending on national, international, and previous extensive DIPAS scientific studies, a formulation (DIP-SIP) has been finalized.

Various types of drinks are available in market such as sports drinks, carbonated drinks, oral rehydration fluids, etc. Key purpose of these drinks is to provide energy mainly for people engaged in sports. DIP-SIP is a replenishment drink, which is formulated for replacement of lost body water along with the electrolytes. DIP-SIP is devised to alleviate heat-related disorders under extreme environmental conditions such as desert operations. Evaluation of DIP-SIP was extensively carried out by the scientists at DIPAS. The study was conducted in three phases. Phase I and phase II were laboratory studies, while phase III was field trial.

Study Protocol

Lab Trials

Phase I: Taste Evaluation Study

The organoleptic tests were carried out on male and female volunteers alongside army personnel for taste evaluation. Four flavors were tested: lemon, orange, litchi, and mango. Two flavors were recommended by all the participants, viz. orange and litchi.

Phase II: Efficacy Evaluation Study (Simulation Study)

DIP-SIP was further evaluated under simulated heat stress conditions. This study was carried out using "standard work-heat test" (SWHT) protocol, established by this laboratory. These subjects were acclimatized in a human climatic chamber maintained at 45 °C and 30 % RH. Immediately after acclimatization, each subject was exposed to simulated heat stress. They underwent 2 % hypohydration and then supplied with DIP-SIP. The findings consuming DIP-SIP were compared with when replenishment is done with normal water.

Fig. 1 Field study at Thar desert, Rajasthan- in progress

Field Trials

Phase III: Performance Evaluation of DIP-SIP (Replenishment Mix) Under Field Condition

This study was conducted at Pokhran, Jaisalmer, Rajasthan (Fig. 1). The study was conducted on BSF (Border Security Force) personnel at two locations: (a) 6BN Pokhran and (b) Observatory Post at LOC, during peak summer months. The environmental physical factors recorded were maximum globe temperature (GT) 59 °C, dry-bulb temperature (DB) 44.3 °C, and wet-bulb temperature (WB) 28.5 °C. The overall heat stress index (WBGT) recorded was 36.18 °C. Volunteers performed varied intensities of physical activities under heat stress condition. They were categorized into light work, moderate work, and heavy work depending on their intensity.

Observation and Discussion

From the results it became clear that physiological responses in volunteers consuming DIP-SIP remained at lower values compared to the volunteers not consuming DIP-SIP replenishment drink. Heart rate remained at basal level in light working group compared to the control batch (volunteers not drinking DIP-SIP) doing light work. Rise in heart rate was seen at the end of exposure but it was lower in experimental group (Fig. 2). This difference in heart rate was even aggravated in the heavy work group. The experimental batch showed much lesser rise in heart rate than the control group volunteers. When comparing thermophysiological parameters, core body temperature and mean skin temperature showed less value as compared to control batch. Core temperature remained near basal values in light work group. Skin temperature showed lesser increase in heavy work group. Among light and moderate work groups, onset of sweating helped cool the skin better as compared to the control batch. Blood sodium content was measured in all three working groups. In all the cases there was depletion of sodium ions from the body. Among volunteers consuming DIP-SIP, depletion of sodium was less as compared to control batch. Normal range of sodium in body ranges from 135 to 145 mEqiv/l. In all working groups and batches, content of sodium remained within permissible physiological limits. When DIP-SIP was used, sodium content showed a rise (Fig. 3). This rise was within the physiological permissible limit. Thus, it can be stated that consumption of DIP-SIP would be effective and beneficial in counteracting blood sodium depletion due to sweating. When measuring oxygen consumption, it was observed that in volunteers drinking replenishment fluid DIP-SIP, oxygen consumption reduced. Similarly there was a reduction in pulmonary ventilation. From earlier DIPAS studies and scientific findings, it is known that performance is dependent on oxygen consumption and pulmonary ventilation. Among the experimental (with DIP-SIP) batch, both oxygen consumption and pulmonary ventilation reduced. This reduction in oxygen requirement and pulmonary ventilation signifies improvement in physical performance. From this finding it can be stated that DIP-SIP may have some beneficial effects in improving physical performance. This aspect gets further supported by the findings from energy expenditure. It is seen that after consuming DIP-SIP, requirement of energy to do the same intensity of work was reduced. From observations it can be quantified that for carrying out heavy work, volunteer not using DIP-SIP formulation utilized more than 400 kcal energy, whereas to do the same activity, volunteer drinking DIP-SIP utilized much lesser amount of energy (330 kcal) (Fig. 4). These findings again indicate the beneficial effects of DIP-SIP in improving physical performance.

DIP-SIP formulation is in dry and powdered form, packaged in a sachet. One sachet is to be

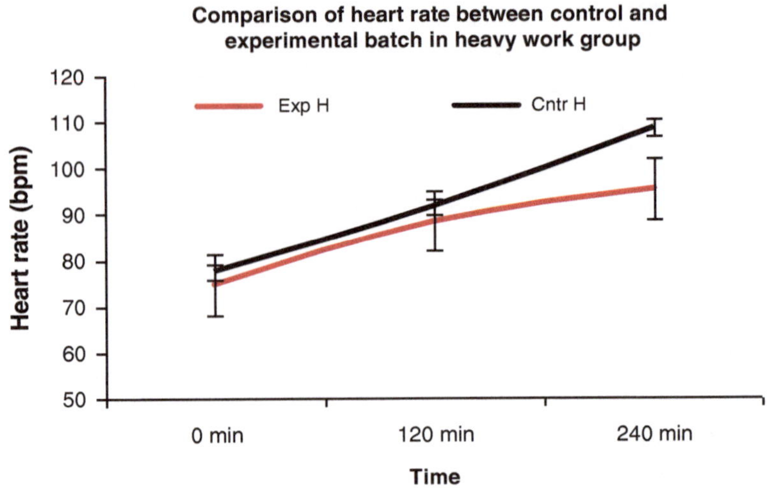

Fig. 2 Comparison of heart rate between control and experiment batch in heavy work group. Data is Mean ± SEM

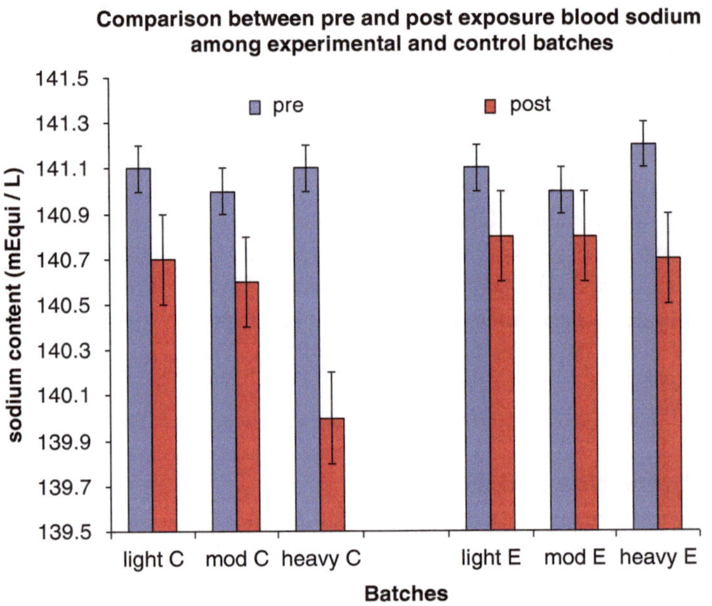

Fig. 3 Comparison between pre and post exposure blood sodium among experimental and control batches. Data is Mean ± SEM

dissolved in one glass of water (250 ml). Different flavors were tested for palatability from which orange and litchi were found most acceptable. DIP-SIP is meant to replenish the electrolyte and fluid loss of individuals exposed to heat stress. Apart from replenishment, an effort is also made to help delay and decrease loss in physical performance of the individual under such stress. The major components of DIP-SIP are electrolytes (viz. sodium, potassium, calcium, iron, phosphorous, selenium, copper, etc.) and carbohydrate.

DIP-SIP can be applicable to a number of circumstances in the military as well as civilian working situations under heat stress, which are as follows:

- Maintaining adequate fluid intake prior to operations, during which voluntary dehydration is probable
- Providing fluid, electrolyte, and carbohydrate replacement during physical work in a variety of environmental conditions where heat stress is imperative

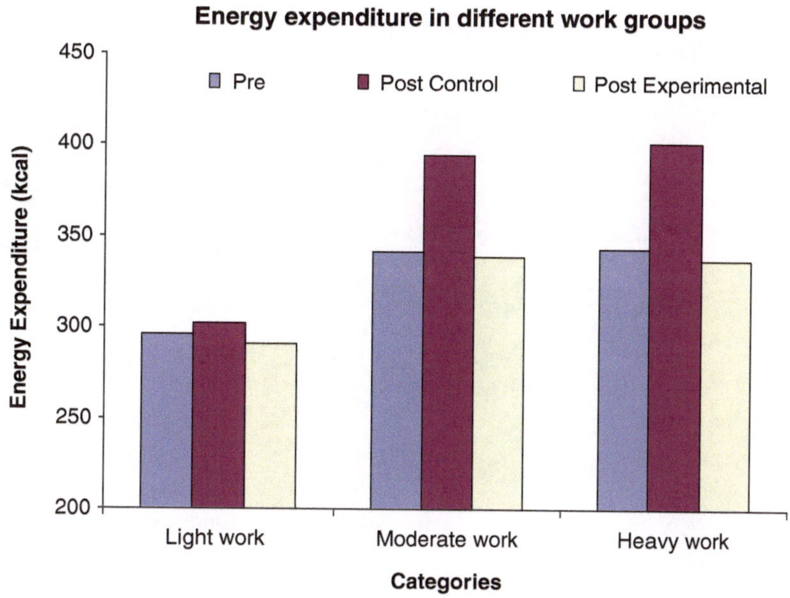

Fig. 4 Energy expenditure in different work groups

- Providing rapid rehydration following heavy or prolonged physical work, thereby facilitating recovery from heat injury
- Additional supplementation of various ions (like copper, calcium, selenium, manganese, magnesium, etc.) to compensate for electrolyte loss during hypohydration
- Encouraging fluid intake due to extra palatability and flavor
- Beneficial in delaying decrement of physical performance.

Auxiliary Cooling Systems

Maintaining body temperature when exposed to high ambient environment is a daunting task for the human body. The innate physiological response to thermoregulate body's core temperature is sweating. Sweating is essential for body's well-being, but not without comprises. Excessive sweating would lead to body dehydration and ultimately heat-related ailments. Hence, it would be much better if the rise in body temperature could be kept under check so as to delay the onset of thermoregulatory responses of body. In such a scenario, auxiliary cooling systems are beneficial, delaying the rise of body temperature. There are several auxiliary cooling methods and devices. The most common in use is phase change. Apart from phase change, technologies like vortex and cooling garments are also in practice [6, 9, 20, 37].

Phase-Change Material (PCM) Cooling Systems

In military, soldiers are exposed to unusual thermal stress during summer months. In peak summer, soldiers from armored corps are particularly affected as the crew compartment (tank turret) temperature rises by 8–10 °C above ambient temperature. Diurnal temperature often exceeds 50 °C and crew members experience heat stress, both due to ambient temperature and also while operating inside the tank. This extreme stress is difficult to compensate by normal physiological heat exchange process. Studies revealed that exposure to hot environment triggers thermoregulatory mechanism with an increase in sweat rate to transfer excess body heat to environment [5]. If such condition persists, the critical problems of heat disorder first manifest as dehydration. High level of dehydration cannot be sustained for a long time because it leads to increase in the body heat storage

Fig. 5 Phase-change material ensemble

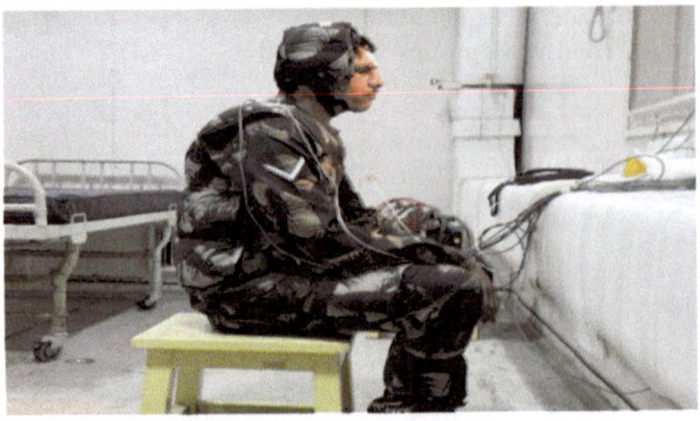

Fig. 6 Laboratory experiments wearing PCM ensemble

drastically. This leads to hyperthermia and related pathological consequences. The main concern in this regard is "microenvironment." Thus, the focus of study related to auxiliary cooling became studies in microenvironments [13]. Earlier studies on phase-change material (PCM) clothing were aimed at protecting troops from heat exposure while working under hot environments either in desert operations or microenvironments. PCM based on hybrid salts binary system of calcium and zinc has been found effective for operations in hot environment. It functions by absorbing heat from surrounding thereby making microenvironment cooler. PCM composition optimizing studies were carried out to render maximum comfort to human volunteers. This PCM material is a combination of three eutectic compositions of temperature: 5–7, 14–16, and 24–26 °C. PCM ensemble has been fabricated to extract heat from microenvironment while a human is exposed to hot environment [4, 5, 13] (Fig. 6).

The combinations of salts are made into pouches and are strategically placed in PCM ensemble. PCM ensemble consists of a sleeveless vest with pockets to hold the pouches (Fig. 5). Along with vest the ensemble also consists of a thigh and head protector.

On performance evaluation of PCM ensemble under simulated human climatic chamber (HCC), the following observations were made: PCM jacket with 10 mm inside insulation was able to combat heat stress at 45 °C for duration of 1 h 30 min under sedentary condition. While doing submaximal exercise, it offered 30 min of cooling comfort. The PCM ensemble was also tested with NBC suit [2, 7, 11, 12, 14]. Cooling comfort duration of 1 h 35 min under sedentary condition was noted under this condition.

Vortex Cooling System (VCS)

Vortex cooling system (VCS) is designed for defense personnel serving under heat stress environments. Vortex tube works under Bernoulli's principle. Rapid swirling motion around a center is called vortex (Fig. 7). Air exhibits vortex and dissipates energy from the core of vortex to periphery. Therefore, cold air gets dissipated from the core of vortex tube and warm air is discharged from periphery as a result of rotational forced vortex. Vortex tube as a cooling device is prevalent in industrial sector, where factory ambient temperature is very high, viz. glass and iron industry, whereas usage of vortex as an integral part of man-mounted cooling system for (NBC) protective ensemble is scarce. To address this area, optimization of vortex cooling was carried out using compressed air. It was achieved with an air compressor along with air filter system used for moisture and

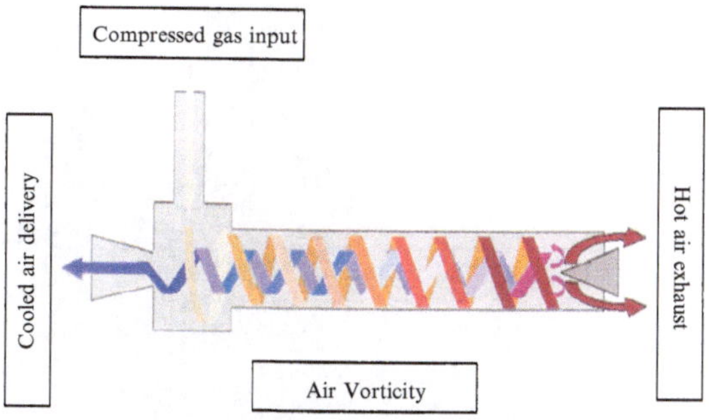

Fig. 7 Schematic diagram of vortex tube

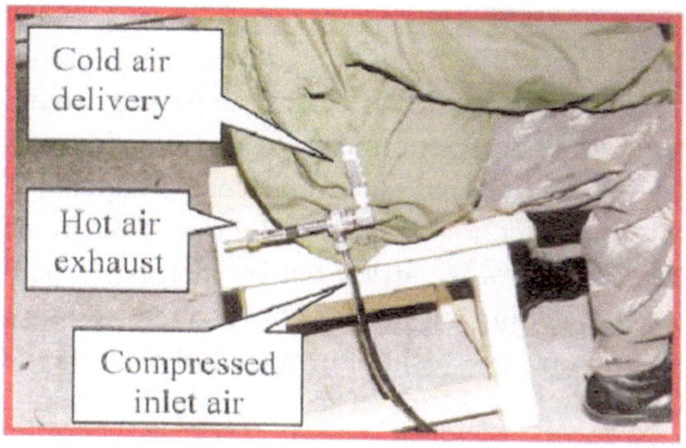

Fig. 8 Vortex cooling system

oil-free nonpulsating compressed air. This compressed air (Fig. 8) was supplied at 2–5 kg sq. cm with airflow of 340–380 l/min (as per requirement). Input air supply to the vortex tube was optimized with respect to cooling air yield. This vortex tube was proficient to provide 19 °C cooled air with 60–80 l/min for circulation and exhausted out hot air at 89 ± 3 °C [3].

Integration of Vortex Tube with NBC Ensemble

Vortex cooling system (VCS) ensemble was taken up for performance evaluation while integrated with NBC suit. Outlet of the cooled vortex air was placed on dorsal and ventral side of NBC jacket towards inner lining to circulate air in microclimate of the wearer. The extra tubing showed a reduction in cooling efficiency. On investigating it was seen that cooled air was mixing with hot air exhaust within microclimate of the ensemble. Therefore, redesigning of vortex tube fitting to NBC suit was carried out to improve

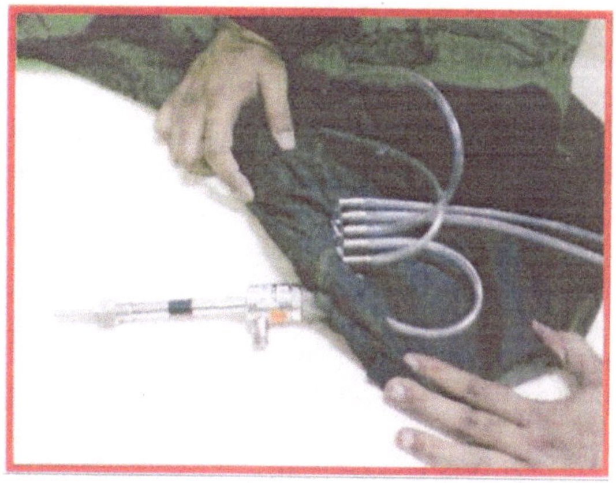

Fig. 9 Integration with NBC suit

its cool air circulation. Vortex tube was attached to NBC jacket with a provision to flush out hot air, thus just allowing cool air to circulate into the microclimate. This was achieved by keeping half of vortex tube (hot air exhaust) towards outside and other half (cold air providing end) within the jacket. This facilitated cooled air circulation within the microclimate of suit (space between NBC suit and combat uniform) through an open tube circulation. The vortex tube was located at lateral side of NBC jacket. This provided comfort and freedom of movement to the user. Placement of the tubes ensured easy exhaust of hot air through dorsal side of NBC jacket [3] (Fig. 9).

Performance evaluation regarding effectiveness of vortex cooling system was conducted in simulated human climatic chamber. This study considered 60 min exposure as safe thermal exposure limits (STEL) for NBC protective ensemble [40]. Duration of exposure and heat stress simulation for 45 °C was chosen for effective performance evaluation of VCS. The exposure limit was worked out as per occupational safety health administration (OSHA), Tech manual III, Chapter 4. Physiological comfort was measured based on core temperature, which should not cross 38 °C apart from other variables. "The American Conference of Governmental Industrial Hygienists (1992)" also stated that workers should not be permitted to work when their deep body temperature exceeds 39 °C. During exposure, volunteers remained sedentary in human climatic chamber to simulate passive thermal stress. Objective indicators of thermal sensation and thermal comfort were taken as mean skin temperature, mean body temperature, and sweat rate.

In this study, all physiological variables of experimental group were found within permissible physiological limits. Oral temperature and mean skin temperature were within the physiologically permissible range. From the parameters it can be said that experimental group had received benefit of cooling using VCS. Therefore, VCS is very effective to provide comfort in an environment of heat stress index equivalent to 34 °C WBGT for duration of 60 min. Thus, vortex cooling system has been found efficient enough to reduce heat stress.

Solid-State Cooling Units

Whole-body microclimate cooling by liquid cooling garments (LCG) has been found to be very effective in alleviating effect of high temperature [6]. For armored corps personnel, microclimate cooling using liquid cooling garment is very effective to counter effects of hot environment. System provides comfortable thermal conditions close to the body surface when surrounding is otherwise thermally uncomfortable. Solid-state cooling system developed on principles of thermoelectric effect has the potential to be used as microclimate cooler. The unit works on principle of thermoelectric cooling based on phenomenon known as "Peltier effect." This effect is observed when direct current is allowed to flow across the junction of two dissimilar metal conductors. The junction region is found to either absorb or release heat, depending on the direction of current. Physiological optimization of required cooling for miniaturization of prototype was carried out in human climatic chamber (HCC), and the volunteers were exposed in HCC while wearing liquid cooling garment (LCG). The LCG was connected to solid-state cooling unit with quick coupling system. Studies went on for 60 to 120 min. The simulation varied from thermoneutral temperature increasing up to 50 °C. It was found that heat extraction of around 250 to 300 W would protect the tank crew from undue body heating at 50 °C.

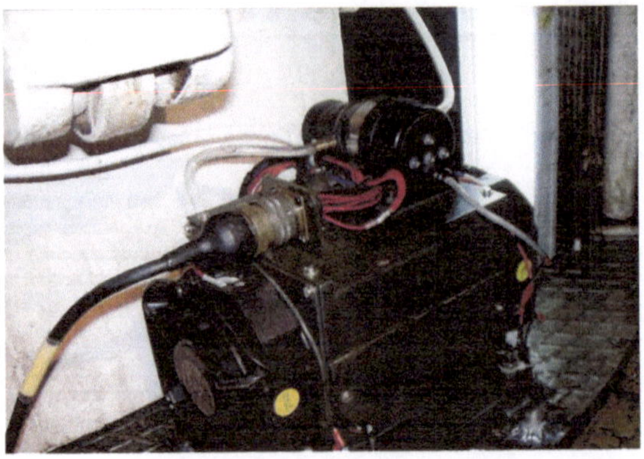

Fig. 10 Solid-state cooling unit fitted in the tank

Thermoelectric cooling is based on "Peltier effect" in which when current is passed around a circuit of different materials, one junction gets heated, while other junction is cooled depending on the direction of current flow. Design of components was carried out by thermal impedance matching principle. The cooling systems, employing this technology, are solid state and CFC gas-free device. These are highly reliable, low power-consuming, and easily maintainable devices. Performance of cooling system has been ensured by innovative design made possible by software developed from mathematical modeling and thermal impedance matching for all the components. Discrete thermal modeling has ensured liquid cooling garments worn under protective clothing (microclimate) that absorb excess metabolic heat from the body. Using thermoelectrically chilled fluid such as water in this unit, cooling is transferred by conduction through silicon tubes carrying cold-water circulation [9].

Technology Translation

The thermoelectric cooler (TE cooler) assembly consists of a thermoelectric module (TEM) core, one liquid positive displacement pump, ambient air fan, and liquid lines. When cooler assembly is turned on, water is pumped through the liquid channels of TE cooler core where it is cooled. The water is then pumped through the internal liquid lines leading to the outlet port. From the outlet port, external liquid lines are connected to LCG trunk, which directs water to the liquid circulating garments (Fig. 10). Chilled liquid

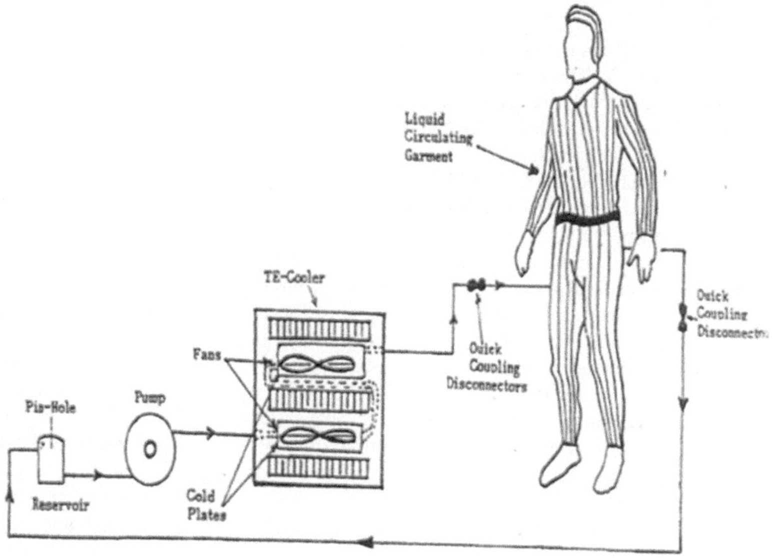

Fig. 11 Schematic diagram of solid-state cooling unit

passing through garment provides body and head cooling to the crew members. During the cooling process, heat is transferred to hot side of TE cooler and this is dissipated to the ambient air. Fans are used for forced dissipation of heat through the hot side of TE cooler. Present solid-state cooling unit has been miniaturized. Dimension and weight has been reduced considerably to 9.5 kg from 24 kg maintaining the efficiency of the cooling.

The use of solid-state thermoelectric cooling is recommended for tank crew during operations and also other vehicles operating in desert environments (Fig. 11). The system may be of use in several industrial environments where workers are exposed to high ambient temperature without any protective devices. Efforts are under way to further miniaturize the cooling system and make it into a portable wearable unit. This portable unit would be meant for individual soldier use. The unit would be wearable on the body and the user will be able to carry the unit comfortably (*man-mounted air conditioning system (MMACS)*).

Operation of the system is reversible, as cooling or heating is dependent on direction of current flow. Thus the system can be used as microclimate heating device for soldiers working in sub-zero ambient by changing the direction of current flow.

References

1. Adolph EF (1947) Physiology of man in the desert. Interscience, New York
2. Baburaj TP, Bharadwaj A, Gupta R, Bajaj AC, Singh G. Physiological evaluation of NBC protective Clothing under heat stress. Report No. DRDO-DIPAS-DLS-TR-01-2012. DIPAS, Delhi
3. Baburaj TP, Bharadwaj A, Gupta R, Bajaj AC, Singh G. (2010) Physiological evaluation of Vortex cooling system. Report no: DIPAS/4/2010. DIPAS, Delhi
4. Baburaj TP, Bharadwaj A, Gupta R, Bajaj AC, Singh G, Kumar R, Meena RR, Sharma PK, Kumar R, Majumdar D. Physiological evaluation of PCM based Cooling suit (Mark –III) under heat stress. Report no: DRDO-DIPAS-DLS-TR00/2/2011. DIPAS, Delhi
5. Baburaj TP, Bharadwaj A, Singh G, Bajaj AC, Banerjee PK. Physiological evaluation of PCM ensemble under heat stress. Report no: DIPAS/3/2006. DIPAS, Delhi
6. Baburaj TP, Bharadwaj A, Singh G, Bajaj AC, Ray US. Heat stress evaluation of tank crew in desert environment. Report no: DIPAS/4/2008. DIPAS, Delhi
7. Baburaj TP, Himashree G, Singh G, Bajaj AC, Banerjee PK. Physiological evaluation of NBC suits (2 Ensembles). Report no: DIPAS/6/2002. DIPAS, Delhi
8. Baburaj TP, Ray US, Ilavazhan G, Bajaj AC, Singh G, Bharadwaj A. Heat stress problems in desert operation and its managements. Report no: DIPAS/17/2006. DIPAS, Delhi
9. Baburaj TP, Ray US, Ilavazhan G, Bajaj AC, Singh G, Bharadwaj A, Chatterjee S. Miniaturization of the cooling unit for solid state cooling garment for tank crew. Report no: DIPAS/10/2006. DIPAS, Delhi
10. Baburaj TP, Singh G, Bajaj AC, Pichan G, Selvamurthy W. Potential status of ergogenic herbal

fluid replenishment on work during heat stress. Report no: DIPAS/06/2000. DIPAS, Delhi

11. Baburaj TP, Singh G, Bajaj AC, Banerjee PK, Selvamurthy W. Thermoregulatory effects of tropical man in Antarctica with reference to special clothing. Report no: DIPAS/3/2001. DIPAS, Delhi
12. Baburaj TP, Gulab Singh, Bajaj AC., Deepk Das, Harish Kumar, Hegde K Himashree G. Physiological evaluation of combat free fall clothing in cold chamber. Report no: DIPAS/11/2004. DIPAS, Delhi
13. Baburaj TP, Bharadwaj A, Gupta R, Bajaj AC, Singh G, Majumdar D, Ravinder Kumar, Rohitash Kumar, Meena RR, Sharma PK. Physiological evaluation of cooling ensemble made up of Phase Change Material (PCM) for hot environment conditions. Report no: DIPAS/2/2010. DIPAS, Delhi
14. Baburaj TP, Gulab Singh, Bajaj AC, Banerjee PK, Selvamurthy W. Physiological evaluation of NBC protective ensembles (4 ensembles). Report no: DIPAS/6/2001. DIPAS, Delhi
15. Bhambhani Y, Buckley S, Maikala R (1997) Physiological and biomechanical responses during treadmill walking with graded loads. Eur J Appl Physiol Occup Physiol 76:544–551
16. Burgess ML, Robertson RJ, Davis JM, Norris JM (1991) RPE, blood glucose, and carbohydrate oxidation during exercise: effects of glucose feedings. Med Sci Sports Exerc 23:353–359
17. Cade R, Spooner G, Schlein E, Pickering M, Dean R (1972) Effect of fluid, electrolyte, and glucose replacement during exercise on performance, body temperature, rate of sweat loss, and compositional changes of extracellular fluid. J Sports Med Phys Fitness 12:150–156
18. Castell LM, Poortmans JR, Leclercq R, Brasseur M, Duchateau J, Newsholme EA (1997) Some aspects of the acute phase response after a marathon race, and the effects of glutamine supplementation. Eur J Appl Physiol Occup Physiol 75:47–53
19. Costill DL, Sparks KE (1973) Rapid fluid replacement following thermal dehydration. J Appl Physiol 34:299–303
20. Cotter JD, Patterson MJ, Taylor NA (1997) Sweat distribution before and after repeated heat exposure. Eur J Appl Physiol Occup Physiol 76:181–186
21. Coyle EF, Coggan AR, Hemmert MK, Ivy JL (1986) Muscle glycogen utilization during prolonged strenuous exercise when fed carbohydrate. J Appl Physiol 61:165–172
22. Gisolfi CV, Duchman SM (1992) Guidelines for optimal replacement beverages for different athletic events. Med Sci Sports Exerc 24:679–687
23. Goldfinch J, Mc Naughton L, Davies P (1988) Induced metabolic alkalosis and its effects on 400-m racing time. Eur J Appl Physiol Occup Physiol 57:45–48
24. Harrison MH (1985) Effects on thermal stress and exercise on blood volume in humans. Physiol Rev 65:149–209
25. Jimenez C, Melin B, Koulmann N, Charpenet A, Cottet-Emard JM, Péquignot JM, Savourey G, Bittel J (1997) Effects of various beverages on the hormones involved in energy metabolism during exercise in the heat in previously dehydrated subjects. Eur J Appl Physiol Occup Physiol 76:504–509
26. Latzka WA, Sawka MN, Montain SJ, Skrinar GS, Fielding RA, Matott RP, Pandolf KB (1997) Hyperhydration: thermoregulatory effects during compensable exercise-heat stress. J Appl Physiol 83:860–866
27. Leithhead CS, Lind AR (1964) Heat stress and heat disorders. Cassel, London
28. Lyons TP, Riedesel ML, Meuli LE, Chick TW (1990) Effects of glycerol- induced hyperhydration prior to exercise in the heat on sweating and core temperature. Med Sci Sports Exerc 22:477–483
29. Maughan RJ, Leiper JB, Shirreffs SM (1996) Restoration of fluid balance after exercise-induced dehydration: effects of food and fluid intake. Eur J Appl Physiol Occup Physiol 73:317–325
30. McNaughton L, Dalton B, Palmer G (1999) Sodium bicarbonate can be used as an ergogenic aid in high-intensity, competitive cycle ergometry of 1 h duration. Eur J Appl Physiol Occup Physiol 80:64–69
31. Mudambo KS, Leese GP, Rennie MJ (1997) Dehydration in soldiers during walking/running exercise in the heat and the effects of fluid ingestion during and after exercise. Eur J Appl Physiol Occup Physiol 76:517–524
32. Pichan G, Baburaj TP, Singh G, Bajaj AC, Sreedhran K (1996) Human performance under various stress conditions. DIPAS Report no: 13/96. DIPAS, Delhi
33. Ramanathan NL (1964) A new weighting system for mean surface temperature of the human body. J Appl Physiol 19:531–533
34. Rastogi RP, Mehrotra BN (1993) Compendium of Indian medicinal plants, vol 2. Central Drug Research Institute, Lucknow; Publications & Information Directorate, New Delhi
35. Riebe D, Maresh CM, Armstrong LE, Kenefick RW, Castellani JW, Echegaray ME, Clark BA, Camaione DN (1997) Effects of oral and intravenous rehydration on ratings of perceived exertion and thirst. Med Sci Sports Exerc 29:117–124
36. Shirreffs SM, Maughan RJ (1997) Restoration of fluid balance after exercise-induced dehydration: effects of alcohol consumption. J Appl Physiol 83:1152–1158
37. Singh AP, Bhatia MR, Baburaj TP, Singh G, Bajaj AC, Chaterjee S, Pandey KG, Selvamurthy W. Development of solid-state cooling Garments for the tank crew. Report no: DIPAS/13/2001. DIPAS, Delhi
38. Singh LP, Kapoor M, Singh SB (2013) Heat: not black, not white. It's gray!!! J Basic Clin Physiol Pharmacol 24:209–224
39. Utter AC, Kang J, Nieman DC, Williams F, Robertson RJ, Henson DA, Davis JM, Butterworth DE (1999) Effect of carbohydrate ingestion and hormonal responses on ratings of perceived exertion during prolonged cycling and running. Eur J Appl Physiol Occup Physiol 80:92–99
40. Van Dokkum W, Van Boxtel LB, Van Dijk MJ, Boer LC, Van der Beek EJ (1996) Influence of a carbohydrate drink on performance of military personnel in NBC protective clothing. Aviat Space Environ Med 67:819–826

Improved Habitability Under Extreme Environments at High Attitude

Ashok Salhan, Sanjeev K. Sharma, Satish Chauhan, and Manan Oza

Abstract

Indian Armed Forces protect the boundaries of our country under diversified climatic conditions like the hot and dry deserts of Rajasthan, the humid forests of the northeast, coastal regions, cold deserts with snowbound areas with extremely low temperatures, and hypoxic conditions at high altitudes and underwater. The hypoxic and cold conditions at high and extreme altitudes lead to many physiological and psychological problems. The inclement weather conditions and the lack of proper facilities further exaggerate the problem. These soldiers fight more against harsh climatic conditions rather than the enemy. The survivability aspects of the soldiers have already been addressed by the Defence Institute of Physiology & Allied Sciences (DIPAS). Problems related to high-altitude exposure have largely been addressed by providing an acclimatization schedule, which is currently followed by the Armed Forces.

Sustainability is now the main priority, so that soldiers can perform better under harsh environmental conditions. It has been observed that most of the physiological problems are due to the uncomfortable living conditions, and if these problems are solved and living conditions are improved, then many physiological problems can be addressed. So the ideas generated by observing these problems were translated into technological solutions by developing products that can minimize these effects, and the lives of soldiers can be made comfortable. Various products developed that utilize nonconventional energy sources have been discussed.

Introduction

Indian Armed Forces protect the boundaries of our country under diversified climatic conditions like hot and dry deserts of Rajasthan, humid forests of northeast, coastal regions, cold deserts with snow-bound areas with extremely low temperature,

A. Salhan (✉) • S.K. Sharma • S. Chauhan • M. Oza
Defence Institute of Physiology & Allied Sciences (DIPAS), Lucknow Road, Timarpur, New Delhi, Delhi 110 054, India
e-mail: ashoksalhan@yahoo.com; sksgeeta27@yahoo.com; satish.dipas@gmail.com; mananoza@gmail.com

S.B. Singh et al. (eds.), *Translational Research in Environmental and Occupational Stress*, DOI 10.1007/978-81-322-1928-6_16,

hypoxic conditions at high altitude and under water. The hypoxic and cold conditions at high and extreme altitude lead to health problems like acute mountain sickness, high-altitude pulmonary and cerebral oedema, deep vein thrombosis, hypertension, renal disorders, frostbite and cold injuries and uncomfortable daily living. The inclement weather conditions, lack of proper facilities and isolation from the families lead to psychological problems like monotony-induced adjustment disorders, post-traumatic stress disorder (PTSD) due to causalities of colleagues, social detachment-induced mood disorders due to lack of communication and stress due to inadequate medical aid, which lead to several health problems. The frequent movement of service personnel further exaggerates the problem. Besides this, there are other logistic problems like delayed casualty evacuation due to inclement weather, avalanches en route, lack of fresh food and limited recreational activities, generators not working efficiently because of extreme low temperature and emission of carbon monoxide (CO) by bukharis leading to CO poisoning. These soldiers fight more with harsh climatic conditions rather than the enemy. The survivability aspects of the soldiers have already been addressed by DIPAS. Problems related to high-altitude exposure have largely been addressed by providing an acclimatization schedule which is currently followed by the Armed Forces.

Sustainability is now the main priority so that soldiers could perform better under the harsh environmental conditions. It has been observed that most of the physiological problems are due to the uncomfortable living conditions and if these problems are solved and living conditions are improved, then many physiological problems can be addressed. So the ideas generated observing these problems were translated into technological solutions by developing products so that the effect of these problems could be minimized and the life of the soldier could be made comfortable. For example, hypoxic problems which are due to lack of sufficient oxygen can be addressed by giving oxygen-enriched environment especially under the conditions of rapid induction during emergency conditions. Heating devices like bukharis and comfortable solar shelters can solve the problem of extreme cold conditions. Further, life can be made comfortable if sufficient power supply is available which can be harnessed from solar and wind energies which are plentiful in these environments.

In this chapter the following products that can help a soldier at extreme conditions have been discussed:

1. Solar shelter 'Sourja' for cold conditions
2. Improved space heating device (bukhari)
3. Compact bukhari
4. Oxygen-enriched shelters at high and extreme altitudes to combat hypoxia
5. Solar snow melter

Solar Shelter 'Sourja' for Cold Conditions

Keeping the living space warm is an essential requirement at high altitude because of extreme cold conditions which requires electric power or kerosene oil. Both these things are scarce in remote mountain areas. Generators do not work efficiently because of oxygen deficiency. So, lack of power makes the cold problem even more severe. Besides this, wires and connections of solar panels become brittle. Batteries get drained out early. On the other hand, natural power sources like wind and solar energy are available in plenty. The concept of using solar energy for heating the living space using the greenhouse effect was used in this shelter. This technology has been in use in greenhouses for the past many years but was never tried for a dwelling unit. Only roof of the dwelling unit was made transparent using polycarbonate sheet to trap the heat. For power generation, nonconventional sources like sunlight and wind energy have been used so as to make the shelter self-sustaining and no additional space heating device like bukhari or heater is required during daytime, thus saving large quantities of kerosene oil for power.

The shelter has been constructed with prefabricated smugly fitted insulated panels. The size of the shelter is 4 mts × 3 mts and can be used as a dwelling unit for two persons. It is a modular structure and can be extended according to requirement. The shelter traps solar heat

Fig. 1 Solar shelter along with vertical axis wind turbine installed at high altitude (Siachen Base Camp)

Fig. 2 Solar shelter along with vertical axis wind turbine installed at high altitude (Leh, Ladakh)

through transparent double-walled polycarbonate sheets on the roof. The heat thus generated in the daytime is retained up to late evening hours due to the insulated material used in the construction of the shelters. The polycarbonate sheet has been fitted at steep angle to prevent snow accumulation, obviating obstruction to heat and light in inclement weather.

Power is generated using solar panels fitted below the transparent sheets. It helps to prolong their life and provide easy maintenance. Additional power is generated from a 1 KW vertical axis wind turbine (VAWT) that provides power backup in the night as well as during days with low sunlight. The power thus generated is stored in four heavy-duty batteries (12 V 100 Ah). The output of these batteries can be used to DC-DC charged smaller batteries of the electrically heated garments (gloves, socks, jackets, etc.) that the soldiers will be provided on field duties in inclement weather, charging of mobile phones, communication system and LED lighting system. An inverter has been provided which supplies 220/230 V AC. Provision has been made for using inverter output for running energy-efficient lighting systems and other devices such as oxygen generator/concentrator etc. The 4 batteries in the shelter can provide about 4,000 W power. The batteries have been kept at room temperature for their better performance.

Vertical axis wind turbine has the advantages of low detectability, ease of installation and utilization of omnidirectional surface winds. These are important considerations from the defence point.

Three prototype shelters have been installed, one at the base camp (11,000 ft) of Siachen Glacier (Fig. 1) and two at Leh (Fig. 2) along with the solar panels and vertical axis wind turbine. Temperature data loggers were fitted inside the shelter to record the hourly temperature

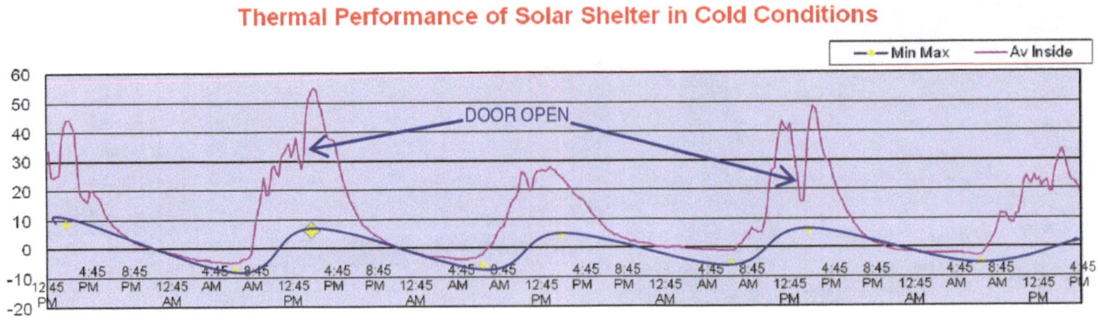

Fig. 3 Thermal performance of the shelter at Leh under cold conditions

for complete 1 year. It was found that comfortable temperature is maintained inside the shelter even when the outside temperature is subzero. In closed experimental shelter the temperature even touched 55 °C, and whenever doors were opened, the temperature dropped, but improved immediately as shown in Fig. 3.

Efforts are being made to use the phase-changing materials to trap the extra heat during daytime which can be used to keep the shelter warm during the night when the temperature falls to subzero.

Improved Space Heating Device (Bukhari) for Cold Conditions

Bukhari or the space heating device is the basic requirement in the cold conditions at high altitude. Army has provided kerosene-based bukharis to soldiers deployed in cold areas. These bukharis consist of a barrel and a burner but have low operating efficiency and have safety hazards. The burner is only a metal plate, and the fuel is gravity fed to the burner by a tank placed at a height. The fuel falls on the preheated plate, vaporizes and is burnt. Because of the improper fuel combustion, there is generation and accumulation of deadly carbon monoxide (CO) gas in the room. The exhaust system is usually a simple pipe which opens outside the roof bent at right angle. The winds are intense in these locations and suddenly change direction. The high wind velocity-induced backdraught extinguishes the flame in the bukhari. The flow of the fuel continues and vapour generated by red hotplate results in reignition and a sudden blast. These bukharis give out only radiative heat and a lot of heat goes waste in the form of exhausted hot flue gases.

In some of the Army units, imported kero-heaters have also been provided. These kero-heaters do not have any exhaust pipe. Any CO produced by these heaters remains in the room, and ultimately CO level in the room can go considerably high.

With the aim of designing and developing safe and efficient system, an improved space heating device (bukhari) and burner have been developed taking into consideration all the problems mentioned above (Fig. 4). The salient features of these devices are:

1. Bukhari is two chambered: one for heating air and other for flue gases so that there is no mixing of harmful flue gases in the air circulated in the room. The CO level in the room remains below the detectable level.
2. It has been designed for maximum extraction of heat so that there is minimum wastage at the exhaust level.
3. There are three level protections from the backdraught, so there is no chance of blast or fire.
4. For better efficiency, it is having both convective and radiative type of heating. For convective heating, airflow is maintained with the help of a DC fan resulting in uniform heating of room for better comfort.
5. To operate the fan, an AC/DC adapter of 6 V 500 mA has been provided. In remote areas where power supply is not available, a charge controller system having one 12 V 7.4 Ah battery has been provided which is charged

Fig. 4 Improved bukhari with backdraught-proof oxygen vent system and triple-burner stove

with a 12 V 10 W solar panel. The charged battery provides backup for about 8 h after the sunset.

6. An improved backdraught-proof exhaust system has been developed as an additional safety measure. It works on Bernoulli's principle and does not allow the air to come down. It is not having any moving part and is thus maintenance-free.
7. The stove is a multiple-wick triple burner that is separated from the shell of the bukhari. The stove burns fuel by wicking action; hence once extinguished, there is no spontaneous reignition and no chance of explosion ever.
8. The tank of the burner has been specially designed to keep the oil temperature low.
9. The burner is efficient and consumes only 500 ml/h of kerosene oil. Being separate from bukhari shell, it can be also used for heating food.

Most of the works reported in the literature [1–3] are on the heat exchangers for different industrial purposes using coal or other types of fuels. In our case, besides a good exchanger, safety was the biggest concern. The system has been provided with multiple levels of fire hazard protections. Secondly, the rarefied air at high altitude heightens the inefficiency of burners for fuel combustion and leads to the accumulation of harmful gases in the room because of incomplete combustion and mixing of flue gases with the room air. This system allows for efficient exhaust of flue gases with no build-up of such gases in the room. Existing systems radiate heat that excessively heats the surfaces facing it while other things in shadow region remain cold. This system gently heats the room air, thus making the whole room comfortable. Maximum heat is extracted from the flue gases so that the exhaust is at less than 50 °C, making it an efficient system.

Structural Details

The final prototype of the bukhari is a double-chambered shell with 30 in. height and 14 in. diameter. The inner chamber is 9 in. in diameter. There are 3 outlets for the hot air, each 3 in. in

Table 1 Results of field trials at high and extreme altitude

Location	Ambient outside temp. (°C)	Size of the room (ft)	Ambient room temp. (°C)	Initial CO level (ppm)	Ambient room temp. after 20 min (°C)	CO level after 20 min (ppm)
Leh (11,000 ft)	−6	18 × 14 × 10	3	0	11	0
South Pullu (14,500 ft)	−10	10 × 8 × 7	7	0	16.8	0
Khardung La 1 (8,380 ft)	−7	10 × 8 × 7	11	0	26.6	0

diameter, which are connecting inner chamber to outside. The legs are 9 in. in height and are detachable.

For extracting the maximum heat, three specially cut horizontal plates have been fitted between the outer and inner chambers. These plates also help to break the impact of the wind. Additional cylinder of 4 in. diameter and 13 in. height has been fitted in the inner chamber to extract additional heat. A conical cap has been given on the top of this cylinder for spreading the fan air uniformly in the inner chamber. The outer chamber provides the radiative heat, while the convective heat from the inner chamber and baseplate is extracted by the downward moving air from a fan fitted on the top of the inner cylinder. The exhaust pipe is connected to the outer chamber for exhaust of flue gases.

As an extra precaution, a backdraught-proof exhaust vent system has been developed and integrated with the improved bukhari which gives further protection and does not allow the wind to come down to the burner. The system is based on Bernoulli's principle. The negative pressure created by the strong winds sucks the air from the lower side; thus the flow of flue gases increases upwards rather than downwards. So the strong winds do not extinguish the burner and thus reduces the chances of blast. It is not having any moving part which further prevents its wear and tear and makes it maintenance-free (Fig. 4). This system can be used even with the existing bukharis and can make them safe by preventing the backdraught-mediated blasts.

Lab and Field Trials

Efficiency of the bukhari was tested both in the lab conditions and in the field condition at high and extreme altitude. Parameters that were taken into consideration were the rise in room temperature in relation to ambient temperature and outside temperature at that time, size of the room, CO level in the room initially and after the specified duration. The room temperature was recorded with the help of four aluminium strips fitted on wooden blocks and kept on four corners of the room. The temperature was recorded with the help of an IR thermometer (Fluke). The average temperature of the four strips was taken as the room temperature. CO, CO_2 and O_2 levels in the room were checked with the help of a Gas Alert Micro 5 IR gas detector (BW Technologies Canada).

Most of the field trials were done at Leh (11,000 ft), South Pullu (14,500 ft) and Khardung La (18,380 ft). In the trial rooms where some heating device like bukhari or kero-heater was already in use, that device was removed, and the room was kept open for some time to bring down the temperature and level of gases close to ambient. Outside temperature and room temperature were recorded initially and after 20 min. The CO level was also recorded initially and after 20 min. The quality of the burner flame and the oil consumption was also recorded. Table 1 shows the results of testing at all the three locations. Figure 5 shows the temperature profile of the room at Leh after different durations. It was observed that comfortable temperature of about 23–24 °C reached within 20 min and this temperature was then maintained for the rest of the experiment for 8 h even when the outside temperature dropped to subzero. Figure 6 shows the temperature profile of the room at different distances from bukhari. In this case it took comparatively more time to reach the comfortable temperature because the outside temperature was subzero from the beginning.

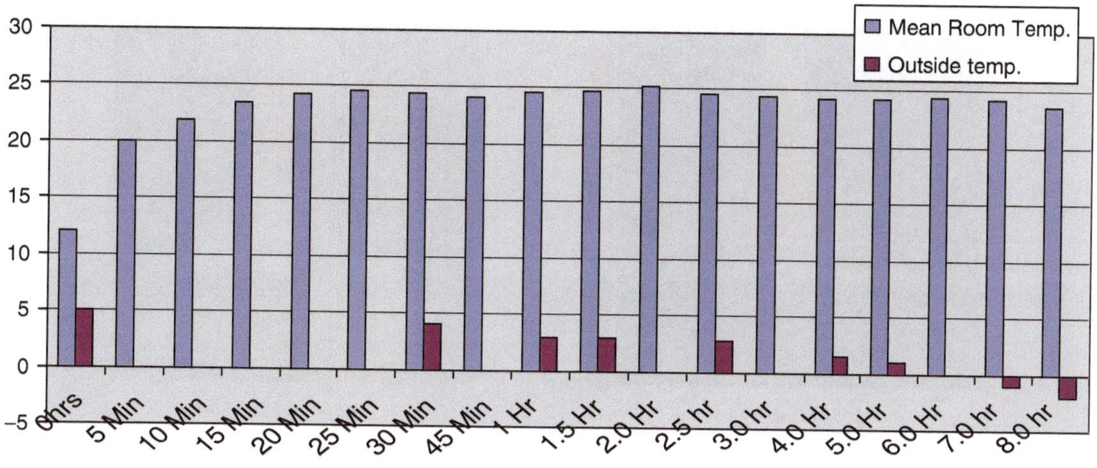

Fig. 5 Temperature profile of the room with bukhari (at Leh)

Fig. 6 Temperature profile of the room at different distances from bukhari (at Leh)

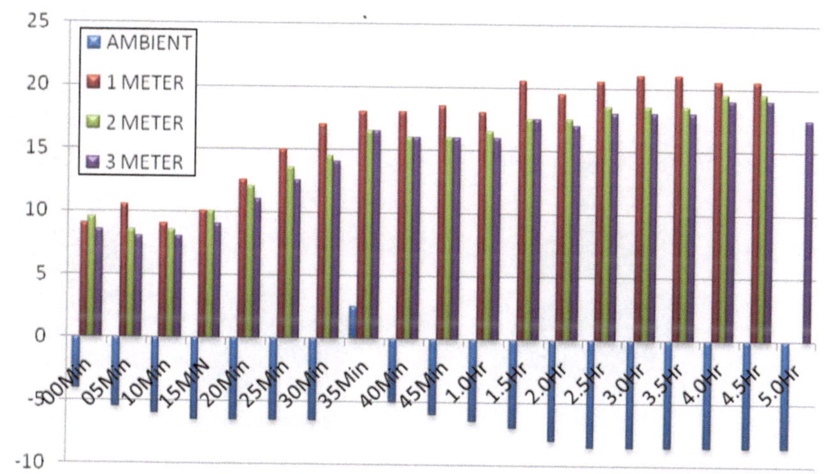

IR Thermography

IR thermography was done to check the thermal performance and the wastage of the heat from the exhaust in the improved bukhari in comparison to the existing bukhari. It was found that the temperature of the outer shell varies from 144 to 192 °C in the existing bukhari, while in improved bukhari, it was around 212 °C in the upper part and around 350 °C near the basal plate. The exhaust pipe temperature was 186.35 ° C in existing bukhari as compared to 38.1 °C in improved bukhari showing that maximum heat had been extracted and there was very little wastage of heat in flue gases in improved bukhari (Fig. 7).

So the improved Bukhari is totally safe for extraction of harmful gases out of the room, and there is no chance of backdraught-induced fire or blast. One bukhari is best suited for a volume of about 1,000–1,200 cubic ft, i.e. for a room size of 10–12 ft × 10 ft × 10 ft in the cold or extreme cold conditions for maintaining a comfortable temperature of about 20–25 °C. For a bigger room, more number of bukharis should be used accordingly.

Compact Bukhari

The above-mentioned improved bukhari saves lot of energy as most of the heat is extracted

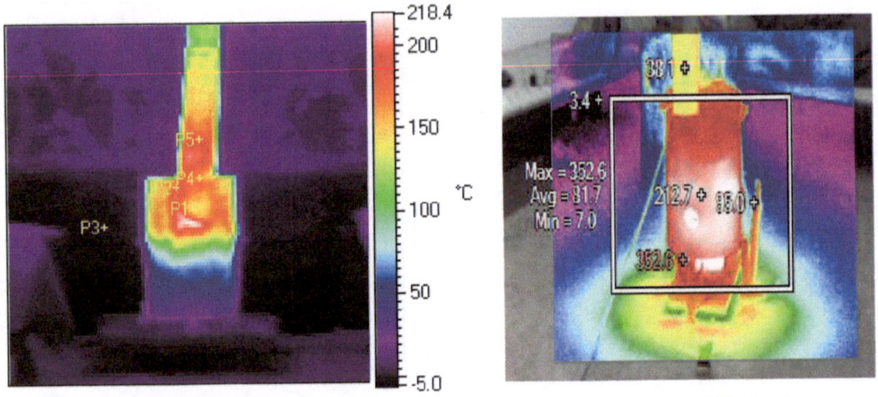

Fig. 7 IR thermograms of existing bukhari and improved bukhari

Fig. 8 Compact bukhari and snow melting attachment (separate and assembled)

by the improved design as shown by the IR thermographs. Still a little heat is wasted through the exhaust pipe.

A compact bukhari has been developed which utilizes the heat of the exhaust pipe to melt the snow or to heat the water. For this an additional attachment has been fabricated which can be fitted on the compact bukhari.

The bukhari has been made compact and lighter in weight. Fan is fitted on one side, and hot air comes out from two openings on the opposite side (Fig. 8).

Oxygen-Enriched Shelters for High Altitude

Indian Army soldiers deployed at high-altitude areas spend 6–14 days for acclimatization depending upon the altitude of their deployment. Similarly, Air Force pilots regularly perform duties at high altitude. An acclimatization of few days is not convenient for a stay of few hours that they spend at high altitude during loading/unloading of aircraft or short-duration landings undertaking sorties.

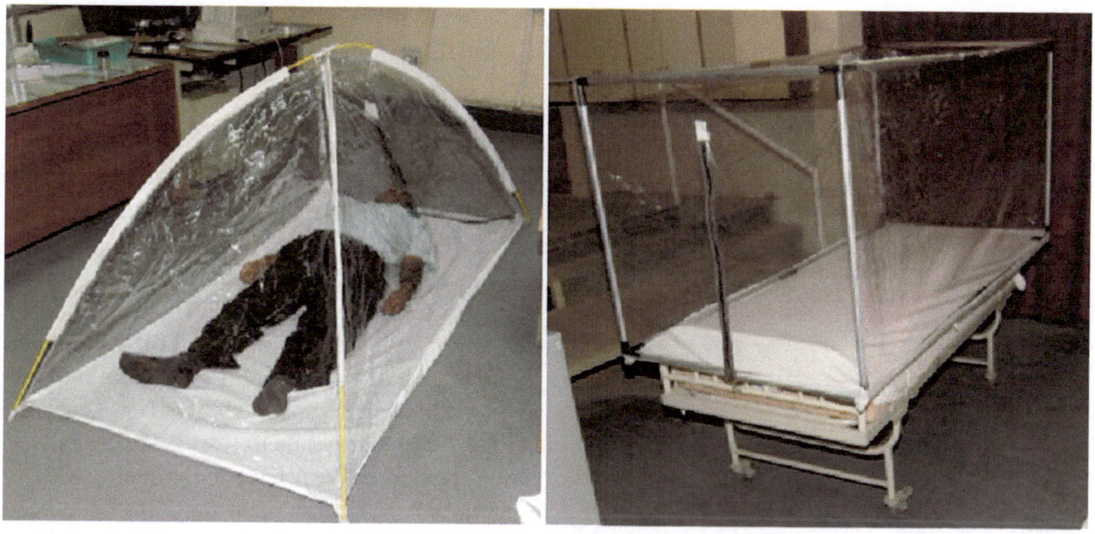

Fig. 9 Oxygen tents for single person

Most of the work by West, Gerard et al., McElroy et al. and Luks et al. [4–12] has shown that oxygen enrichment improves the performance at high altitude. It also improves sleep and subsequent daytime performance at high altitude. It has been reported that increasing the oxygen concentration by 1 % reduces the effective altitude by 300 mts. The highest recommended safe limit for oxygen saturation is around 30 %.

Our experiments also confirmed that inhaling 27–30 % oxygen is sufficient to maintain a SpO_2 level at 92–94 %. Thus, oxygen enrichment of the ambient air to 27–30 % can take care of a large number of high-altitude maladies associated with rapid induction to high altitude. Such oxygen-enriched shelters/devices can be of great help to the pilots staying at high altitude in emergency situations and also to the VIPs on short-duration visit to such areas. These shelters can be of great use in case of exigencies or warlike scenario requiring emergent deployment of forces at high altitude. Three types of oxygen enrichment systems have been developed.

Oxygen Tents for a Single Person

These tents are meant for a single person and consist of a collapsible frame and a transparent tent which can be used on ground or on the bed. Two prototype shells of single bed tents have been developed (Fig. 9). These are 2 mts long, 1 mt wide and 1.5 mts in height. Since the volume is quite less, minimum oxygen will be required to enrich this type of tent. A sensor cum controller has been developed to maintain the desired level of oxygen in the tent.

Personalized Oxygen Inhalation System

Personalized oxygen inhalation system can be used to perform short-duration emergent operations to minimize the effect of hypobaric hypoxia. It is meant for a single person on the move. It has a lightweight carbon composite cylinder of 2 l water capacity filled with medical grade oxygen at 300 bar pressure (600 l of oxygen gas). Flow regulator demand valve assembly adjusts the flow of oxygen in steps of 0.5 l (range 0.5–2 l) and 1 l steps from 2 to 6 l/min. The flow of oxygen occurs only during inspiration using silicone mask or nasal prongs, thus saving lot of oxygen (Fig. 10). Personalized oxygen inhalation system will be of great use during the ascent of troops to the forward posts and will help in minimizing the morbidity, reducing the time taken for ascent and improving the load carrying capacity.

It was observed that at an of altitude of 11,000 ft at Leh, a flow rate of 0.5 l/min is sufficient to maintain SpO_2 of about 95 % and at 18,000 ft at Khardung La top, a flow rate of 1–1.5 l/min is sufficient to maintain SpO_2 of around 90 % in unacclimatized individuals. The flow rate can be adjusted manually by the user to maintain his/her SpO_2 level above 90 %. Efforts are being made to control the flow rate automatically by the feedback from SpO_2 level.

Fig. 10 Personalized oxygen inhalation system

Larger Oxygen-Enriched Shelters for Multiple Number of Persons

These are useful for more number of persons as the complete room is enriched. These can be used by Air Force pilots during their short stay at high altitude so that there is no ill effect of hypobaric hypoxia on their cognitive ability. These can also be used in the hospitals to treat AMS patients to bring them to lower effective altitude.

Four oxygen enrichment facilities have been installed: three at General Hospital, Leh, and one at Air Crew Resting Room (ACRR), Leh (Fig. 11). Oxygen generator of 40 l/min capacity has been installed at General Hospital. Ducting of the rooms was done for the supply of the oxygen. Oxygen sensors and controllers were installed to maintain the set value of the oxygen in the rooms. In ACRR supply of oxygen was maintained with the help of 4 cylinders of 40 l water capacity as shown in Fig. 11.

Solar Snow Melter

The Army posts located in remote mountains do not have electricity. Kerosene stoves are used to melt the snow as the liquid water is mostly flowing in the valleys far below. Fuel has to be airlifted most of the time adding to the cost and

Fig. 11 Oxygen enrichment facility at Air Crew Resting Room, Leh Airport

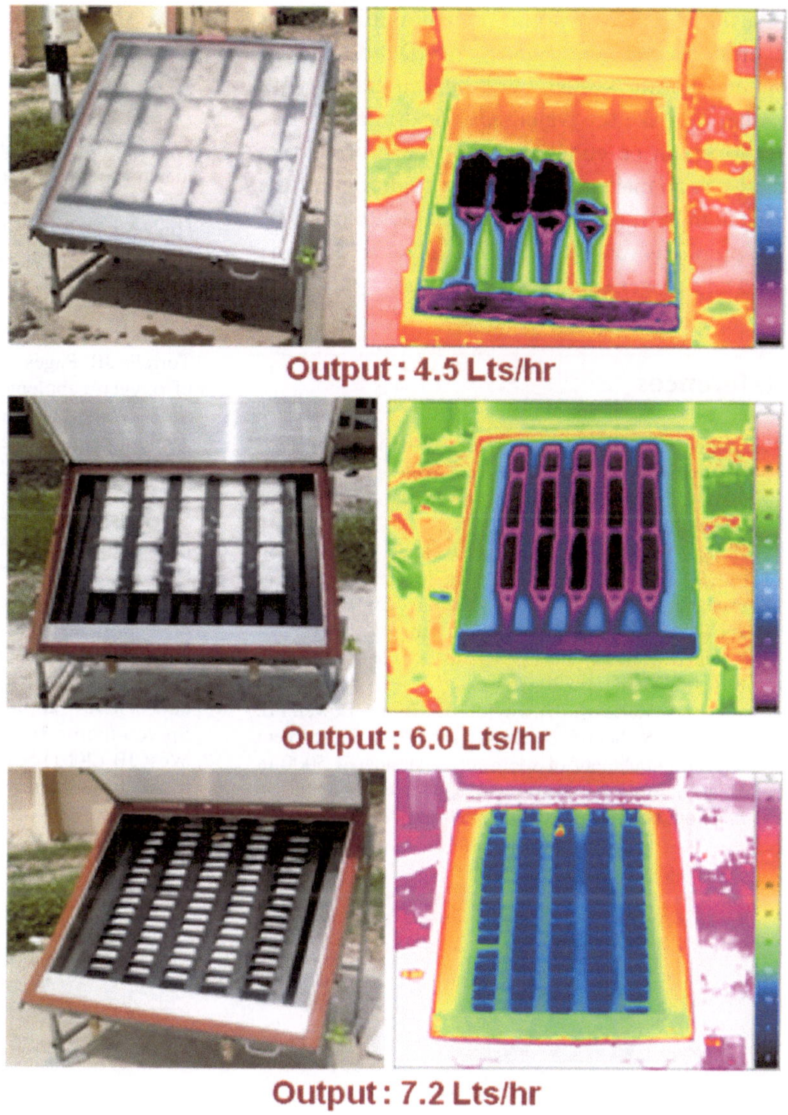

Fig. 12 Three prototypes of solar snow melter and their IR images

logistic problems. To overcome these problems, a solar snow melter has been developed which will be able to meet requirement of water to a great extent at snowbound pickets (>18,000 ft) in high-altitude locations, thus saving cost. The snow melter lets the smaller wavelength IR (infrared) rays in and then converts them to longer wavelength IR rays that cannot escape. Infrared radiation has the right energy to melt the snow. The melter consists of an outer box made of GI. The inner container is made of SS to prevent any rusting in the drinking water. The heating sheet is made up of copper which is covered by polycarbonate sheet. The polycarbonate sheet helps in trapping the heat by the black copper sheet. The experiments done at Delhi using ice shaved by ice shaving machine showed that the temperature inside went up to 119 °C in 30 min. The temperature data obtained by IR thermographs helped to improve the designs of the inner copper sheet for better performance.

Three prototypes were developed and tested in the lab (Fig. 12). The final prototype which generates about 7.2 l/h was tested at high altitude (Khardung La, 18,380 ft). Experiments showed that the melter is working and providing water even in cloudy conditions. A DRDO-developed water filtering system will be integrated with melter to make the water safe for drinking.

So it can be concluded that these protective devices can make the life of the soldier comfortable at high altitude and can help tackle many of the problems related with high altitude, extreme cold and power supply requirements.

References

1. Bingwen Zhang, Yingjin Zhang (2010) A new type heat exchanger for coal burning boilers. In: International conference on future power and energy engineering, pp 11–14
2. Zhang Jie, Ren Yan, Zhang Li Hong, Zhang Kang (2010) Performance evaluation of heat pipe heat exchanger. In: Third international conference on knowledge discovery and data mining, pp 11–15
3. Kevin D, Rafferty PE (1998) Heat exchangers. GHC Bulletin, March, pp 20–26
4. Gerard A, McElroy MK, Taylor MJ, Grant I, Powell FL, Holverda S, Sentse N, West JB (2000) Six percent oxygen enrichment of room air at simulated 5000 m altitude improves some aspects of neuropsychologic function. High Alt Med Biol 1:51–61
5. Luks AM, van Melick H, Batarse RR, Grant I, West JB (1998) Room oxygen enrichment improves sleep and subsequent day-time performance at high altitude. Respir Physiol 113:247–258
6. McElroy MK, Gerard A, Powell FL, Prisk GK, Sentse N, Holverda S, West JB (2000) Nocturnal O_2 enrichment of room air at high altitude increases daytime O2 saturation without changing control of ventilation. High Alt Med Biol 1:197–206
7. National Fire Protection Association (1999) Standard for hypobaric facilities, 1999 edn. National Fire Protection Association, NFPA 99B, Quincy
8. Silva-Urra JA, Urizar C, Basualto-Alarcón C, Torrella JR, Pagés T, Behn C, Viscor G (2011) Effects of oxygen supplementation on acute mountain sickness symptoms and functional capacity during a 2-kilometer walk test on Chajnantor plateau (5050 meters, Northern Chile). Wilderness Environ Med 22(3):250–256
9. West JB (1995) Oxygen enrichment of room air to relieve the hypoxia of high altitude. Respir Physiol 99:225–232
10. West JB (1996) Prediction of barometric pressures at high altitudes with the use of model atmospheres. J Appl Physiol 81:1850–1854
11. West JB (1997) Fire hazard in oxygen-enriched atmospheres at low barometric pressures. Aviat Space Environ Med 68:159–162
12. West JB (2001) Safe upper limits for oxygen enrichment of room air at high altitude. High Alt Med Biol 2:47–51

Inhaled Nitric Oxide Therapy for Treatment of High-Altitude Pulmonary Edema

G. Himashree, R.K. Gupta, B.A.K. Prasad, and Shashi Bala Singh

Abstract

High-altitude pulmonary edema (HAPE) is a severe form of acute mountain sickness. It may be fatal if not diagnosed well in time. Treatment includes descent to low altitude, supplemental oxygen, and calcium channel blocker. Endogenous production of nitric oxide in the lungs has been found to be reduced in individuals susceptible to HAPE. Administration of inhaled nitric oxide and oxygen combination has been found to be more effective in improving tissue oxygenation than giving oxygen or nifedipine. Additional benefits are less toxicity and rapid action due to inhalation mode of therapy. Commercially available inhaled nitric oxide delivery systems are ventilator-based, designed for patients of respiratory failure who require ventilatory support. However, HAPE patients breathe spontaneously and do not require ventilatory support. Hence there was a need for an indigenous NO delivery system tailor-made for HAPE patients. Two prototypes have been developed and currently undergoing clinical trials.

Pathophysiology of High-Altitude Pulmonary Edema (HAPE)

HAPE is a noncardiogenic high-permeability edema characterized by alveolar fluid with a high concentration of protein [67] that develops in otherwise healthy individual after 24–72 h of exposure to altitude above 2,400 m. HAPE can be diagnosed clinically and usually confirmed by radiographic demonstration of patchy distribution of edema in peripheral regions of the lung [37]. Incidence depends on rate of ascent and altitude reached which varies from 2 to 15 % [14, 26, 61]. Mountain climbers and skiers who have a previous history of HAPE are susceptible to unpredictable recurrence when exposed to high altitude [41, 49]. Active young men are more susceptible to HAPE compared to women. Other risk factors include strenuous exercise, cold weather, and recent respiratory tract infection. The symptoms and signs are cough, tachypnea, tachycardia, orthopnea, cyanosis, rales, and frothy pink sputum [28, 29, 62].

G. Himashree (✉) • R.K. Gupta • B.A.K. Prasad
High Altitude Physiology Group, Defence Institute of Physiology and Allied Sciences (DIPAS), Timarpur, Delhi
e-mail: himashree@yahoo.com

S.B. Singh
Defence Institute of Physiology and Allied Sciences (DIPAS), New Delhi, India
e-mail: drshashisingh@gmail.com

S.B. Singh et al. (eds.), *Translational Research in Environmental and Occupational Stress*, DOI 10.1007/978-81-322-1928-6_17, © Springer India 2014

Symptoms usually begin two to four days after a rapid ascent, often during nighttime rest.

An excessive rise in pulmonary artery pressure has been demonstrated by invasive [36, 38, 52] and noninvasive [7, 59] measurements at high altitude possibly due to high sympathetic activity [57] in HAPE patients. Increased concentration of RBC and protein in bronchoalveolar lavage fluid are observed in the early stages of HAPE before the inflammatory cytokines are present [60, 67] suggesting that mechanical injury to the pulmonary capillary bed (stress failure) may be important in development of HAPE. Because hypoxic vasoconstriction takes place in the precapillary arterioles, it is uncertain how capillaries might be exposed to high pressure. A mechanism initially proposed by Visscher [71] and later modified by Hultgren [37] is that hypoxic pulmonary vasoconstriction is uneven in HAPE [37]. If this were true, some parts of the capillary bed would be protected by upstream vasoconstriction, whereas the portion of the capillaries that did not have upstream vasoconstriction would be exposed to high pressure and mechanical stress injury [39] thus may be responsible for occurrence of HAPE. Animal studies have shown that hypoxia results in increased spatial heterogeneity of pulmonary perfusion, suggesting that hypoxic pulmonary vasoconstriction is inherently uneven in the mammalian lung [33]. Functional magnetic resonance imaging of HAPE susceptible has shown increased pulmonary blood flow heterogeneity in acute hypoxia, consistent with uneven hypoxic pulmonary vasoconstriction [35]. There is evidence that some degree of asymptomatic alveolar fluid accumulation may represent a normal phenomenon in healthy humans shortly after arrival at high altitude. Whether this fluid accumulation is cleared or whether it progresses to HAPE depends on dynamics of the quantity of liquid escaping from the pulmonary vasculature and the rate of its clearance by the alveolar respiratory epithelium. The former is directly related to the degree of hypoxia-induced pulmonary hypertension, whereas the latter is determined by the alveolar epithelial sodium transport. This is further determined by the presence of impaired pulmonary endothelial and epithelial NO synthesis and/or bioavailability which causes exaggerated hypoxic pulmonary vasoconstriction and, in turn, capillary stress failure and alveolar fluid flooding. Acute pulmonary hypertension therefore appears to be the sine qua non but may not always be sufficient to induce HAPE, and defective alveolar fluid clearance may represent a second important pathogenic mechanism [58]. The role of inflammation being one of the causative mechanisms has not been confirmed in any of the current studies. Presence of inflammatory markers is more an effect than cause of the edema fluid.

Nitric Oxide (NO) and Respiratory Physiology at High Altitude (HA)

Exposure to high altitude initiates various physiological adjustments required for acclimatization. When oxygen availability is reduced at high altitude, an increase in blood flow could potentially improve oxygen delivery to tissue. However, the pulmonary vasoconstriction response to hypoxia decreases blood flow in the lungs. This response probably evolved at sea level to maintain gas exchange by redistribution of blood flow from temporary small, poorly oxygenated to better oxygenated areas of the lung. At high altitude, the entire lung is always hypoxic, and the resulting general vasoconstriction does not redistribute blood flow; instead, it increases pulmonary arterial pressure, sometimes causing pathological remodeling of the heart and lung. However, many people live at high altitude without pulmonary hypertension, which suggests that another factor may intervene to maintain blood flow when the blood carries less oxygen and the usual vasoconstriction response increases pulmonary resistance. This factor may be NO, a vasodilator found in high concentrations in the lungs of high-altitude natives, particularly Tibetans. A wealth of studies support a key role for NO in determining basal pulmonary vascular tone at sea level and in affecting the hypoxic vasoconstriction response. Nitric oxide (NO) is now known to be a ubiquitous biomolecule having multiple vital roles to play in the body. NO is synthesized from the amino acid arginine by the enzyme nitric oxide synthase (NOS) of which three distinct isoforms exist, i.e., eNOS found in endothelial cell, nNOS in neurons, and iNOS induced in different cells as a response to endotoxins or mediator of inflammation. All the three isoforms of NOS are present in the respiratory tract [44, 70]. NO has been identified in the transduction mechanism for the soluble guanylate cyclase with potential role in a number of tissues [46]. In the respiratory tract NO plays an important signaling role in the physiological control of airway function and in the pathophysiology of airway diseases [2, 6, 42, 53, 77]. Studies of human are consistent and have demonstrated that (1) NO is critical in regulating basal pulmonary vascular tone [13, 66]. (2) Inhaling gas mixture with high concentrations of NO diminishes hypoxic pulmonary vasoconstriction at sea level [9, 22] and lowers pulmonary artery systolic pressure at high altitude. (3) Inhibiting NO synthesis exacerbates hypoxic pulmonary vasoconstriction and polymorphisms of the eNOS gene have been associated with HAPE [19].

Exhaled Nitric Oxide

NO is produced endogenously within the upper and lower respiratory tract and can be measured in exhaled air by chemiluminescence analyzer [25]. The measurement of exhaled NO is critically determined by ventilation and blood flow [54]. Studies have shown that HA decreases exhaled NO level in normal subjects and reduced levels are found in those who are susceptible to HAPE compared to control group under hypoxic condition [8]. Study on Tibetans has shown higher exhaled NO compared to lowlanders at same altitude suggesting a higher pulmonary blood flow in them [34].

Indian High-Altitude Natives (Ladakhi Population)

Study on Ladakhi population has shown higher levels of exhaled NO compared to lowlanders at same altitude [24]. A significantly lower end-tidal carbon dioxide level has been observed in these high-altitude natives (HAN) compared to lowlanders (LL) at 3,200 m (Fig. 1) indicating their higher pulmonary ventilation.

Significantly higher pulmonary ventilation in HAN shows their better adaptation to hypoxia compared to LL. A significantly higher level of exhaled NO is maintained despite higher ventilation in HAN (Fig. 2) indicating a greater endogenous production of pulmonary nitric oxide in HAN as compared to LL which is in accordance with a previous study on HAN of different geographical region [8]. HAN showed a better acclimatization status as they maintain lower pulse rate (Fig. 3), and despite lower end-tidal carbon dioxide levels, their exhaled NO levels

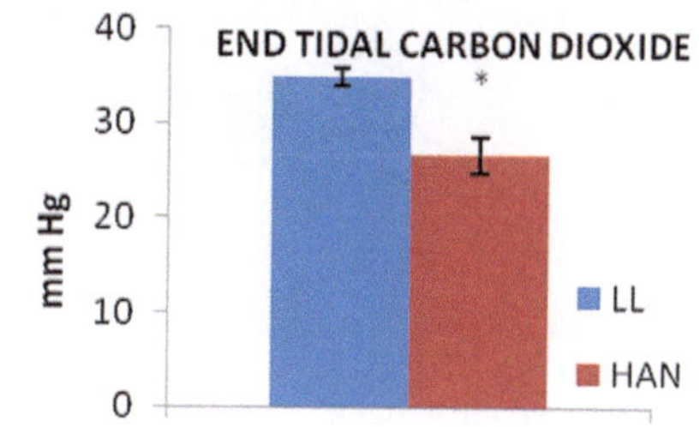

Fig. 1 Comparison of End Tidal Carbon-dioxide (*$ETCO_2$*) levels of lowlanders (*LL*) with high altitude natives (*HAN*) at 3,200 m. Values are presented as Mean ± SD. *$P < 0.05$

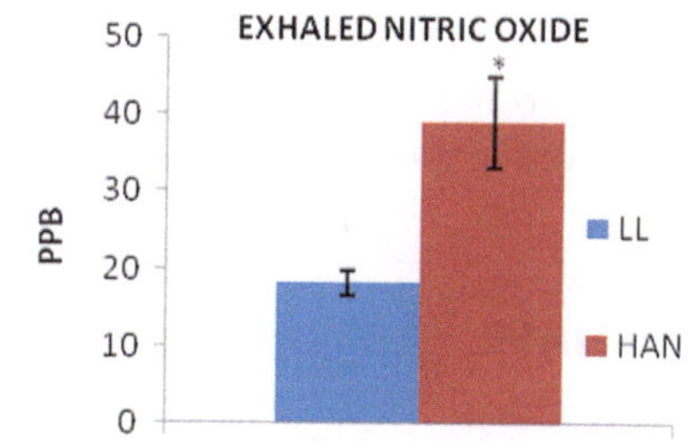

Fig. 2 Comparison of Exhaled Nitric Oxide (*ENO*) levels of lowlanders (*LL*) with high altitude natives (*HAN*) at 3,200 m. Values are presented as Mean ± SD. *$P < 0.05$

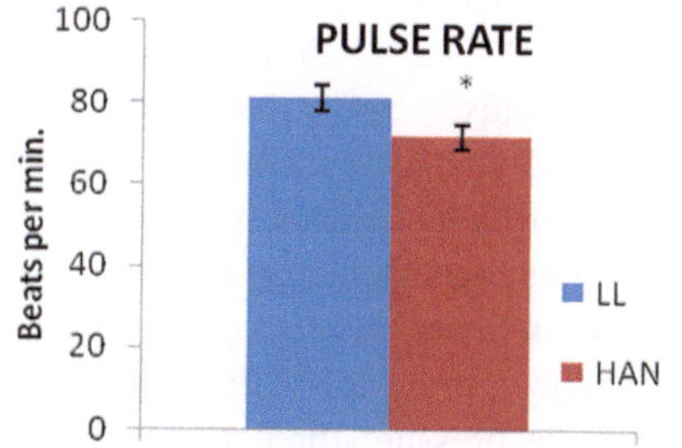

Fig. 3 Comparison of Pulse rate (*PR*) of lowlanders (*LL*) with high altitude natives (*HAN*) at 3,200 m. Values are presented as Mean ± SD. *$P < 0.05$

were higher compared to LL at same altitude. Since in response to hypoxic stress, the rise in sympathetic activity is less and pulmonary ventilation is more in HAN, while LL showed more rise in compensatory sympathetic activity in response to similar hypoxia indicating less cardiorespiratory reserve.

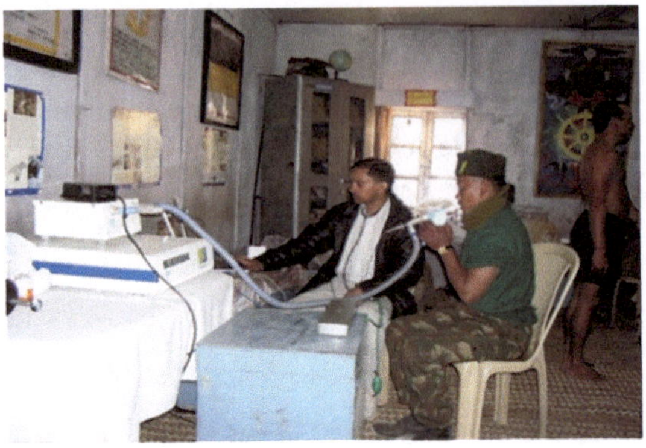

Measuring exhaled nitric oxide

Reduced exhaled NO may be related to altered pulmonary epithelial NO synthesis [19]. Pulmonary hypertension is associated with diminished expression of endothelial nitric oxide synthase. It is possible that decrease expression of nitric oxide synthase may contribute to pulmonary vasoconstriction and to the excessive growth of tunica media observed in this disease [23].

NO and HAPE

An animal study using NOS gene transfer to the airway demonstrates elegantly that increasing NO decreases hypoxic pulmonary vasoconstriction [10]. Moreover, sea-level natives exposed to acute hypoxia (who have relatively high levels of exhaled NO) have less hypoxic pulmonary vasoconstriction [11, 20]. Administration of the NOS antagonist NG-monomethyl-L-arginine (L-NMMA) during hypoxia increases pulmonary artery pressure and vascular resistance [9]. Furthermore, it has been demonstrated that the exogenous administration of 40 ppm (ppm) NO in hypoxic subjects prone to HAPE evokes a decrease in pulmonary pressure three times greater than the decrease in HAPE-resistant subjects. These findings suggest that reduced endogenous NO synthesis in HAPE-susceptible individuals may contribute to their heightened hypoxic pulmonary vascular response. Scherrer in 1996 reported that inhaled NO decreased pulmonary artery pressure and improved ventilation-perfusion mismatch in HAPE-prone subjects exposed to high altitude [59]. There is evidence that oxygen and NO cause pulmonary vasodilatation through the activation of different K + channels in the pulmonary artery smooth muscle. Therefore, there are theoretical grounds to believe that the vasodilatory effect of oxygen and NO could be additive [1]. Further studies also have shown that inhaled NO with oxygen gives immense therapeutic benefit in patients of HAPE [17, 21, 22, 45, 56, 59, 72]. Based on these facts NO and oxygen delivery system has been developed by DIPAS and R&D Engineers, Pune, for treatment of HAPE.

Inhaled Nitric Oxide (iNO) for Treatment of HAPE

Currently HAPE is treated with supplementary oxygen and descent to lower altitudes as soon as feasible or, when descent is not possible, simulated descent with the use of a portable hyperbaric chamber. When neither descent nor simulated descent is possible, administration of calcium channel blockers like nifedipine may help [7, 49]. Attenuating pulmonary hypertension with nifedipine is an effective prevention and carries little risk [27, 50]. The administration of the carbonic anhydrase inhibitor acetazolamide may be useful since it causes bicarbonate diuresis and respiratory stimulation [49]. Without the administration of oxygen, rest, and descent to a lower altitude, death may result.

At this juncture inhaled NO has emerged to be a new, effective, and safe modality of treatment for HAPE. The role of endothelial dysfunction to explain the excessive pulmonary hypertension in HAPE is of burgeoning interest. The decreased nitrates and nitrites in bronchial lavage fluid of HAPE patients reinforce the notion of decreased NO production, possibly because of reduced pulmonary NOS [11]. The focus is on preventing or reversing increased hydrostatic pressure in the pulmonary circulation. Lung scans before and after the inhalation of NO showed that the gas diverted the pulmonary blood flow from edematous regions of the lung to nonedematous regions in patients of HAPE.

It is likely that the INO had a greater vasodilating effect in the nonedematous regions of the lung, where it could more easily enter the alveoli. In contrast, no such redistribution of perfusion in response to NO was seen in subjects resistant to HAPE. Thus, exaggerated hypoxic pulmonary vasoconstriction resulting in ventilation-perfusion mismatch appears to be an important mechanism in the pathophysiology of HAPE [11]. It has also been suggested that HAPE develops because of stress failure in capillaries [75]. HAPE results from the stress failure of overdistended, relatively thin-walled pulmonary arteries [64, 76].

Two sites where NO may exert its beneficial effect are as follows: (a) the muscular pulmonary arterial vessels, where persons prone to HAPE may have defects in NO-mediated vasodilatation, and (b) capillary beds, where a defect in its synthesis may cause increased leaking of water, protein, and cells rather than from capillary rupture [59]. Low-concentration NO inhalation when compared to the prevailing conventional and nifedipine therapies was very effective in the treatment of HAPE [72].

It is advisable to continue administration of combination therapy of NO and oxygen (15 ppm NO + 50 % O_2) to HAPE patient till peripheral oxygen saturation reaches above 90 %.

Indian Scenario

Presently the Indian army is deployed at heights up to 6,700 m (22,000 ft). The incidence of high-altitude (HA) maladies are on the rise because of increasing number of troops being deployed under the prevailing geopolitical situation. HAPE incidence which was 0.3–0.4 % is presently also at the same rate, but the total number of patients is almost threefold. Currently the treatment modality of HAPE is aimed at immediate relief of symptoms and evacuation of patient to a lower altitude. This includes oxygen administration, aspirin, acetazolamide, recompression in a pressurized chamber, and de-induction to lower altitude. However, the therapy of iNO would reduce the morbidity and mortality due to HAPE and also help the organization to re-induct the acclimatized and trained manpower back to the place of duty thereby saving wastage of trained manpower. iNO is a selective pulmonary vasodilator for which the mechanism of action involves guanylate cyclase activation leading to production of cyclic guanosine monophosphate and subsequent smooth muscle relaxation [40, 51, 70]. Several studies have suggested that iNO improves oxygenation.

A study was done to assess the safety and physiological and clinical effects of various doses of NO over a 28-day period [18]. An acute response to NO was defined as a 20 % increase in PaO_2 in the first 4 h. The study group considered days off the ventilator as the primary long-term outcome. 24 % of patients in the placebo group had showed a 20 % increase in PaO_2 during the first 4 h, compared to 60 % of patients receiving NO. Virtually every study found that inhaled NO (1) improves ventilation, (2) reduced the blood pressure in the arteries surrounding the lungs, and (3) improved oxygen levels in the blood.

Experience in the armed forces: A milestone study was conducted where the acute effects of inhaling two gases in isolation and combination (iNO, 50 % oxygen, and a mixture of NO plus 50 % oxygen) on hemodynamics and gas exchange in 14 HAPE patients [1]. Each gas mixture was given in random order for 30 min followed by 30 min washout with room air. All patients had severe HAPE as judged by Lake Louise score (6.460.7), PaO2 (35.63 mmHg), and alveolar to arterial oxygen tension difference (AaDO2) (26.63 mmHg). NO had a selective effect on the pulmonary vasculature and did not alter systemic hemodynamics. Compared with room air, pulmonary vascular resistance declined 36 % with NO ($P < 0.001$), 23 % with oxygen (P, 0.001 vs. air; P, 0.05 vs. NO alone), and 54 % with NO plus 50 % oxygen (P, 0.001 vs. air; P, 0.005 vs. oxygen and NO). NO alone improved PaO2 (114 %) and AaDO2 (231 %). Compared with 50 % oxygen alone, NO plus 50 % oxygen had a greater effect on AaDO2 (218 %) and PaO2 (121 %). This study reconfirmed the therapeutic role of iNO in the management of HAPE. The combined use of inhaled NO and oxygen has additive effects on pulmonary hemodynamics and even greater effects on gas exchange. This study suggested that oxygen and NO may act on separate but interactive mechanisms in the pulmonary vasculature.

Another study was done in which iNO and oxygen were administered to HAPE patients [31, 32]. The study compared the disease progress and outcome of 27 HAPE patients on iNO, 20 HAPE patients on oxygen, and 50 HA stay-matched HAPE-resistant individuals. Significant findings of this study were that compared to the oxygen group; in the iNO group the HR was lower (74.2 ± 7.8 vs. 87.8 ± 16.8), SaO_2 was higher (94 % vs. 90 %), PaO2 was higher (78.1 ± 4.1 vs. 73.1 ± 6.1), pulmonary artery pressure was lower (20.1 ± 3.8 vs. 33.2 ± 4.1), Vock's score on day 2 was lower (3.8 vs. 11.6), and the hospital stay was lower. Complete radiological clearance occurred in 3.8 days in iNO group when compared to the oxygen group which took 5.6 days. All the iNO patients returned to the place of duty in HA and two of the oxygen groups were sent on leave for recuperation. The highlight of this study was that the iNO was administered by an indigenously developed iNO delivery system.

iNO Therapy Unit for HAPE

Indigenous Gas Delivery Systems

Soldiers deployed at HA are prone to develop acute mountain sickness (AMS), HAPE, and other diseases due to malacclimatization. iNO is the treatment of choice in patients of HAPE. Patients suffering from HAPE are spontaneously breathing, whereas commercially available systems work in conjunction with a ventilator [48, 59]. Hence there was a need to develop an indigenous NO delivery system for treatment of HAPE. Thus, an attempt has been made to design and develop an indigenous NO delivery system which may reduce the morbidity and mortality due to HAPE.

Why Do We Need Another iNO Therapy Unit?

We need an iNO therapy unit (ITU) tailor-made for the patients of HAPE because of the number of successful studies establishing the therapeutic value of iNO. Another advantage of INO is its low toxicity and very low incidence of side effects, which makes it a safe treatment modality.

What Is Available?

In the market there are a number of NO delivery systems available. They all are sensitive and accurate electronic flow meters, which deliver a specific dose of NO into the ventilator. However, dedicated NO-equipped ventilators are also available commercially but are not yet common in clinical practice. With other ventilators, there is no standardized procedure for the administration or monitoring of NO. The use of NO in conjunction with a simple time cycled, pressure regulated, flow-generating ventilator attached to a model infant-sized lung has been described. It is recommended that, when used as therapy, NO levels in inspired gases should always be measured [42]. It is also suggested that all compressed air methods using tap water have charcoal filters at the compression site and the gases be assessed periodically for oxidants [69]. These are designed for acutely ill moribund patients who are on ventilatory support. Contrasting to this, patients of HAPE are spontaneously breathing individuals who do not require ventilatory support. Hence such systems are not suitable in their present form to deliver NO.

Designing a Tailor-Made iNO Therapy Unit

The basic design and goal of delivery system is to provide a system for safe gas delivery and precision gas analysis. While delivering the gas through a ventilator, either a continuous or intermittent flow of NO is fed into the inspiratory limb of the ventilator tubing. The rate of NO gas flow is controlled to maintain the desired concentration. Prior to the patient connection of the ventilator tubing, a sensor or sample line is connected to an analyzer that displays NO, NO_2, and possibly oxygen levels, usually the displayed NO and NO_2 reading is measured in ppm.

Initial attempt has been made to modify ventilator-based delivery system available in the market to deliver NO to HAPE patient. This system is coupled with a source of compressed air and ventilator has been removed. However, the whole system is cumbersome, complicated, multimodular, not portable, and with very high chance of developing mechanical faults. This prompted us to design and develop iNO therapy unit tailor-made to suit the needs of HAPE patients. The design aimed at compactness, portability, nonventilator dependent, mask delivery system, simple to operate, accurate delivery of NO + O_2 mixtures, real-time online therapeutic gas mixture monitoring of NO + O_2, safe to use, and nonpolluting to the environment.

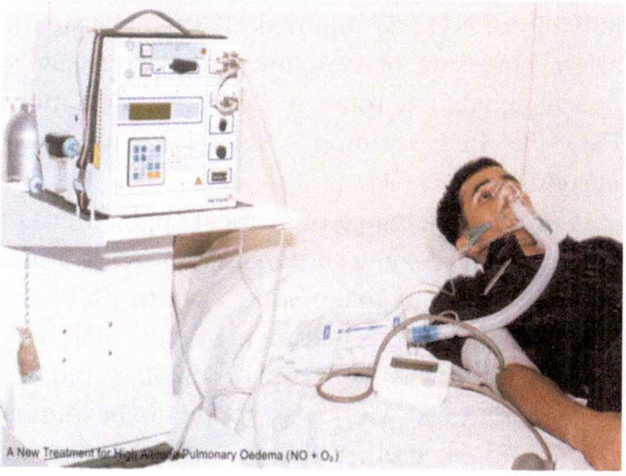

Imported system modified by DIPAS

In addition to the above, special attention was given to ascertain that the delivery system has to be a single module instrument, accurate delivery of 15 ppm NO + 50 % O_2, and minimize formation of NO_2 (permissive limits of $NO_2 < 2$ ppm) and to ensure that these gases are mixed in close proximity (<than 150 cm) to the delivery mask.

Online real-time monitoring of the therapeutic gas mixture (NO + O_2) in which the sample is drawn just before the mask to monitor the levels of NO and NO_2 and a desirable monitoring facility of the level of pollution, i.e., the rise in levels of NO and NO_2 in the vicinity of the machine, has been provided for in the design.

Prototype I (Mark I) of this system has already been developed in collaboration with R&D Engineers. Prototype II (Mark II) has been incorporated with a new design mass flow controller for precise delivery of the gas mixture and has been installed at HAMRC, Leh. The system is to be used in large scale for therapy of HAPE amongst Indian soldiers at various heights, and in the future miniaturized versions of the system would be made available even at the remote posts where the nursing staff would be trained to administer this to patients and save precious life.

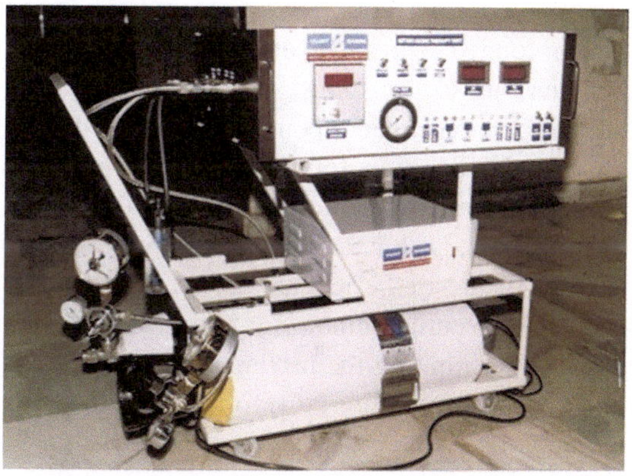

NO and oxygen delivery system *Mark I*

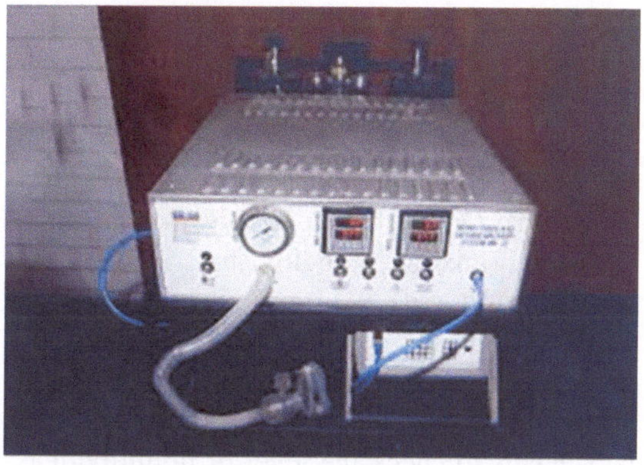

NO and oxygen delivery system *Mark II*

Safety Concerns

As with any drug, there are legitimate safety and toxicity concerns regarding the use of inhaled NO. Inhaling very high levels of NO (5,000–20,000 ppm) can be lethal causing a severe and acute accumulation of fluid in the lungs (pulmonary edema) and methemoglobinemia. However, there is little evidence of such toxicity when the concentration is kept in the normal therapeutic concentration range (1–80 ppm). According to a recent study, animals could withstand inhaling the gas in concentrations of 10–40 ppm, for 6 days to 6 months, without evidence of toxicity. Virtually all patients receiving NO will also be receiving oxygen. ARDS patients usually require high levels of O_2. The by-product of NO and O_2 is nitrogen dioxide (NO_2), which is a highly toxic gas. Although the Occupational Safety and Health Administration (OSHA) has set the safety limit for NO_2 at 5 ppm, some investigators have found that prolonged exposure to even 2 ppm of NO_2 can be injurious to the lungs. The amount of NO_2 produced is dependent upon the levels of NO and O_2 and the duration for which they are mixed together prior to inhalation. Therefore, the lowest dose of NO and lowest concentration of O_2 that achieve the desired effect are used. NO is usually fed into the ventilator tubing as close to the patient as possible, limiting the mixing time between O_2 and NO thus preventing NO_2 formation. All delivery systems monitor NO_2 levels continuously.

Safety Features

The features that make the unit safe for the patient and the operator are accurate dosing of NO and O_2 with minimal formation of NO_2 (<2 ppm). Online breath by breath analysis of the gas mixture, audiovisual alarms for concentrations of NO and NO_2, audiovisual alarm to indicate 30 min of gas reserve, automatic switchover to pure O_2 when NO concentration increases, monitoring of ambient NO, NO_2 levels to check the level of pollution/leakage if any, all materials used being human grade, gases of high purity and negligible chance of mechanical failure, and power leaks or short circuits also augment the safety of this system.

Advantages of the New Design

The advantages of the present design are single module, compact, portable, easy to use, no danger of overdosage, oxygen delivery along with NO, no mandatory requirement of ventilator, emergent switchover to pure O_2 when NO concentration increases preventing NO_2 toxicity demand valve breathing, 24 h power backup, capable of operating in high altitude, less expensive, and easy to procure. This system can be coupled with a ventilator. It will deliver 50 % of the available NO concentration in the cylinder. Although it is not designed for varying doses, different doses can be given by changing the concentration of the feeder NO cylinder.

Other Medical Applications of Inhaled Nitric Oxide

The vasodilatory action of iNO has found a place in treatment of multitude of diseases. The short half life of NO and minimal effects on systemic arterial pressure or systemic vascular resistance due to inhalation route of administration make iNO the first seemingly selective pulmonary vasodilator [21, 63].

Currently the range of medical applications of iNO spans across the cardiovascular, respiratory, neural, renal, and coagulation systems [30].

In neonates NO is helpful in severe respiratory failure. The need for extracorporeal membrane oxygenation (ECMO) was found to be diminished [48] in another disease where satisfying results comparable to ECMO have been found [16, 55, 68]. A number of studies have confirmed the efficiency of iNO in improving oxygenation in patients of acute respiratory distress syndrome (ARDS) in both pediatric and adult age groups. iNO is used with increasing frequency in infants and children with right heart failure and pulmonary hypertension following repairs of congenital heart defects [5, 56, 73]. iNO is found to be beneficial in pulmonary hypertension [30].

It has a life-saving role in patients with primary pulmonary hypertension [43, 74] and is also used in patients of primary pulmonary hypertension (PPH) as a screening tool to predict response to conventional long-term treatment with calcium channel antagonists and prostacyclin [12]. iNO has been shown to reduce pulmonary arterial pressure (PAP), pulmonary vascular resistance, and intrapulmonary shunt fraction following lung transplantation [4, 15, 47]. Cardiac transplant patients who develop right ventricular dysfunction also benefit from the pulmonary vasodilatory properties of iNO [3, 65]. iNO is found to be beneficial in congenital heart diseases pre- and postsurgery, necrotizing enterocolitis in neonates, any other disease whose pathophysiology rests on pulmonary hypertension and hypoxemia.

Conclusion

It may be concluded that acclimatization to high altitude would be indicated by higher pulmonary endogenous production of NO and ventilation as observed in HAN. Reduced endogenous NO synthesis may contribute to HAPE susceptibility and their heightened hypoxic pulmonary vascular response to hypoxia. iNO decreased pulmonary artery pressure and improved ventilation-perfusion mismatch in HAPE-prone subjects exposed to high altitude. iNO therapy has also emerged as a new modality of treatment for a multitude of cardiorespiratory diseases. Currently available NO delivery systems are ventilator-based for patients of respiratory failure who require mechanical ventilatory support. HAPE is a life-threatening disease in which inhaled nitric oxide is of immense therapeutic value. The presently available delivery systems are unsuitable as these patients are spontaneously breathing and cannot be coupled with a mechanical ventilator. In this article we have designed a tailor-made, easy to operate, safe, simple, compact, fixed dose NO and O_2 delivery system that would be especially helpful in the armed forces. The instrument awaits patenting and extensive clinical trials.

References

1. Anand IS, Prasad BA, Chugh SS, Rao KR, Cornfield DN, Milla CE, Singh N, Singh S, Selvamurthy W (1998) Effects of inhaled nitric oxide and oxygen in high-altitude pulmonary edema. Circulation 98:2441–2445
2. Anggard E (1994) Nitric oxide: mediator, murderer, and medicine. Lancet 343:1199–1206
3. Auler AO Jr, Carmona MJ, Bochii EA, Bacal F, Firelli AI, Stoff NA, Jatene AD (1996) Low doses of inhaled nitric oxide in heart transplant recipients. J Heart Lung Transplant 15:443–450
4. Bacha EA, Herve P, Murakami S, Chapelier A, Maxmanian GM, de Montpreville V (1996) Lasting beneficial effect of short term inhaled nitric oxide on graft function after lung transplantation. Paris-Sud University Lung Transplantation Group. J Thorac Cardiovasc Surg 112:590–598
5. Barefield E, Karle VA, Phillips JB III, Carlo WA (1996) Inhaled nitric oxide in term infants with hypoxemic respiratory failure. J Pediatr 129:279–286
6. Barnes PJ (1995) Nitric oxide and airway disease. Ann Med 27:389–393
7. Bartsch P, Maggiorini M, Ritter M, Noti C, Vock P, Oelz O (1991) Prevention of high-altitude pulmonary edema by nifedipine. N Engl J Med 325:1284–1289
8. Beall CM, Laskowski D, Strohl KP, Soria R, Villena M, Vargas E, Alarcon AM, Gonzales C, Erzurum SC (2001) Pulmonary nitric oxide in mountain dwellers. Nature 414:411–412
9. Blitzer ML, Loh E, Roddy MA, Stamler JS, Creager MA (1996) Endothelium derived nitric oxide regulates systemic and pulmonary vascular resistance during acute hypoxia in human. J Am Coll Cardiol 28:591–596
10. Budts W, Pokreisz P, Nong Z, Pelt V, Gillijns H, Gerard R, Lyons R, Collen D, Bloch KD, Janssens S (2000) Aerosol gene transfer with inducible nitric oxide synthase reduces hypoxic pulmonary hypertension and pulmonary vascular remodeling in rats. Circulation 102:2880–2885
11. Busch T, Bartsch P, Pappert D, Grunig E, Hilderbrandt W, Elser H, Falke KJ, Swenson ER (2001) Hypoxia decreases exhaled nitric oxide in mountaineers susceptible to high-altitude pulmonary edema. Am J Respir Crit Care Med 163:368–373
12. Cockrill BA, Kaemarek R, Thompson BT et al (1995) Response to inhaled nitric oxide identifies a subset of patients more likely to respond to high dose calcium channel blockers. Am J Respir Crit Care Med 151:728
13. Cooper CJ, Landzberg MJ, Anderson TJ, Charbonneau F, Creager MA, Ganz P, Selwyn AP (1996) Role of nitric oxide in the local regulation of pulmonary vascular resistance in humans. Circulation 93:266–271
14. Cremona G, Asnaghi R, Baderna P, Brunetto A, Brutsaert T, Cavallaro C, Clark TM, Cogo A, Donis R, Lanfranchi P et al (2002) Pulmonary extra-vascular fluid accumulation in recreational climbers: a prospective study. Lancet 359:303–309

15. Date H, Triantafillou AN, Trulock EP, Pohl MS, Cooper JD, Patterson GA (1996) Inhaled nitric oxide reduces human lung allograft dysfunction. J Thorac Cardiovasc Surg 111:913–919
16. Davidson D, Barefield ES, Kattwinkel J et al (1998) Inhaled nitric oxide for the early treatment of persistent pulmonary hypertension of the term newborn: a randomized, double-masked, placebo-controlled, dose-response, multicenter study. Pediatrics 101:325–334
17. Day RW, Lynch JM, White KS, Ward RM (1996) Acute response to inhaled nitric oxide in newborns in respiratory failure and pulmonary hypertension. Pediatrics 98:698–705
18. Dellinger RP, Zimmerman JL, Taylor RW, Straube RC, Hauser DL, Criner GJ, Davis K, Hyers TM, Papadakos P (1998) Effects of inhaled nitric oxide in patients with acute respiratory distress syndrome: results of a randomized phase trial. Crit Care Med 26:15–23
19. Droma Y, Hanaoka M, Ota M, Katsuyama Y, Koizumi T, Fujimoto K, Kobayashi T, Kubo K (2002) Positive association of endothelial nitric oxide synthase gene polymorphisms with high altitude pulmonary edema. Circulation 106:826–830
20. Duplain H, Sartori C, Lepori M, Egli M, Allemann Y, Nicod P, Scherrer U (2000) Exhaled nitric oxide in high altitude pulmonary edema. Am J Respir Crit Care Med 162:221–224
21. Frostell C, Fratacci MD, Wain JC, Jones R, Zapol WM (1991) Inhaled nitric oxide. A selective pulmonary vasodilator reversing hypoxic pulmonary vasoconstriction. Circulation 83:2038–2047
22. Frostell CG, Blomqvist H, Hedenstrierna G, Lundberg J, Zapol WM (1993) Inhaled nitric oxide selectively reverses human hypoxic pulmonary vasoconstriction without causing systemic vasodilation. Anesthesiology 78:427–435
23. Adel G, Saleh D (1995) Reduced expression of endothelial nitric oxide synthase in the lung of patients with pulmonary hypertension. N Engl J Med 333:214–221
24. Gupta RK, Selvaraj PT, Kumar H, Kain TC, Ray US (2008) Application of nitric oxide in treatment of high altitude pulmonary edema. DRDO project report no. DIPAS/03/2008
25. Gustafsson LE, Leone AM, Persson MG, Wiklund NP, Moncada S (1991) Endogenous nitric oxide is present in exhaled air of rabbits, guinea pig and humans. Biochem Biophys Res Commun 181:852–857
26. Hackett PH, Rennie D (1976) The incidence, importance and prophylaxis of acute mountain sickness. Lancet 2:1149–1155
27. Hackett PH, Roach RC, Hartig GS, Greene ER, Levine BD (1993) The effect of vasodilators on pulmonary hemodynamics in high altitude pulmonary edema: a comparison. Md Med J 42:641–645
28. Hackett P, Rennie D (2002) High altitude pulmonary edema. JAMA 287:2275–2278
29. Hackett P, Roach RC (2001) High altitude illness. N Engl J Med 345:107–114
30. Dalton HJ, (1998) Inhaled nitric oxide: medical miracle or passing fad? Respir Care 43:978–987
31. Himashree G, Chattopdhya PK, Selvamurthy W (2005) Indigensisation of nitric oxide therapy for treatment of HAPE. Published in trends in respiratory diseases, The fourth Sir Dorabji Tata symposium, Tata McGraw Hill, India, pp 131–149
32. Himashree G (2007) Use of nitric oxide at high altitude – advantage and challenges. Proceedings of 10th annual conference of Indian Association of Cardiovascular and Thoracic Anaesthesiologists, p 86
33. Hlastala MP, Lamm WJ, Karp A, Polissar NL, Starr IR, Glenny RW (2003) Spatial distribution of hypoxic pulmonary vasoconstriction in the supine pig. J Appl Physiol 96:1589–1599
34. Hoit BD, Dalton ND, Erzurum SC, Laskowski D, Strohl KP, Beall CM (2005) Nitric oxide and cardiopulmonary hemodynamics in Tibetan highlanders. J Appl Physiol 99:1796–1801
35. Hopkins SR, Garg J, Bolar DS, Balouch J, Levin DL (2005) Pulmonary blood flow heterogeneity during hypoxia and high altitude pulmonary edema. Am J Respir Crit Care Med 171:83–87
36. Hultgren HN, Groover RF, Hartley LH (1971) Abnormal circulatory responses to high altitude in subjects with a previous history of high altitude pulmonary edema. Circulation 44:759–770
37. Hultgren HN (1997) High altitude medicine. Hultgren Publications, Stanford
38. Hultgren HN (1995) High altitude pulmonary edema: current concepts. Adv Exp Med Biol 381:15–25
39. Hultgren HN (1982) Pulmonary hypertension and pulmonary edema. In: Loeppky JA, Riedesel ML (eds) Oxygen transport to human tissue. Elsevier/North Holland, New York, pp 243–254
40. Ignarro LJ, Buga GM, Wood KD, Byrns RE, Chaudhuri G (1987) Endothelium derived relaxing factor produced and released from artery and vein is nitric oxide. Proc Natl Acad Sci U S A 84:9265–9269
41. Jerome SJW (1996) High altitude pulmonary edema. N Engl J Med 334:662
42. Jones PR, Hooper RH, Stopp N, Moore NA (2000) Nitric oxide delivery to the lung: a model. Eur J Anaesthesiol 17:160–167
43. Kinsella JP, Truog WE, Walsh WF et al (1997) Randomized, multicenter trial of inhaled nitric oxide and high-frequency oscillatory ventilation in severe, persistent pulmonary hypertension of the newborn. J Pediatr 131:55–62
44. Kobzik L, Bredt DS, Lowenstein CJ, Drazen J, Gaston B, Sugarbaker D, Stamler JS (1993) Nitric oxide synthase in human and rat lung: Immunocytochemical and histochemical localisation. Am J Respir Cell Mol Biol 9:371–377
45. Lunn R (1995) Subspecialty clinics: anesthesiology; inhaled nitric oxide therapy. Mayo Clin Proc 70:247–255
46. Moncada S, Palmer RMJ, Higgs EA (1991) NO: physiology, pathophysiology & pharmacology. Pharmacol Rev 43:109–142

47. Myles PS, Venema HR (1995) Avoidance of cardiopulmonary bypass during bilateral sequential lung transplantation using inhaled nitric oxide. J Cardiothorac Vasc Anesth 9:571–574
48. Neonatal Inhaled Nitric Oxide Study Group (NINOS) (1997) Inhaled nitric oxide in full-term and nearly full-term infants with hypoxic respiratory failure. N Engl J Med 336:597–604
49. Voelkel NF (2002) High altitude pulmonary edema. N Engl J Med 346:1606–1607
50. Oelz O, Maggiorini M, Ritter M, Noti C, Waber U, Vock P, Bartsch P (1992) Pathophysiology, prevention and therapy of altitude pulmonary edema. Schweiz Med Wochenschr 122:151–1158
51. Palmer RM, Ferrige AG, Moncada S (1987) Nitric oxide release accounts for the biological activity of endothelium,-derived relaxing factor. Nature 327:524–526
52. Penaloza D, Sime F (1969) Circulatory dynamics during high altitude pulmonary edema. Am J Cardiol 23:369–378
53. Persson MG, Gustaffson LE, Wiklund NP, Moncada S, Hedquist P (1990) Regulation of pulmonary circulation by nitric oxide. Acta Physiol Scand 140:449–457
54. Phillips CR, Giraud GD, Holden WE (1996) Exhaled nitric oxide during exercise: site of release and modulation by ventilation and blood flow. J Appl Physiol 80:1865–1871
55. Roberts JD Jr, Fineman JR, Morin FC III et al (1997) Inhaled nitric oxide and persistent pulmonary hypertension of the newborn. N Engl J Med 336:605–610
56. Rossaint R, Falke KJ, Lopedz F, Salma K, Pison U, Zapol WM (1993) Inhaled nitric oxide for the adult respiratory distress syndrome. N Engl J Med 328:399–405
57. Sartori C, Vollenweider L, Loffler BM, Delabays A, Nicod P, Bartsch P, Scherrer U (1999) Exaggerated endothelin release in high altitude pulmonary edema. Circulation 99:2665–2668
58. Scherrer U, Rexhaj E, Jayet P, Allemann Y, Sartori C (2010) New insights in the pathogenesis of high-altitude pulmonary edema. Prog Cardiovasc Dis 52:485–492
59. Scherrer U, Vollenweider L, Delabays A, Savcic M, Eichenberger U, Kleger G, Fikrle A, Ballmer PE, Nicod P, Bartsch P (1996) Inhaled nitric oxide for high altitude pulmonary edema. N Engl J Med 334:624–629
60. Schoene RB, Hackett PH, Henderson WR, Sage EH, Chow M, Roach RC, Mills WJ, Martin TR (1986) High altitude pulmonary edema: characteristic of lung lavage fluid. JAMA 256:63–69
61. Schoene R, Swenson E, Hultgren H (2001) High altitude pulmonary edema. In: Hornbein TF, Schoene RB (eds) High altitude. An exploration of human adaptation. Lung biology in health and disease, vol 161. Marcel Dekker, New York, pp 777–814
62. Selvamurthy W, Dass D, Himashree G (2005) Pathophysiology of respiratory maladies at high altitude. Published in trends in respiratory diseases, The fourth Sir Dorabji Tata symposium, Tata McGraw Hill, India, pp111–119
63. Semigram MJ, Cockrill BA, Kaemarek R, Thompson BT, Zapol WM, Dee GW, Fifer MA (1994) Hemodynamic effects of inhaled nitric oxide in heart failure. J Am Coll Cardiol 24:982–988
64. Severinghaus JW (1971) Transarterial leakage: a possible mechanism of high altitude pulmonary edema. In: Porter R, Knight J (eds) High altitude physiology: cardiac and respiratory aspects. Churchill Livingstone, London, pp 61–77
65. Snell GI, Salmonsen RF, Bergin P, Esmore DS, Khan S, William TJ (1995) Inhaled nitric oxide used as a bridge to heart lung transplantation in a patient with end stage pulmonary hypertension. Am J Respir Crit Care Med 151:1263–1266
66. Stamler JS, Loh E, Roddy M, Currie KE, Creager MA (1994) Nitric oxide regulates basal systemic and pulmonary vascular resistance in healthy humans. Circulation 89:2035–2040
67. Swenson ER, Maggiorini M, Mongovin S, Gibbs JS, Greve I, Mairbauri H, Bartsch P (2002) Pathogenesis of high altitude pulmonary edema: inflammation is not an etiologic factor. JAMA 287:2228–2235
68. The Franco-Belgium Collaborative NO Trial Group (1999) Early compared with delayed inhaled nitric oxide in moderately hypoxaemic neonates with respiratory failure: a randomized controlled trial. Lancet 354:1066–1071
69. Thibeault DW, Rexaiekhaligh MH, Ekekezie I, Truog WE (1999) Compressed air as a source of inhaled oxidants in intensive care units. Am J Perinatol 16:497–501
70. Tracey WR, Xue C, Klinghoffer V, Barlow J, Pollock JS, Forstermann U, Johns RA (1994) Immunocytochemical detection of inducible NO synthase in human lung. Am J Physiol 266:722–727
71. Visscher MB (1962) Studies on embolization of lung vessel. Med Thorac 19:334–340
72. Wang W, Zhang X, Ma Y (1998) Low-concentration nitrous oxide inhalation in the treatment of high altitude pulmonary edema. Zhonghua Jie He He Hu Xi Za Zhi 21:12–14
73. Wessel DL, Adatia I, Giglia TM, Thomson KTJ (1993) Use of inhaled nitric oxide and acetylcholine in evaluation of pulmonary hypertension and endothelial function after cardiopulmonary bypass. Circulation 88:2128–2138
74. Wessel DL, Adatia I, Van Marter LJ et al (1997) Improved oxygenation in randomized trials of inhaled nitric oxide for persistent pulmonary hypertension of the newborn. Pediatrics 100:E7
75. West JB, Tsukimoto K, Mathieu-Costello O, Prediletto R (1991) Stress failure in pulmonary capillaries. J Appl Physiol 70:1731–1742
76. Whayne TF Jr, Severinghaus JW (1968) Experimental hypoxic pulmonary edema in the rat. J Appl Physiol 25:729–737
77. Xue C, Johns RA (1995) Endothelial nitric oxide synthase in the lungs of patients with pulmonary hypertension. N Engl J Med 333:1642–1644

High-Altitude Medicine: The Path from Genomic Insight to Clinical Applications

Soma Sarkar

Abstract

Translation of research findings into clinical practice is an important aspect of medical progress. In context of high-altitude medical research, understanding the functions of all genes and their regulation under high-altitude environment is far from complete, and translation of genomics research findings into clinical practice is yet an unexplored domain. With advancement and accessibility to genomic and genomewide data sets, understanding of the role of genes and environment under high altitude will get accelerated which can potentially get converted to translational research.

Introduction

Translational genomics research is a relatively new field which aims to translate research findings into clinical practice by way of diagnostics, prognostics, and therapies which are important aspects of medical progress by applying the innovative advances from the Human Genome Project. A significant proportion of genomics research is still at an early stage in terms of clinical outcome; nevertheless, research endeavors which aim to expand understanding of underlying mechanisms of disease and usefulness of such research in medical treatment and public health are going to be of prime importance. Primarily based on the Watson and Crick's [83] model of the first proposed DNA structure, genomics in current context encompasses a wide array of technologies like DNA sequencing, microarrays, and polymerase chain reaction (PCR) technology including real-time and quantitative PCR (qPCR) that permit amplification of genetic material within a very short period, thereby advancing greater uses for the technology. Genomics has enabled pharmaceutical companies to undertake genetic diagnostics, and improved information technologies have allowed biological information to be better managed and analyzed. Current research prospects involve crossing the barrier of sequencing the human genome for identifying novel genes and therapeutic targets. Combination of automated technologies such as robotics and microarrays has further expanded the scope for genomics. One of the technologies which is now being widely used in clinical diagnostics is single nucleotide polymorphism (SNP)-based DNA

S. Sarkar (✉)
Defence Institute of Physiology and Allied Sciences (DIPAS), Lucknow Road, New Delhi 110 054, India
e-mail: soma_sarkar2000@yahoo.com

S.B. Singh et al. (eds.), *Translational Research in Environmental and Occupational Stress*, DOI 10.1007/978-81-322-1928-6_18,

fingerprinting. The industries have diversified applications of genomics in various areas such as molecular diagnostics, pharmacogenomics, personalized medicine, and drug discovery.

Mapping of the human genes was an important step in the development of medicines and other aspects of health care. Human Genome Project (HGP), which begun in 1990, was an international scientific research project with a primary goal to determine the DNA sequence and identify and map the approximately 25,000–30,000 genes of the human genome from both physical and functional standpoint. Widespread international cooperation and advances in sequence analysis as well as major advances in computing technology enabled completion of the "rough draft" of the genome in 2000, and announcement of the essentially complete human genome was made in April 2003, 2 years earlier than planned. Key findings of the Human Genome Project were as follows: (a) there are approximately 25,000 genes in human beings, which is the same range as in mice and twice that of roundworms. Understanding how these genes express themselves will provide clues to how diseases are caused (http://www.genome.gov/12011238); (b) the human genome has significantly more segmental duplications (newly identified repeated segments of DNA) than other mammalian genomes; (c) fewer than 796 of protein families appeared to be vertebrate specific [17].

Although most of the human genome (about 92.3 %) was completed by the end of 2003, a number of regions like centromeres, telomeres, and loci that contain members of multigene families remained unfinished. The centromeres, central regions of each chromosome, which are possibly tens of millions of base pair long and have highly repetitive DNA sequences, are difficult to sequence using current technology and still remain to be sequenced. The telomeres, the ends of the chromosomes, are also highly repetitive, and for most of the 46 chromosome ends, sequencing is incomplete. Presently it is not known how much sequence remains before the telomeres of each chromosome are reached. As with the centromeres, current technological restraints for sequencing the telomeres are also prohibitive. Several loci in each individual's genome that contain members of multigene families that often encode proteins important for immune function are similarly difficult to sequence. There also remain many gaps scattered around the genome which still remain to be sequenced. The centromeres and telomeres will remain unsequenced until new sequencing technologies develop. Much of the remaining unsequenced highly repetitive DNA is unlikely to contain genes, but this cannot be ascertained till it is entirely sequenced. The role of junk DNA, the evolution of the genome, differences between individuals, and many other questions still remain to be answered. In context of high-altitude medical research, understanding the functions of all genes and their regulation under high-altitude environment is far from complete, and translation of genomics research findings into clinical practice is yet an unexplored domain.

Medical Problems at High Altitude

More than 140 million people are permanently settled on high-altitude regions, on continents ranging from Africa and Asia to South America, while a substantial number of people visit high altitude for various reasons. The medical problems at high altitude largely stem from hypoxia along with risks of dehydration, accidental injury, cold injury, weight loss, and psychological stress under such environment in lowland sojourners who travel to high altitude. At high and extreme altitudes (>3,000 and 5,500 m above sea level, respectively), hypobaric hypoxia induces a range of normal or adverse responses at molecular, cellular, and systemic levels. At the systems level, the major physiological responses include increased heart and ventilation rate and rapid erythrocyte expansion [13, 35]. Symptoms of increased heart rate and hyperventilation appear within the first few hours when ascending to high altitude, due to stimulation of the peripheral chemoreceptors by hypoxia. In the following days, increased hemoglobin concentration and erythrocyte numbers are observed in the blood, which are more likely to be a gene level regulation of the body. At the cellular level, increased expression of genes that participate in anaerobic energy supply and decreased expression of those involved in ATP consumption processes are noted [20]. These signs along with other responses of different systems of the body comprise altitude acclimatization. When acclimatization does not happen, such individuals get maladapted and develop high-altitude illnesses which may be classified as (1) *acute* such as acute mountain sickness (AMS), high-altitude cerebral edema (HACE), and high-altitude pulmonary edema (HAPE) and (2) *subacute* such as subacute mountain sickness and chronic mountain sickness (CMS) which happen after sustained periods at high altitude. AMS occurs in 50–90 % of the people who travel to high altitude; HACE and HAPE have lower frequencies than AMS but are fatal if not treated on time [8, 35]. Inhabitants of Andes mountain region of South America suffer CMS [59]. CMS is common in Andeans, occasionally found in Tibetans and absent from Ethiopians living on the East African high-altitude plateau. The disease is characterized by an array of neurologic symptoms, including headache, fatigue, sleepiness, and depression. Individuals with CMS often suffer from stroke or heart attacks in early adulthood due to increased blood viscosity.

Field studies on the physiological aspects of high-altitude hypoxia have generated a great deal of valuable data including ECG measurements [66], arterial blood gases, and oxygen content [32] for understanding the associated processes. Studies on molecular mechanisms of high-altitude hypoxia on human system as well as mechanisms of human acclimatization/adaptation to high altitude are limited [42, 69, 76]. Some studies on the role of hypoxia-induced factors (HIFs) and erythropoietin response in the high altitude are available [13, 20, 34]. The signaling pathways and molecular functions that mediate such responses (especially those that modulate or are independent of HIFs and erythropoietin) are largely unknown. Genes are undoubtedly the major determinants of susceptibility to diseases, clinical outcome, as well as response to therapy. It is challenging to reliably identify the underlying specific genes and gene mutations. Clinical studies have implicated genetic background in susceptibility to HAPE [35]. Those with higher pulmonary artery pressure at rest or an exaggerated response to hypoxia or exercise are considered to be at higher risk of developing HAPE [39], and thus genetic variants that affect pulmonary arterial pressure may serve as susceptibility alleles for HAPE.

Genetic Basis for Altitude Illness

Evolution of complex physiological systems that underlie acclimatization/adaptation to high-altitude stress requires coordinated evolution of many interacting genes [24, 81]. Susceptibility to high-altitude maladaptation may be evidenced at the individual level, familial level, or population level. At the individual level, some persons repeatedly develop altitude illness (recurrent incidence) and may be considered as the true susceptibles. Identifying the potential candidate genes in such individuals is one of the ways of establishing a reliable risk predictor of altitude susceptibility which can have clinical translational potential. Some research laboratories are actively engaged in such identification process through candidate gene association studies [21, 22, 44, 61, 63, 71, 75, 79, 82]. Very little evidence is available for familial pattern of genetic transmission of susceptibility [39, 47, 55]. Highlander population has developed different adaptive strategies for coping with high-altitude environment [10, 11]. Lower frequency of AMS is reported from Tibetans [84, 85]. In a study on Andean Quechua, G allele at the G894T polymorphism in *NOS3*, suggested to be beneficial in hypoxic environments, was more frequent compared to lowland Amerindians [80], while the frequency of the same allele was low in Nepalese population who suffered from AMS [79]. In the Himalayan Ladakhi population, allele 5160A of *CYP11B2* was less frequent compared to Indian lowlanders [65] and the same allele was also shown to be associated with HAPE [61]. *EPAS1* along with other genes in HIF pathway was studied in CMS in Andeans although no association was reported [51]. Genetic contribution to susceptibility to altitude illness as well as contribution of specific genes as potential contributors to predisposition have been recently reviewed [48]. Understanding of genetic basic and molecular mechanisms leading to altitude illness is critical for designing of preventive or prophylactic strategies. A number of genomic approaches that have provided insight into the genetic basis of high-altitude acclimatization and/or adaptive traits are discussed in the following sections.

Genomewide Screening of Gene and Protein Transcript Variations Under High Altitude Stress

Gene expression profiles, apart from identifying genes which cope with physiological challenges, also provide insight into the putative pathways present at the system level. Oligonucleotide and cDNA microarrays have revolutionized the study of differential gene expression in different cells and tissues, enabling analysis of various disease processes. Microarray techniques are sensitive enough to detect expression of a target gene among 50,000 to 300,000 transcripts [46]. Similar array techniques are also being developed to analyze proteins and their variants [58]. mRNA expression with DNA microarrays has proven useful in generating hypotheses concerning the mechanisms of acclimatization and adaptation [28, 29]. Our studies on global gene expression profiling in individuals who acclimatized to high altitude showed downregulation of more gene transcripts than upregulation, and these transcripts were implicated in the pathways like vascular smooth muscle contraction (hsa04270), regulation of actin cytoskeleton (hsa04810), calcium signaling pathway (hsa04020), ubiquitin-mediated proteolysis (hsa04120), and cytokine-cytokine receptor interaction (hsa04060) [70]. Global gene expression profiling in individuals of established HAPE compared to acclimatized controls provides evidence for concurrent modulation of multiple transcripts representative of different classes of factors that regulate lung fluid homeostasis as well as pathways that regulate vasoconstriction through smooth muscle contraction and cellular actin cytoskeleton rearrangement during HAPE. Several regulators of systemic/pulmonary hypertension including *ADRA1D*, *ECE1*, and *EDNRA* were observed to be differentially expressed. Other notable genes included Rho family members, myosin light chain 2, focal adhesion molecules, cadherins, claudins, paxillins, and *VCAM1*. Pathway-specific transcripts which could potentially culminate in perturbed endothelial cell permeability (dysfunction) and edema formation were seen to be modulated (communicated) [73]. These studies provide evidence for hypoxic signature of high-altitude exposure in sea-level sojourners, and it may be hypothesized that designing drugs that would downregulate those genes which show upregulated expression during HAPE or upregulating those which show downregulation could be beneficial in coping with hypoxic stress. As an example, designing drugs which downregulate expression of *ADRA1*, *ECE1*, and *EDNRA* would be good candidates for translational purposes.

A proteomic study of native Tibetans showed overexpression of glutathione-S-transferase (*GSTP1-1*), Δ^2-enoyl-CoA hydratase (*ECH*), phosphoglycerate mutase (muscle form, PGA), NADH-ubiquinone oxidoreductase (*NUGM*), and myoglobin and downregulation of glyceraldehyde-3-phosphate dehydrogenase (*GAPDH*) and lactate dehydrogenase (*LDH*) compared to lowlander Nepalese population indicating thereby that Tibetans may be protected from ROS-induced tissue damage and may also possess specific metabolic adaptations [31]. Scope exists for translating these and related research findings for prevention and treatment of inhabitants of high altitude as well as sojourners who travel to high altitude.

Genome-Wide Association Studies for High-Altitude Adaptation/ Acclimatization

The availability of high-density single nucleotide polymorphism (SNP) microarrays in recent years has proven to be a great step forward in the context of global analysis of genomic abnormalities in disease. SNP arrays offer great robustness, high resolution, and the possibility to detect a variety of different genomic copy number variations such as submicroscopic deletions, amplifications, loss of heterozygosity, and uniparental disomy. Moreover, they can be used to perform genome-wide association studies. Therefore, SNP arrays harbor several advancements over traditional molecular methods such as cytogenetic analyses, fluorescence in situ hybridization, or comparative genomic hybridization methods to analyze genomic aberrations. Several recent genomic studies of high-altitude Andean and Tibetan population samples have identified a set of candidate genes (including *EPAS1*, *EGLN1*, and *PPARA*) thought to contain variants that play a role in physiological adaptation to high altitude [1, 12, 14, 15, 74, 86, 87]. High-altitude Amhara individuals (living at 3,202 m above sea level) and low-altitude Aari and Hamer individuals (living at <1,500 m above sea level) residing in Ethiopia showed that there are significantly higher hemoglobin levels in high-altitude (3,200 m) relative to low-altitude (<1,500 m) Ethiopian residents [72]. Genome-wide analysis of over 1 million SNPs identified several candidates for involvement in high-altitude adaptation in the Ethiopians, including *CBARA1*, *VAV3*, *ARNT2*, and *THRB* which were also the strongest candidates for selection for genotype/phenotype associations with hemoglobin levels [72]. HAPE is a multifactorial disorder resulting from the interaction of genetic and environmental factors. The combined study designs of genome-wide association and epigenetic analysis should be undertaken to elucidate the complex interactions between the genome and hypoxic environment, pathogenesis of HAPE, and appropriate drug designing both for prophylactic and prevention purposes.

Whole Genome Sequencing

Genetic mechanisms underlying high-altitude adaptation were recently studied using whole genome sequencing and comparing genetic variation between Peruvian individuals from the Andes region with CMS and those without CMS. Two genes, *ANP32D* and *SENP1*, with significantly increased expression in the CMS individuals compared to non-CMS individuals were identified. It was hypothesized that downregulating these genes could be beneficial in coping with hypoxia [88]. This finding may have important implications not only for those who live at high altitudes but also in treating certain cardiovascular and brain diseases related to low oxygen levels in individuals living at any altitude. These researchers are planning whole genome sequencing for the almost 100 remaining patient samples to test if biomarkers exist to predict CMS.

Polymorphism Profiling for High-Altitude Acclimatization/Adaptation/Maladaptation

Much of what is known about the genetic basis of high-altitude acclimatization/adaptation comes from studies using "candidate gene" approach; a total of 58 genes have so far been tested for association with acute mountain sickness, high-altitude pulmonary edema, chronic mountain sickness, and high-altitude pulmonary hypertension, and out of these, 17 genes with at least one variant were found to be associated with one or more altitude illness [reviewed in 48]. A number of genes encoding components of the renin-angiotensin-aldosterone system (RAAS) have been investigated in altitude illness, and *ACE*, *AGTR1*, and *CYP11B2* were found to be associated with HAPE. A significant amount of work has been undertaken to investigate the insertion (I)/deletion (D) polymorphism of the angiotensin-converting enzyme (*ACE*) gene (*ACE* I/D polymorphism) in relation to human adaptation to high altitude [4, 60, 67]. The *ACE* I allele has been shown to be associated with Caucasian mountain climbers who ascended to extremely high altitudes without supplemental oxygen [53]. An overrepresentation of *ACE* D allele was shown in HAPE-susceptible individuals, while *ACE* I/I genotype in conjunction with specific *EDN1* genotype was more prevalent in HAPE-resistant individuals [19]. A meta-analysis of association of *ACE I/D* polymorphism with HAPE showed that *ACE D* allele carriers were at significant increased risk of developing HAPE [64]. Variants of *CYP11B2* gene polymorphisms are also associated with HAPE [2, 61] although more studies may be required to unravel those variants that may be functionally linked to HAPE. Single locus analysis showed that the polymorphisms C-344T and K173R in the cytochrome P450 family protein *CYP11B2* and the A-240T polymorphism in the angiotensin I-converting enzyme (*ACE*) protein were significantly associated with HAPE. Gene-gene interaction analysis found that the *ACE* A-240T, A2350G, and *CYP11B2* C-344T polymorphisms had a strong synergistic effect on HAPE. Homozygous genotype combination of -240AA, 2350GG, and -344TT, in particular, conferred high genetic susceptibility to HAPE providing evidence for synergistic effect of RAAS gene polymorphisms on HAPE susceptibility [62, 64, 68]. Mass spectroscopy-based genotyping assay of polymorphisms in *EGLN1* identified seven SNPs to be significantly different in individuals of HAPE compared to HAPE-free controls, and genotypes AA of rs1538664, TT of rs479200, AA of rs2486729, GG of rs2790879, CC of rs480902, AA of rs2486736, and GG of rs973252 were identified as risk genotypes [52]. Upregulated expression of *EGLN1* was also noted in HAPE [52]. In our gene expression experiments, we noted upregulated expression of *EGLN3* in HAPE compared to acclimatized controls (communicated) [73]. Designing drugs/inhibitors which downregulate expression of *EGLN* in maladapted individuals will have translational application.

Familial character and heritability studies have suggested that genetic factors could make a contribution to the pathogenesis of chronic mountain sickness (CMS) and high-altitude pulmonary hypertension (HAPH) [42]. The *ACE* I allele and I/I genotype were shown to be significantly overrepresented in high-altitude pulmonary hypertension (HAPH) in Kyrgyz highlanders compared to unaffected individuals [4] as was I-G-A haplotype [3]. In Tibetan population the *AGT 235M* allele was shown to be associated with CMS. Even though some alleles are more prevalent (G allele of *eNOS* polymorphism, Glu298Asp in Sherpas, and *ACE* I allele in HAPH Kyrgyz) or less prevalent (*ACE* D allele in Andeans) in the different high-altitude populations, data so far are still insufficient to rigorously test any hypothesis regarding the implications of these gene polymorphisms in CMS or HAPH.

Medication for Prevention and Treatment of High-Altitude Illness

Much of the prevailing medications for high altitude were developed as approved medications for other purposes. As these drugs confer some advantage in amelioration of high-altitude adverse effects and also in absence of other drugs at the moment, these are continued to be used. These medications do not restore sea-level physical performance at high altitude and/or may compromise natural acclimatization processes. The prevailing medications are discussed in the following section.

Acetazolamide

[N-(5-Sulfamoyl-1,3,4-thiadiazol-2-yl)acetamide] (trade name, Diamox): The role of acetazolamide in prevention of AMS has been established through multiple trials [9, 18, 25, 78]. Acetazolamide is a carbonic anhydrase inhibitor wherein carbonic anhydrase is inhibited and carbonic acid levels build up. It works by interfering with bicarbonate (HCO_3^-) and forces the kidneys to excrete the bicarbonate through the urine thereby reacidifying the blood [26]. Inhibition of carbonic anhydrase leads to a slowing of the reverse reaction which causes an increase in respiration and thereby aids in acclimatization. It helps some people in rapid ascent to sleeping altitude above 2,700 m (9,000 ft) and allows one to breathe faster and utilize more oxygen, thereby minimizing the symptoms caused by poor oxygenation. The drug is especially helpful at night when respiratory drive is decreased. It takes a while for acetazolamide to have an effect, and thus it is advised to start the drug 24 h before travel to altitude and continue for at least 5 days at the high altitude. It is taken prophylactically starting a few days before going to high attitude [35]. The standard dose recommendation of the Himalayan Rescue Association Medical Clinic is 125 mg twice a day (morning and night) although some individuals may need 250 mg. The Everest Base Camp Medical Clinic (Everest ER), a project of the Himalayan Rescue Association-USA (US-based nonprofit charity organization) and Himalayan Rescue Association (a nonprofit Nepali NGO) established during the 2003, cautions against routine use of acetazolamide as a substitute for a reasonable ascent schedule, except where rapid ascent is forced by flying into high-altitude locations or due to terrain considerations. The center suggests a dosage of 125–250 mg twice daily for prophylaxis, starting from 24 h before ascending until a few days at the highest altitude or on descending with 250 mg twice daily recommended for treatment of AMS. As acetazolamide is a sulfonamide derivative, it is not recommended for sulfa-allergic patients. An undesirable side effect of acetazolamide is a reduction in aerobic endurance performance. Dosage of 1,000 mg/day produces a 25 % decrease in performance over and above reduction in performance due to high-altitude exposure [54]. Other possible side effects include tingling of the lips and finger tips, blurring of vision, and alteration of taste. These side effects may be reduced with the 125 mg. A study by the Denali Medical Research Project concluded that in established cases of acute mountain sickness, treatment with acetazolamide relieves symptoms, improves arterial oxygenation, and prevents further impairment of pulmonary gas exchange [33]. A single study in human showed blunt hypoxic pulmonary vasoconstriction with acetazolamide [77], but any other data supporting its role in HAPE prevention is not available. Interestingly in our gene expression studies, we noted upregulation of carbonic anhydrase IX (*CA9*) gene transcript in HAPE compared to acclimatized controls (fold change 2.52, p-value 2.00E-004) and suggest that the role of this drug or some other moiety designed through in silico approaches should be further probed in HAPE.

Dexamethasone

[8S,9R,10S,11S,13S,14S,16R,17R)-9-Fluoro-11,17-dihydroxy-17-(2-hydroxyacetyl)-10,13,16-trimethyl-6,7,8,9,10,11,12,13,14,15,16,17-dodecahydro-3H-cyclopenta[a]phenanthren-3-one] is a prescription drug that is very effective in treatment of AMS [23, 36, 43]. It decreases brain and other swellings reversing the effects of AMS. Extensive clinical experience supports the use of dexamethasone in patients with HACE. It is a steroid and is administered as an 8 mg dose (intramuscularly, intravenously, or orally) followed by four every 6 h until symptoms resolve. A dose of 4 mg twice a day for a few days starting with ascent prevents most symptoms of altitude illness. It should be used with caution and only on the advice of a physician because of possible serious side effects. It may be combined with Diamox. No other medications have been proven valuable for preventing AMS. The Centers for Disease Control and Prevention (CDC), the national public health institute of the United States, advises that dexamethasone be reserved for treatment of AMS and HACE during descents [37].

Nifedipine

(3,5-Dimethyl 2,6-dimethyl-4-(2-nitrophenyl)-1,4-dihydropyridine-3,5-dicarboxylate) is a dihydropyridine calcium channel blocker. Nifedipine (20 mg sustained release version) over 8 h was shown to be effective at preventing HAPE in climbers/trekkers with a history of recurrent episodes of HAPE [6]. Climbers/trekkers with history of recurrent episodes of HAPE often consider this drug for prophylaxis and are advised to carry nifedipine when at altitude and use with the first sign of HAPE. Nifedipine rapidly decreases pulmonary artery pressure and relieves HAPE. A single, nonrandomized, unblended study demonstrated utility of nifedipine in treatment of HAPE when oxygen or descent was not available [56].

Other Medications for Altitude Illnesses

Ibuprofen [(RS)-2-(4-(2-methylpropyl)phenyl) propanoic acid], a nonsteroidal anti-inflammatory drug, has been found to be effective at relieving altitude headache. Following the onset of altitude sickness, ibuprofen is suggested as anti-inflammatory and painkiller that helps alleviate both headache and nausea associated with AMS, as well as in combating cerebral edema (swelling of the brain) associated with extreme symptoms of AMS [45]. It works by inhibiting the enzyme cyclooxygenase (COX), which converts arachidonic acid to prostaglandin H_2 (PGH_2). PGH_2, in turn, is converted by other enzymes to several other prostaglandins (which are mediators of pain, inflammation, and fever) and to thromboxane A_2 (which stimulates platelet aggregation leading to the formation of blood clots). Two recent studies showed that ibuprofen 600 mg three times daily was effective at preventing AMS. But it was not clear if this affected HAPE or HACE [27, 30]. A single, randomized, controlled trial found that **sumatriptan** may help prevent altitude sickness [40]. Despite their popularity, antioxidant treatments have not been found to be effective medications for prevention of AMS [5]. Phosphodiesterase-5 inhibitors such as **sildenafil** can also selectively lower pulmonary artery pressure, with less effect on systemic blood pressure. Sildenafil significantly improved the cardiovascular and exercise performance measures of trained cyclists at high altitude, and the drug helped some participants improve performance up to 45 %, while others showed little change although it provided no benefit at sea level [38]. Interest in sildenafil has been, however, limited by the possibility that these drugs might worsen the headache of mountain sickness [7]. In a single, randomized, placebo-controlled trial, **tadalafil**, also a phosphodiesterase type 5 enzyme inhibitor, 10 mg twice a day, was found to be effective in preventing HAPE in susceptible individuals during ascent [49] and is being studied for treatment. Another promising possible preventative for altitude sickness may be **myo-inositol trispyrophosphate** (ITPP), a novel membrane-permeant allosteric effector of hemoglobin (Hb) which increases the amount of oxygen released by hemoglobin [16].

Conclusion

With advancement and accessibility to genomic and genome-wide data sets, our understanding of the role of genes and environment in high-altitude environment will get accelerated which can potentially get converted to translational research. Regulating key pathway target genes will be important for management of physiological changes due to environmental stress of high altitude which may be beneficial for rapid acclimatization as well as prevention of maladaptation. One approach which could be promising is tailoring function of hemoglobin-based oxygen carriers (HBOCs) in a specific fashion in disease states and conditions by modulating activity of HIF (hypoxia-inducible factor) which is a master regulator of oxygen homoeostasis and has important functional role in the transcriptional regulation of major pathway genes that mediate the adaptive responses to hypoxia.

The opportunities for using modern research advances in genomics and proteomics and other novel strategies to bring new insights into the study of disease and human populations have never been greater. It will be advantageous to utilize these opportunities and transform medical practice by expanding vision beyond cure and intervening earlier in the treatment process. In current times, great technological potentials are available to stop diseases before they occur by predictive, personalized, preemptive, and participatory medicine. Practicing medicine in this way will help us in understanding the fundamental causes of diseases at their earliest molecular stages so that one can reliably predict how and when a disease will develop and in whom. Utilization of biomarkers, which are the molecular, biological, or physical characteristics that indicate a specific underlying physiological state, is showing impact upon prevention and treatment of diseases like blood pressure [57]. Cholesterol biomarkers have enabled diagnostics and therapies contributing to a nearly 50 % decrease in cardiovascular mortality in the United States over the past 30 years. New powerful scientific resources like genotype-phenotype database for NIH-supported genome-wide association studies are being created through which it is envisioned to establish a common data repository to promote and facilitate widespread sharing of genotype and phenotype data as well as the significant associations between them.

Many challenges, however, remain as one seeks to obtain complete descriptions of the susceptibility architecture of biomedical traits of interest and translate the information into improvements in clinical management. A major challenge is that the risk marker alleles confer very small relative risks (over a range of 1.1–1.5) [50]. When alleles that are associated with a modest increase in risk are combined, they show low discriminatory and predictive ability [41]. In coming years many more common variants conferring a risk of disease or maladaptation to environmental stress may be identified paving way for stability of individual risk estimates. Translational research in high-altitude medicine has just arrived, and a major challenge in coming years will be determining when and how high-altitude genomic information can be usefully applied in clinical health practice. Future research in high-altitude genomics will definitely identify candidate processes at the systemic level and drug candidates which will be suitable for development of new drug and other novel products for translational purposes. The ultimate goal would lie in making genomic information readily accessible for more informed and better management of high-altitude environmental stress.

References

1. Aggarwal S, Negi S, Jha P, Singh PK, Stobdan T, Pasha MAQ, Ghosh S, Agrawal A, Praasher B, Mukerji M (2010) EGLN1 involvement in high-altitude adaptation revealed through genetic analysis of extreme constitution types defined in Ayurveda. Proc Natl Acad Sci U S A 107:18961–18966
2. Ahsan A, Charu R, Pasha MA, Norboo T, Charu R, Afrin F, Ahsan A, Baig MA (2004) eNOS allelic variants at the same locus associate with HAPE and adaptation. Thorax 59:1000–1002
3. Aldashev AA (2007) Gene polymorphisms and high altitude pulmonary hypertension. In: Aldashev AA, Naeije R (eds) Problems of high altitude medicine and biology. Springer, New York, pp 151–168
4. Aldashev AA, Sarybaev AS, Sydykov AS et al (2002) Characterization of high-altitude pulmonary hypertension in the Kyrgyz: association with angiotensin-converting enzyme genotype. Am J Respir Crit Care Med 166:1396–1402
5. Baillie JK, Thompson AA, Irving JB, Bates MG, Sutherland AI, Macnee W, Maxwell SR, Webb DJ (2009) Oral antioxidant supplementation does not prevent acute mountain sickness: double blind randomized placebo-controlled trial. QJM 102:341–348
6. Bartsch P, Maggiorini M, Schobersberger W, Shaw S, Rascher W, Girard J, Weidmann P, Oelz O (1991) Enhanced exercise induced rise of aldosterone and vasopressin preceding mountain sickness. J Appl Physiol 71:136–143
7. Bates MG, Thompson AA, Baillie JK (2007) Phosphodiesterase type 5 inhibitors in the treatment and prevention of high altitude pulmonary edema. Curr Open Invest Drugs 8:226–231
8. Basnyat B, Murdoch DR (2003) High-altitude illness. Lancet 361:1967–1974
9. Basnyat B, Gertsch JH, Holck PS et al (2006) Acetazolamide 125 mg BD is not significantly different from 375 mg BD in the prevention of acute mountain sickness: the Prophylactic Acetazolamide Dosage Comparison for Efficacy (PACE) trial. High Alt Med Biol 7:17–27
10. Beall CM (2000) Tibetan and Andean patterns of adaptation to high-altitude hypoxia. Hum Biol 72:201–228
11. Beall CM, Decker MJ, Brittenham GM, Kushner I, Gebremedhin A, Strohl KP (2002) An Ethiopian pattern of human adaptation to high-altitude hypoxia. Proc Natl Acad Sci U S A 99:17215–17218
12. Beall CM, Cavalleri GL, Deng L, Elston RC, Gao Y, Knight J, Li C et al (2010) Natural selection on EPAS1 (HIF_2alpha) associated with low haemoglobin concentration in Tibetan highlanders. Proc Natl Acad Sci U S A 107:11459–11464
13. Berglund B (1992) High-altitude training aspects of haematological adaptation. Sports Med 14:289–303
14. Bigham A, Mao X, Mei R, Brutsaert T, Wilson MJ, Julian CG, Parra EJ, Akey JM, Moore LG, Shriver

MD (2009) Identifying positive selection candidate loci for high-altitude adaptation in Andean populations. Hum Genomics 4:79–90
15. Bigham A, Bauchet M, Pinto D, Mao X, Akey JM, Mei R, Scherer SW, Julian CG, Wilson MJ, López Herráez D, Brutsaert T, Parra EJ, Moore LG, Shriver MD (2010) Identifying signatures of natural selection in Tibetan and Andean populations using dense genome scan data. PLoS Genet 6:e100111
16. Biolo A, Greferath R, Siwik DA, Qin F, Valsky E, Fylaktakidou KC, Pothukanuri S, Duarte CD, Schwarz RP, Lehn JM, Nicolau C, Colucci WS (2009) Enhanced exercise capacity in mice with severe heart failure treated with an allosteric effector of hemoglobin, myo-inositol trispyrophosphate. Proc Natl Acad Sci U S A 106:1926–1929
17. Bryant JA, Atherton MA, Collins MW (2007) Design and Information in Biology. From Molecules to System. WIT Press, Southampton, p 108
18. Cain SM, Dunn JE (1966) Low doses of acetazolamide to aid accommodation of men to altitude. J Appl Physiol 21:1195–1200
19. Charu R, Stobdan T, Ram RB, Khan AP, Pash MAQ, Norboo T, Afrin F (2006) Susceptibility to high altitude pulmonary oedema: role of ACE and ET-1 polymorphisms. Thorax 61:1011–1012
20. Clerici C, Planes C (2009) Gene regulation in the adaptive process to hypoxia in long epithelial call. Am J Physiol Lung Cell Mol Physiol 296:L267–L274
21. Droma Y, Hanaoka M, Hotta J, Katsuyama Y, Ota M, Kobayashi T, Kubo K (2003) The r506 Q mutation of coagulation factor V gene in high altitude pulmonary edema susceptible subjects. High Alt Med Biol 4:497–498
22. Denhert C, Grunig E, Mereles D, von Lennep N, Bartsch P (2005) Identification of individuals susceptible to high-altitude pulmonary oedema at low altitude. Eur Respir J 25:545–551
23. Ferrazzini G, Maggiorini M, Kriemler S, Bartsch P, Oelz O (1987) Successful treatment of acute mountain sickness with dexamethasone. Br Med J (Clin Res Ed) 294:1380–1382
24. Flowers J, Sezgin E, Kumagai S, Duvernell D, Matzkin L, Schmidt P et al (2007) Adaptive evolution of metabolic pathways in Drosophila. Mol Biol Evol 24:1347–1354
25. Forwand SA, Landowne M, Follansbee JN, Hansen JE (1968) Effect of acetazolamide on acute mountain sickness. N Engl J Med 279(16):839–845
26. Fulco CS, Ditzler D, Soares R, Lammi E, Muza SR, DW Degroot (2002) Effect of acetazolamide on isolated quadriceps muscle endurance performance at sea level and during acute altitude exposure. US Army Research Institute of Environmental Medicine Technical Report no T 02/9
27. Grant KP, Nicholas PH, Constance B, Gertssch J (2012) Ibuprofen prevents altitude illness a randomized controlled trial for prevention of altitude illness with non-steroidal anti-inflammatory. Ann Emerg Med 59:484–490
28. Gracey AY (2007) Interpreting physiological responses to environmental change through gene expression profiling. J Exp Biol 210:1584–1592
29. Gracey AY, Cossins AR (2003) Application of microarray technology in environmental and comparative physiology. Ann Rev Physiol 65:231–259
30. Gertsch JH, Lipman GS, Kanaan NC, Holck PS (2012) Altitude sickness in climbers and efficacy of NSAIDs trial (ASCENT) randomized controlled trial of Ibuprofen versus placebo for prevention of altitude illness. Wilderness and Environ Med 23:307–315
31. Gelfi C, De Palma S, Ripamonti M, Eberin I, Cerretelli P (2004) New aspects of altitude adaptation in Tibetans: a proteomic approach. FASEB J 18:612–614
32. Grocott MPW, Martin DS, Levett DZH, McMorrow R, Windsor J, Montgomery HE (2009) Arterial blood gases and oxygen content in climbers on Mount Everest. N Engl J Med 360:140–149
33. Grissom CK, Roach RC, Sarnquist FH, Hackett PH (1992) Acetazolamide in the treatment of acute mountain sickness clinical efficacy and effect on gas exchange. Ann Intern Med 116:461–465
34. Hochachka PW, Rupert JL (2003) Fine tuning the HIF-1 global O_2 sensor for hypobaric hypoxia in Andean high-altitude natives. Biol Essays 25:515–519
35. Hackett PH, Roach RC (2001) High altitude illness. N Engl J Med 345:107–114
36. Hackett PH, Roach RC, Wood RA et al (1988) Dexamethasone for prevention and treatment of acute mountain sickness. Aviat Space Environ Med 59:950–954
37. Hackett P, Shlim D (2009) The pre-travel consultation – self-treatable diseases – altitude illness. In: Turell D, Brunette G, Kozarsky P, St. Louis Mosby, Lefor A (eds) CDC health information for international travel 2010 "The Yellow Book". St. Louis. Mosby. ISBN 0-7020-3481-9
38. Hsu AR, Barnholt KE, Grundmann NK, Lin JH, McCallum SW, Friedlander AL (2006) Sildenafil improves cardiac output and exercise performance during acute hypoxia, but not normoxia. J Appl Physiol 100:2031–2040
39. Hultgren HN, Spickard WB, Hellriegel K, Houston CS (1961) High altitude pulmonary edema. Medicine (Baltimore) 40:289–313
40. Jafarian S, Gorouhi F, Salimi S, Lotfi J (2007) Sumatriptan for prevention of acute mountain sickness: randomized clinical trial. Ann Neurol 62:273–277
41. Janssens AC, van Duijin CM (2008) Genome-based prediction of common diseases: advances and prospects. Hum Mol Genet 17:166–173
42. León-Velarde F, Mejía O (2008) Gene expression in chronic high altitude diseases. High Alt Med Biol 9:130–139

43. Levine BD, Yoshimura K, Kobayashi T, Fukushima M, Shibamoto T, Ueda G (1989) Dexamethasone in the treatment of acute mountain sickness. N Engl J Med 321:1707–1713
44. Li FZ, Zhou F, Jiang CZ, Sun SY, He MA, Zhang SY, Liao R, Zeng HS, Wu YM, Wu TC (2004) Relationship between heat stress protein 70 gene polymorphisms and the risk of acute mountain sickness. Zhonghua Lao Dong Wei Sheng Zhi Ye Bing Za Zhi 22:413–415 (article in Chinese)
45. Lipman JS, Kanan NC, Holck PS, Constance BB, Gertsc JH (2012) Ibuprofen prevents altitude illness: a randomized controlled trial for prevention of altitude illness with nonsteroidal anti-inflammatories. Ann Emerg Med 58:484–490
46. Lockhart DJ, Winzeler EA (2000) Genomics, gene expression and DNA arrays. Nature 405:827–836
47. Lorenzo VF, Yang Y, Simonson TS, Nussenzvei R, Jorde LB, Prchal JT, Ge RL (2009) Genetic adaptation to extreme hypoxia: study of high-altitude pulmonary edema in a three generation Han Chinese family. Blood Cells Mol Dis 42:221–225
48. Maclnnis MJ, Michael SK, Jim LR (2010) Evidence for a genetic basis for altitude illness: 2010 update. High Alt Med Biol 11:349–368
49. Maggiorini M, Brunner La Rocca HP, Peth S et al (2006) Both tadalafil and dexamethasone may reduce the incidence of high-altitude pulmonary edema: a randomized trial. Ann Intern Med 145:497–506
50. Manolio TA, Collins FS, Cox NJ et al (2009) Finding the missing heritability of complex diseases. Nature 461:747–753
51. Mejia OM, Prchal JT, Leon-Velarde F, Hurtado A, Stockton DW (2005) Genetic association analysis of chronic mountain sickness in an Andean high-altitude population. Haematologica 90:13–19
52. Mishra A, Mohammad G, Thinlas T, Pasha MAQ (2013) *EGLN1* variants influence expression and SaO_2 levels to associate with high-altitude pulmonary oedema and adaptation. Clin Sci 124:479–489
53. Montgomery HE, Marshall R, Hemingway H et al (1998) Human gene for physical performance. Nature 393:221–222
54. Muza SR, Fulco CS, Cymerman A (2004) Altitude acclimatization guide. USARIEM Technical Note TN 04-05
55. Norboo T, Saiyed HN, Angchuk PT, Tsering P, Angchuk ST, Phuntsog ST, Yahya M, Wood S, Bruce NG, Ball KP (2004) Mini review of high altitude health problems in Ladakh. Biomed. Pharmacotherapy 58:220–225
56. Oelz O, Maggiorini M, Ritter M et al (1989) Nifedipine for high altitude pulmonary oedema. Lancet 2:1241–1244
57. Orringer CE (2013) Non-HDL Cholesterol, ApoB and LDL particle concentration in coronary heart disease risk prediction and treatment. Clin Lipidol 8:69–79
58. Pandey A, Mann M (2000) Proteomics to study genes and genomes. Nature 405:837–846
59. Penaloza D, Arias-Stella J (2007) Healthy highlanders and chronic mountain sickness. Circulation 115:1132–1146
60. Pasha MAQ, Khan AP, Kumar R et al (2001) Angiotensin converting enzyme insertion allele in relation to high altitude adaptation. Ann Hum Genet 65:531–536
61. Pasha MAQ, Charu R, Ahsan A, Norboo T (2005) HAPE susceptibility associates with variants of genes of vascular homeostasis. Human Genome Meeting (HUGO HGM). Poster abstract 310
62. Qi Y, Niu W, Zhu T, Zhou W, Qiu C (2008) Synergistic effect of the genetic polymorphisms of the rennin- angiotensin aldosterone system on high altitude pulmonary edema: a study from Qinghai-Tibet altitude. Eur J Epidemiol 23:143–152
63. Qi Y, Niu WQ, Zhu TC, Liu JL, Dong WY, Xu Y, Ding SQ, Cui CB, Pan YJ, Yu GS et al (2009) Genetic interaction of Hsp70 family genes polymorphisms with high-altitude pulmonary edema among Chinese railway constructors at altitudes exceeding 4000 meters. Clin Chim Acta 405:17–22
64. Qi L, Yang L, Chen H (2011) Detecting and quantitating physiological ER stress in mammals. Methods Enzymol 490:137–146
65. Rajput C, Arif E, Vibhuti A, Stobdan T, Khan AP, Norboo T, Afrin F, Qadar Pasha MA (2006) Predominance of interaction among wild-type alleles of CYP11B2 in Himalayan native associates with high-altitude adaptation. Biochem Biophys Res Commun 348:735–740
66. Richalet JP, Larmiqnat P, Poitrine E, Letournel M, Canoui-Poitrine F (2012) Physiological risk factors for severe high altitude illness: a prospective cohort study. Am J Respir Crit Care Med 185:192–198
67. Rupert JL, Kidd KK, Norman LE, Monsalve MV, Hochachka PW, Devine DV (2003) Genetic polymorphisms in the rennin-angiotensin system in high-altitude and low-altitude native American populations. Ann Hum Genet 67:17–25
68. Stobdan TA, Khan Z, Nejatizadeh AP, Ram A, Thinlas R, Mohammad T, Norboo G, Himashree T, Pasha G, MAQ (2011) Polymorphisms of renin-angiotensin system genes as a risk factor for high-altitude pulmonary oedema. J Renin Angiotensin Aldosterone Syst 12:93–101
69. Sarkar S, Banerjee PK, Selvamurthy W (2003) High altitude hypoxia: an intricate inerplay of oxygen responsive macroevents and micromolecules. Mol Cell Biochem 253:287–305
70. Sarkar S (2012) Hypoxic signature of high altitude acclimatization: a gene expression study. Ind J Aerosp Med 56:1–10
71. Saxena S, Kumar R, Madan T, Gupta V, Muralidhar K, Sarma PU (2005) Association of polymorphisms in pulmonary surfactant protein A1 and A2 gene with high-altitude pulmonary edema. Chest 128:1611–1619
72. Scheinfeldt LB, Soi S, Thompson S, Ranciaro A, Woldemeskel D, Beggs W, Charla Lambert C, Jarvis JP, Abate D, Belay G, Tishkoff SA (2012) Genetic

adaptation to high altitude in the Ethiopian highlands. Genome Biol 13:R1
73. Sharma M, Singh SB, Sarkar S (2014) Genomewide expression analysis suggests perturbation of vascular homeostatis during high attitude pulmonary edema. Plos One 9:e85902
74. Simonson TS, Yang Y, Huff CD, Yun H, Qin G, Witherspoon DJ, Bai Z, Lorenzo FR, Xing J, Jorde LB, Prchal JT, Ge R (2010) Genetic evidence for high-altitude adaptation in Tibet. Science 329:72–75
75. Srivastava S, Bhagi S, Kumari B, Chandra K, Sarkar S, Ashraf MZ (2011) Association of polymorphisms in angiotensin and aldosterone synthase genes of the renin–angiotensin–aldosterone system with high-altitude pulmonary edema. JRAAS 13:155–160
76. Storz JF, Lockhart DJ, Winzeler E, Scott GR, Cheviron ZA (2010) Phenotypic plasticity and genetic adaptation to high-altitude hypoxia in vertebrates. Nature 405:827
77. Teppema LJ, Balanos GM, Steinback CD et al (2007) Effects of acetazolamide on ventilatory, cerebrovascular, and pulmonary vascular responses to hypoxia. Am J Respir Crit Care Med 175:277–281
78. van Patot MC, Leadbetter G, Keyes LE, Maakestad KM, Olson S, Hackett PH (2008) Prophylactic low-dose acetazolamide reduces the incidence and severity of acute mountain sickness. High Alt Med Biol 9:289–293
79. Wang P, Koehle MS, Rupert JL (2009) Genotype at the missense G894T polymorphism (Glu298Asp) in the NOS3 gene is associated with the susceptibility to acute mountain sickness. High Alt Med Biol 10:261–267
80. Wang P, Ha AY, Kidd KK, Koehle MS, Rupert JL (2010) A variant of the endothelial nitric oxide synthase gene (NOS3) associated with AMS susceptibility is less common in the Quechua a high altitude Native population. High Alt Med Biol 11:27–30
81. Wagner A (2009) Evolutionary constraints permeate large metabolic networks. BMC Evol Biol 9:231
82. Wagner DR, Fargo JD, Parker D, Tatsugawa K, Young TA (2006) Variables contributing to acute mountain sickness on the summit of Mt. Whitney. Wilderness Environ Med 17:221–228
83. Watson JD, Crick FHC (1933) A structure of deoxyribose nucleic acid. Nature 171:737–738
84. Wu T, Li S, Ward MP (2005) Tibetans at extreme altitude. Wilderness Environ Med 16:47–54
85. Wu TY, Ding SQ, Liu JL, Yu MT, Jia JH, Duan JQ, Chai ZC, Dai RC, Zhang SL, Liang BZ et al (2009) Reduced incidence and severity of acute mountain sickness in Qinghai- Tibet railroad construction workers after repeated 7 month exposures despite 5 month low altitude periods. High Alt Med Biol 10:221–232
86. Xu S, Li S, Yang Y, Tan J, Lou H, Jin W, Yang L, Pan X, Wang J, Shen Y, Wu B, Wang H, Jin L (2011) A genome-wide search for signals of high-altitude adaptation in Tibetans. Mol Biol Evol 28:1003–1011
87. Yi X, Liang Y, Huerta-Sanchez E, Jin X, Cuo ZXP, Pool JE, Xu X, Jiang H et al (2010) Sequencing of 50 human exomes reveals adaptation to high altitude. Science 329:75–78
88. Zhou D, Udpa N, Ronen R, Stobdan T, Liang J, Appenzeller O, Zhao HW, Yin Y, Du Y, Guo L, Cao R, Wang Y, Jin X, Huang C, Jia W, Cao D, Guo G, Gamboa JL, Villafuerte F, Callacondo D, Xue J, Liu S, Frazer KA, Li Y, Bafna V, Haddad GG (2013) Whole-genome sequencing uncovers the genetic basis of chronic mountain sickness in Andean highlanders. Am J Hum Genet 93:452–462

Hypoxia in Acute Chemical Emergencies: Toxicity, Mechanism, and Treatment

Rahul Bhattacharya and M.P. Kaushik

Abstract

Hypoxia is a pathological condition in which the whole body or a particular region of the body suffers from oxygen deprivation. There are various kinds of hypoxia viz. hypoxemic hypoxia (e.g., high-altitude pulmonary edema, high-altitude cerebral edema, etc.), anemic hypoxia (e.g., chronic anemia, carbon monoxide (CO) poisoning, etc.), stagnant hypoxia (heart failure, vasodilatory shock, etc.), and histotoxic hypoxia. In histotoxic hypoxia, the quantity of oxygen reaching the cells is normal, but the cells are unable to utilize the oxygen efficiently. This is due to inactivation of enzymes involved in oxidative phosphorylation. This causes profound reduction in adenosine triphosphate production by the mitochondria, leading to various cellular perturbations, and possibly cell death. The gases considered as chemical asphyxiants act either by decreasing the oxygen-carrying capacity of the blood (CO) or by interfering with cellular utilization of oxygen (cyanide and hydrogen sulfide). Regardless of mode of action, chemical asphyxiants can cause severe hypoxia (chemical hypoxia), which is detrimental for the living organisms. Acute chemical emergencies can be encountered during industrial disasters, occupational exposures, accidents, use of chemical warfare agents, and acts of terrorism. This chapter discusses the toxicity, molecular mechanism(s) of action, and treatment modalities of chemical asphyxiants. Possible implication of organophosphorus compounds in producing chemical hypoxia is also addressed. This information will be useful for medical management of hypoxia-related chemical emergencies.

Introduction

Oxygen is a fundamental requirement for animal existence. Complete deficiency of oxygen in the body is referred as anoxia, while deprivation of oxygen is regarded as hypoxia. Hypoxia is a life-threatening condition in which oxygen supply

R. Bhattacharya (✉) • M.P. Kaushik
Defence Research and Development Establishment, Jhansi Road, Gwalior 474 002, Madhya Pradesh, India
e-mail: Rahul@drde.drdo.in; rbhattacharya41@rediffmail.com

S.B. Singh et al. (eds.), *Translational Research in Environmental and Occupational Stress*, DOI 10.1007/978-81-322-1928-6_19,

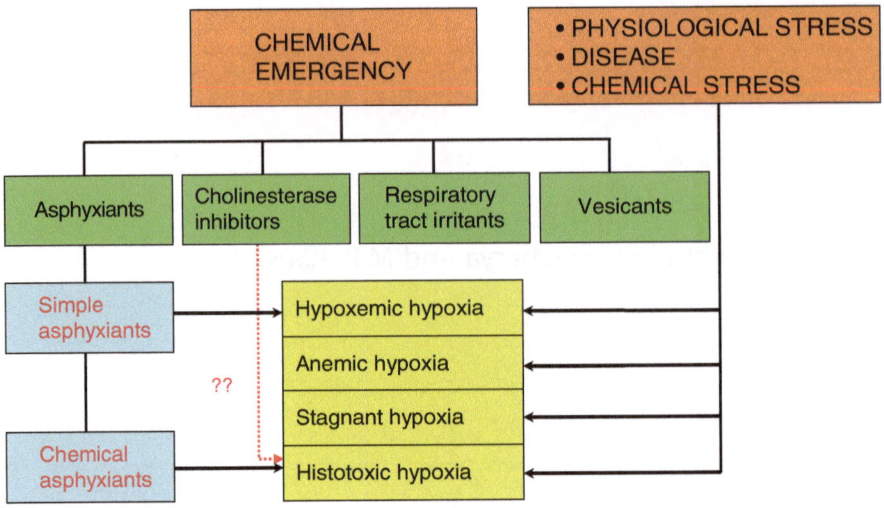

Fig. 1 Factors contributing to different types of hypoxia

in the body is inadequate to meet the metabolic requirements of the tissue [1, 2]. Hypoxia is known to occur after ischemia, hemorrhage, stroke, premature birth, and other cardiovascular difficulties, among which hemorrhagic shock is the leading cause of death and complications. These complications can be experienced in both military and civilian settings [3, 4]. Hypoxia may lead to systemic inflammation response syndrome (SIRS), multiple organ dysfunction (MOD), and multiple organ failure (MOF) [5]. Hypoxia can occur during physiological stress, disease, or chemical stress. There are four major types of hypoxia as shown in Fig. 1: (1) hypoxemic hypoxia is the most common type of hypoxia which is caused by decreased oxygen in air or the inability to diffuse the oxygen across the lungs. In this condition, the person has less than 100 % saturation of the blood in the arteries. This can happen at an altitude where the oxygen content of air is low. This can usually be experienced at an altitude over 12,500 ft, when one suffers from high-altitude pulmonary edema (HAPE) and high-altitude cerebral edema (HACE). Also, in the case of damaged lungs, oxygen transfer is impaired as in the case of smokers; (2) anemic hypoxia is a state in which the arterial partial pressure of oxygen is normal, but total oxygen content of the blood is reduced due to decrease in circulating hemoglobin, which carries the oxygen. This occurs in chronic anemia and carbon monoxide (CO) poisoning, (3) stagnant hypoxia is also known as ischemic hypoxia, which occurs due to decreased blood flow, as observed during heart failure and vasodilatory shock. This usually happens due to reduced cardiac output (impaired pumping) or venous pooling during high G force stress, and (4) histotoxic hypoxia occurs when blood is enriched in oxygen, but cells are unable to utilize the oxygen. This happens due to impaired extraction of oxygen from the circulation by the body tissues, as observed in case of cyanide, CO, and hydrogen sulfide (H_2S) poisoning [1, 2].

A chemical emergency occurs when a hazardous chemical has been released, and the release has the potential for harming people's health. Chemical release can be unintentional or accidental. This includes industrial mishap, occupational exposure, recreational mishap, and natural catastrophe. Intentional exposures include use of chemical warfare in war or terrorist attack with a chemical weapon [6–14]. The major classes of chemicals responsible for chemical emergencies include asphyxiants (e.g., cyanide, CO), cholinesterase inhibitors (e.g., organophosphorus (OP) nerve agents), respiratory tract irritants (e.g., chlorine, phosgene), and vesicants (e.g., mustards, arsenicals) [12]. Various factors contributing to different types of hypoxia are depicted in Fig. 1. Asphyxiants are gases that deprive body tissues of oxygen. They are generally divided into two categories, namely, simple and chemical asphyxiants [15]. Simple asphyxiants usually displace oxygen from ambient air. In a confined space, when the concentration of such gases increases, with concomitant decrease in the fraction of inspired oxygen (FiO_2), it leads to hypoxia. Simple asphyxiants include carbon dioxide (CO_2), nitrogen, and propane. On the other hand, chemical asphyxiants act either by decreasing the oxygen-carrying capacity of the blood (e.g., CO) or by interfering with cellular utilization of oxygen (e.g., cyanide and H_2S). Such agents inactivate the enzymes like cytochrome *c* oxidase (CCO), which is involved in oxidative phosphorylation, leading to histotoxic hypoxia. This results in profound reduction in adenosine triphosphate (ATP) production by mitochondria, leading to various cellular perturbations and possibly cell death [12]. There are several other chemicals, which are also known to interrupt the mitochondrial electron transport chain leading to hypoxia, namely, rotenone, antimycin A, azides, and 1-methyl-4-phenyl-1,2,3,6-tetrahydropyridine (MPTP), etc. [16, 17]. Several reports were published on cellular and molecular events immediately following hypoxia [18–20]. Although, organophosphorus (OP) compounds, including the highly toxic nerve agents, have not been categorized as chemical asphyxiants, they have also been shown to produce effects similar to hypoxia caused by chemical asphyxiants [21–24]. This chapter discusses the toxicity, molecular mechanism(s) of action, and treatment modalities of some important chemical asphyxiants. Possible implications of OP compounds in induction of chemical hypoxia are also discussed. This chapter will be useful for medical management of hypoxia during chemical emergencies.

Biology of Hypoxia

In this section we have discussed the molecular mechanisms underlying generalized hypoxia, without any particular focus on histotoxic hypoxia. Oxygen is essential for cell survival; therefore, tissue oxygenation is essential for all normal physiological functions, predominantly the cell growth, respiration, and metabolism. Oxygen is an essential nutrient for mammalian cells because of its role as the terminal electron acceptor in oxidative phosphorylation. Supplied through breathing, oxygen dissolves in plasma and tissues and is carried throughout the body as oxyhemoglobin (O_2Hb), which releases oxygen when it is needed for mitochondrial respiration. However, in several medical conditions like high-altitude or underwater activities, and rigorous exercise, deprivation of oxygen is experienced leading to hypoxia. Additionally, accidental or deliberate exposure to certain chemicals (asphyxiants) can cause severe hypoxia and tissue pathology, resulting in morbidity or mortality [25]. The molecular biology of hypoxia has been possibly explained by many [3, 4, 26–28]. Three thresholds of hypoxia have been defined [29]. They are as follows: (1) when cellular oxygen decreases but ATP production is maintained at a level sufficient to match ATP demand by metabolic adaptation; (2) when steady state ATP turnover is maintained by the production of ATP from anaerobic glycolysis, generating two molecules of ATP per one molecule of glucose metabolized; this, however, cannot meet the metabolic demands of functionally active brain, kidney, and liver, and they suffer from hypoxia; and (3) when glycolysis becomes insufficient to produce enough ATP to maintain cell function and structural integrity. Many tissues after the second threshold develop ATP depletion. Nerve cells in the brain require high levels of ATP to operate the sodium-potassium pump. At 50–60 % depletion of ATP, depolarization of the membrane and uptake of sodium and water take place [25]. Possible mechanistic pathways of hypoxia-induced cell injury have been depicted in Fig. 2. Hypoxia has been shown to perturb calcium homeostasis. An influx of $[Ca^{2+}]_i$ through voltage-gated calcium channels occur that lead to glutamate neurotransmitter release [30, 31]. Glutamate binds to glutamate receptors, initiating a process of $[Ca^{2+}]_i$ influx and excitatory injury called the glutamate cascade. This results in cascade of complex events leading to cell swelling and lysis,

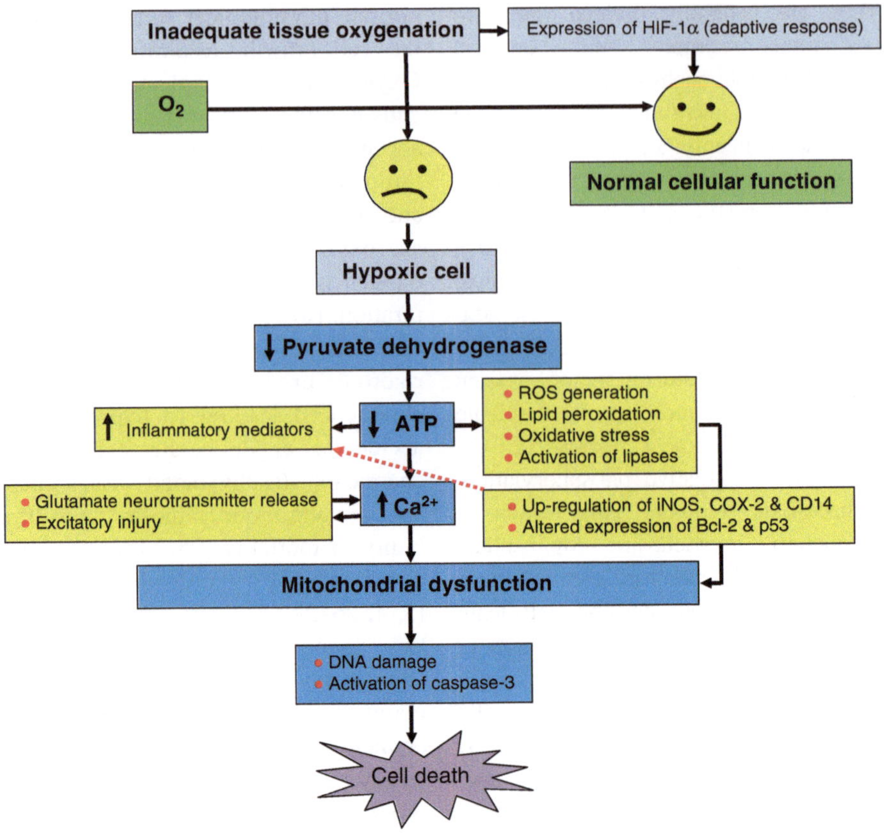

Fig. 2 Possible mechanistic pathway of hypoxia-induced cell injury

characterizing apoptotic type of cell death [32]. Influx of $[Ca^{2+}]_i$ is also known to be activated due to inhibition of Na^+/Ca^{2+} exchanger [33]. It is also believed that cell death following abrupt increase in $[Ca^{2+}]_i$ is mediated by activation of Ca^{2+}-dependent enzymes that trigger the downstream cascade reaction, thereby probably leading to either necrosis or apoptosis [34, 35]. Overexpression of heat-shock protein (HSP-70i) has been shown to reduce the hypoxia-induced increase in intracellular Ca^{2+} and probably increases cell survival [36]. There are several instances of hypoxia, where the induction of inducible nitric oxide synthase (iNOS), along with cyclooxygenase (COX)-2, and CD14 upregulation has been documented [37, 38]. This promotes the inflammatory response by the rapid and excessive production of nitric oxide (NO) and prostaglandins. In addition, hypoxia also alters expression of Bcl-2 and p53. Expression of Bcl-2 is usually accompanied by inactivation of tumor suppressor protein p53, which frequently leads to tumor progression [39]. However, these responses to hypoxia are organ and cell specific [3]. Hypoxia has also been shown to trigger the activity of caspases (executioners of apoptosis), particularly caspase-3, which has been associated with DNA fragmentation [40]. One study elucidated the implication of nuclear Ca^{2+} influx and nuclear Ca^{2+}-mediated mechanisms, including signal transduction, apoptotic gene transcription, caspase activation, and nuclear DNA fragmentation, which resulted in hypoxic neuronal injury in the newborn brain [18]. Further, hypoxia-induced Bax and Bcl-2 protein expression, caspase-9 activation, DNA fragmentation, lipid peroxidation in mitochondria of the cerebral cortex of newborn piglets, and possible role of NO was also reported [19]. Authors concluded that during post-hypoxic reoxygenation, the increase in Bax protein expression and nuclear DNA fragmentation were mediated by NO derived from neuronal nitric oxide synthase (nNOS). It was further proposed that in addition to NO-mediated nuclear DNA damage, the hypoxia-induced increased ratio of Bax/Bcl-2-protein could lead to caspase-activated cascade of hypoxic neuronal death during post-hypoxic reoxygenation [20]. Chemical hypoxia induced by antimycin A, yet another inhibitor of mitochondrial electron transport, showed the involvement of ATP depletion, reactive oxygen species (ROS), alteration in the mitochondrial permeability transition, and increase in intracellular Ca^{2+} in human glioma cells [41].

Whatever may be the sequence of events after hypoxia, there are some neuronal cells which can withstand the hypoxic crisis and survive [31]. These surviving cells develop adaptive responses, and express high levels of hypoxia inducible factor 1α (HIF-1α), which is a transcription factor and a major regulator of the cellular response to hypoxia [42]. In mammalian cells, HIF-1α has been shown to play an essential role in the cellular and systemic oxygen homeostasis [43]. HIF-1α has been considered to be a master transcriptional regulator in hypoxia [44] and be a key modulator for the induction of genes that facilitate adaption and survival of cells [45]. Various genes targeted by HIF-1α include vascular endothelial growth factor (VEGF), insulin-like growth factors (IGF), and transforming growth factors (TGF), which can activate signal transduction pathways, leading to cell proliferation and survival [3].

Cyanide

Source

The use of cyanide in military operations is negligible. However, its possible use in mass terrorism is a matter of concern to civil authorities [13, 46, 47]. Cyanide is widely used in industries including mining operations, photography, tannery, production of plastics, pigments, and dyes and often used as fumigant pesticides [48–50]. Fire victims can also inhale significant amount of cyanide and CO gases, which may cause synergistic toxicity in humans [51]. In a closed space, fire induces a combination of oxygen deprivation directly related to combustion and simultaneous intoxication by asphyxiant and irritant gases [52]. Iatrogenic release of toxic levels of cyanide has been reported after administration of drugs like amygdalin (Laetrile®), sodium nitroprusside (Nipride®), and succinonitrile [53–55]. Cigarette smoke and cassava-based food also contain cyanide [54].

Signs and Symptoms

The maximum permissible concentration for hydrogen cyanide (HCN) in human is 11 mg/m^3, and orally the fatal dose is estimated at 50–100 mg [50]. Inhalation of high concentration of hydrogen cyanide produces reddish pink hue on the skin, accompanied by tachypnea, tachycardia, and nonspecific symptoms originating from central nervous system (CNS). Stupor, coma, and seizure immediately precede respiratory arrest and cardiovascular collapse, culminating in death [15]. Signs and symptoms of acute cyanide poisoning are often nonspecific and vary in both time and intensity depending upon the scale of exposure [56]. Probably the most widespread pathologic condition attributed to chronic cyanide poisoning is tropic ataxic neuropathy (TAN) following cassava consumption [57]. Usually arterial blood gases, serum electrolytes, cyanide, and thiocyanate levels in the blood or urine are measured to confirm cyanide poisoning. "Lee-Jones Test" is a quick bedside test that can qualitatively detect cyanide in gastric aspirate [50].

Mechanism of Action

Cyanide is described as a selective neurotoxin [58]. The cyanide anion (CN^-) rapidly causes a non-competitive inhibition of the mitochondrial metalloenzyme, cytochrome *c* oxidase (CCO). The enzyme is essential for ATP production. Consequently, aerobic oxidative metabolism and phosphorylation are compromised leading to histotoxic hypoxia and lactic acidosis [52]. Various molecular events mediating cyanide toxicity have been illustrated in Fig. 3. Cyanide toxicity cannot be

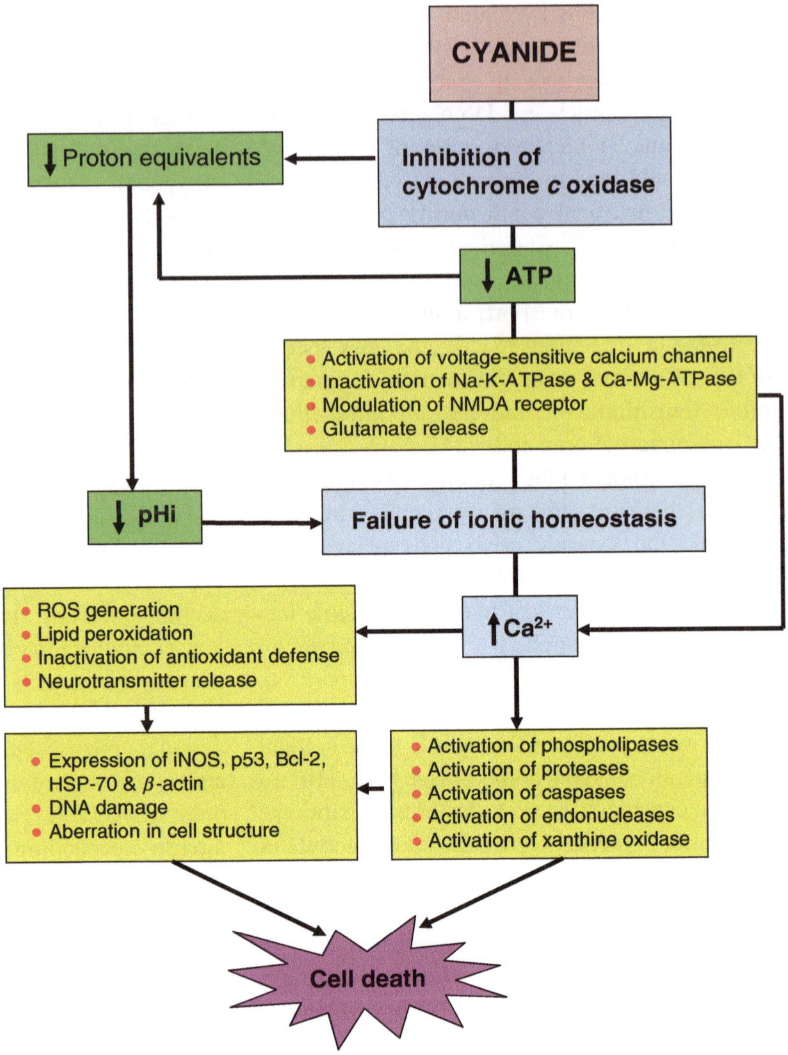

Fig. 3 Various molecular events underlying cyanide toxicity

attributed to a single biochemical lesion but a complex phenomenon. The hypoxia-induced neurotoxicity of cyanide is also known to be mediated through elevated intracellular $[Ca^{2+}]$i levels, inhibition of antioxidant enzymes, generation of ROS, and lipid peroxidation [59–61]. Cyanide-induced chemical hypoxia has been shown to increase the expression of mRNA of iNOS, p53, Bcl-2, heat-shock protein 70 (HSP-70), and β-actin in human cells in vitro [40]. A correlation between chemical hypoxia-induced apoptosis and expression of iNOS and NO production has been shown earlier. Further, inhibition of iNOS limited T-cell apoptosis by decreasing the activity of caspase-3 without affecting the expression of Fas/Apo-1/CD95 on the surface membrane of T cells [38]. Subsequently, cyanide-induced oxidative stress [62] was found to be accompanied by increased expression of HSP-70 activity in vivo, without any concurrent change in the gene expression of antioxidant enzymes [63]. Modulation of *N*-methyl-D-aspartate (NMDA) receptor has been widely implicated in cyanide-induced neurotoxicity [64]. Alterations in the levels of dopamine and protein kinase C, calmodulin, and NO-dependent cyclic guanosine monophosphate (GMP)-dependent enzymes are known to be involved in cyanide-induced convulsions [65, 66]. Most of the biochemical lesions produced by cyanide are emanating from histotoxic hypoxia, lactic acidosis, and altered bioenergetics [67].

Treatment

The onset of cyanide toxicity is very fast, which necessitates immediate and vigorous treatment. The recovery of the victim largely depends on termination of exposure, supportive care, and institution of specific treatment. The supportive therapy includes mechanical airway support, artificial ventilation with 100 % oxygen, possibly delivered through an Ambu bag containing amyl nitrite. Lactic acidosis resulting from anaerobic metabolism and convulsions can be treated with sodium bicarbonate and diazepam, respectively [48, 50]. Critical patients, who are not satisfactorily responding to supportive therapy, should be administered specific cyanide antidotes including sodium nitrite and sodium thiosulfate. Sodium nitrite oxidizes hemoglobin to ferrihemoglobin, which preferentially binds with cyanide to form cyanomethemoglobin. As a result, the inhibited CCO is restored for normal activity. Further, sodium thiosulfate enzymatically eliminates cyanide through urine in the form of thiocyanate. 4-dimethylaminophenol (DMAP), dicobalt edetate, and hydroxocobalamin are also some popular cyanide antidotes used in different countries [50]. There are several other investigational drugs, which have shown promise against experimental cyanide poisoning [50]. One of these is a carbonyl compound viz. α-ketoglutarate, which neutralizes cyanide to form a cyanohydrin complex [68]. Presently, this molecule is undergoing clinical trials in humans as an oral antidote for cyanide [69].

Carbon Monoxide

Source

Yet another classical chemical asphyxiant is CO, which is a colorless, odorless, tasteless, nonirritating gas, and difficult to detect. It is produced due to incomplete combustion of organic matter due to insufficient oxygen supply, which prevents complete oxidation to CO_2. After CO_2, it is the most abundant atmospheric pollutant. CO is the most common cause of fatal occupational inhalation in the United States [70, 71]. Natural sources of CO include forest fires and volcanic eruptions, but mostly anthropogenic activities include escape of carbon monoxide from internal combustion engine and industrial discharges [15, 72]. It is often produced in domestic or industrial settings by old motor vehicles and gasoline-powered engines, heaters, and cooking equipments. Cigarette smoke is also known to produce CO [73].

Signs and Symptoms

Different people and populations may have different CO tolerance levels [74]. On an average, exposure to 100 ppm or greater can be dangerous to human health [73]. During mild CO poisoning, headache, nausea, vomiting, lactic acidosis, dizziness, myalgia, or confusion may be experienced [75]. However, after severe CO poisoning, victims are known to suffer from serious neuropsychiatry abnormality (e.g., depression, memory loss, delirium, hallucination, dizziness, etc.) or cardiovascular instability (e.g., altered sensorium, seizure, coma, syncope, ischemia, etc.). In severe cases, CNS depression, unconsciousness, and respiratory arrest occur, immediately followed by death [76]. A diagnosis of CO poisoning is confirmed by an elevated carboxyhemoglobin (HbCO) level, for which there are specific and rapid tests available [12, 77]. However, low HbCO values may be interpreted with caution, because of possible interference of oxygen therapy [77]. Usually, HbCO is measured by a CO-oximeter and pulse oximeter [78].

Mechanism of Action

When CO is not ventilated, it binds to the ferrous (Fe^{2+}) in the heme prosthetic group of hemeproteins, which includes hemoglobin, myoglobin, and some intracellular enzymes (cytochromes, P-450). The affinity between CO and hemoglobin is approximately 230 times stronger than the affinity between oxygen and hemoglobin. As a result, oxygen fails to dissociate from the HbCO, leading to tissue hypoxia. Hemoglobin is a tetramer with four oxygen binding sites. The binding of CO at one of these site increases the oxygen affinity of the remaining three sites, which refrains the hemoglobin molecule from delivering the oxygen to the tissue, leading to hypoxia [79]. Additionally, HbCO causes a shift of the O_2Hb dissociation curve to the left, reducing delivery of oxygen to already anoxic tissues [78]. The HbCO can revert to hemoglobin, but the recovery takes time because the complex is fairly stable. Also, CO binds to the hemeprotein myoglobin, impairing its ability to utilize oxygen. This may cause reduced cardiac output and hypotension leading to brain ischemia [80]. Similar to cyanide, CO also inhibits the activity of CCO, hampering the ATP synthesis. As a result, cells resort to anaerobic metabolism, causing anoxia, lactic acidosis, and death [81]. Further, CO has also been shown to prompt endothelial cells and platelets to release NO, leading to formation of oxygen free radicals including peroxynitrite [75]. This leads to temporal mitochondrial dysfunction, capillary leakage, leukocyte sequestration, and apoptosis [82]. Lipid peroxidation, demyelinization of white matter in the CNS, edema, various pathological changes in different brain areas, and necrosis are other complications observed after CO poisoning [79].

Treatment

Initial treatment of CO poisoning is to immediately remove the victim from contaminated area. Those who are unconscious may require cardiopulmonary resuscitation (CPR) on site. CO poisoning can be treated with 100 % oxygen. Oxygen reverses hypoxemia and accelerates the elimination of the chemical asphyxiant [12]. Hyperbaric oxygen (HBO) reverses hypoxia, competes with CO for hemoglobin binding, and facilitates HbCO dissociation. It has been shown to shorten HbCO half-life from 4 to 6 h to <30 min [15]. Usually, approximately 30 % of victims with severe CO poisoning are likely to have fatal outcome [83]. Extensive follow-up and supportive treatment is often required to manage delayed neurological damage, neuropsychiatric sequelae, and other complications after CO poisoning. Caution should be observed during treatment of smoke-inhaled victims co-exposed to HCN and CO. In such instances, administration of nitrites to challenge cyanide poisoning is contraindicated. Methemoglobin generated by nitrites may have detrimental effects on the oxygen-carrying capacity of blood, which is already compromised due to HbCO formation following CO poisoning [51, 52].

Hydrogen Sulfide

Source

H_2S is a colorless, flammable, lipid-soluble toxic gas produced in occupational settings and from decaying organic matter. It is particularly produced during the decay of sulfur-containing proteins and is a by-product of human and animal waste [84, 85]. It is naturally present in crude petroleum, natural gas, volcanic gases, and in wetlands. Occupational exposure to H_2S is primarily a problem in the sour gas segment of the natural gas industry, particularly in the Canadian province of Alberta [84, 85]. H_2S is also a product of many industries such as leather tanning, rubber vulcanization, synthetic fabric and paper production, asphalt roofing, etc.

Signs and Symptoms

H_2S is a mucus membrane irritant and has a characteristic pungent odor of rotten eggs [78]. However, olfactory desensitization is known to occur at concentrations above 100 ppm. Thus, at levels above 100 ppm, toxicity can continue without the patient being aware [84, 85]. Also, at lower concentrations, a keratoconjunctivitis or "gas eye" can occur, which is a superficial inflammation of the cornea and conjunctiva [84, 86]. High concentrations (>700 ppm) of H_2S are known to be acutely lethal for humans and animals. At a high concentration (400 ppm), necrosis and ulceration of nasal respiratory and olfactory epithelial cells are observed [87]. H_2S has a unique ability to cause sudden collapse or "knock down" of the victims after breathing contaminated air for a short while. This happens because of its acute effect on the respiratory center [78, 84].

Mechanism of Action

Toxicity of H_2S has often been compared with cyanide. It selectively binds with the Fe^{2+} of CCO enzyme in complex IV, the last step involved in electron transport chain in the mitochondria [88]. This leads to cessation of aerobic metabolism and oxidative phosphorylation and a shift towards anaerobic respiration. Impaired utilization of oxygen by the cells leads to the development of lactic acidosis and histotoxic hypoxia [1, 2, 4]. At higher concentrations of H_2S, death is inevitable due to depression of respiratory center in the brain, and at lower concentrations, death is caused by pulmonary edema and congestion. Survivors with periods of unconsciousness may suffer permanent neurological sequelae such as memory loss. High exposures to near lethal concentrations in animals have shown destruction to nasal epithelium [15]. In one study, effect of H_2S on lung mitochondrial chain enzymes in rats was assessed. The study showed selective inhibition of CCO activity, without any change in the activities of NADH-cytochrome c reductase and succinate-cytochrome c reductase. Such biochemical impairment could lead to functional (histotoxic) hypoxia in the lung tissues [87]. Another study revealed that cytochrome P450 (CYP450)-dependent metabolism of H_2S was responsible for inducing ROS production, which was associated with mitochondrial depolarization [89]. Further study indicated that mechanisms additional to inhibition of CCO in H_2S toxicity could not be excluded [90]. Such mechanisms included electron transport inhibition, uncoupling of mitochondrial respiration, and opening of mitochondrial permeability transition pore (MPTP). Other alternative mechanisms, including activation of ATP-activated potassium channels and cell signaling pathways, were also postulated [91]. Subsequently, H_2S exposure was shown to alter gene expression associated with a variety of biological processes, including cell cycle regulation, protein kinase regulation, and cytoskeletal organization and biogenesis. However, no significant changes in CCO gene expression or bioenergetics were observed [92]. In contrast to its toxic effects, H_2S has also been shown to exert beneficial effects in the body. It is a normal endogenous substance that activates NMDA receptor, induces hippocampal long-term potentiation, increases intracellular calcium in glial cells, protects neurons from oxidative stress by upregulation of glutathione synthesis, and may function as a smooth muscle relaxant [93].

Treatment

Treatment of H_2S poisoning involves immediate evacuation from the contaminated area followed by artificial ventilation. Treatment is currently empirical, with a combination of nitrite therapy and hyperbaric oxygen [84, 85]. Sodium nitrite induces methemoglobin, which in turn binds with HS^- ions to form sulfmethemoglobin. This restores the activity of sulfide-inhibited CCO enzyme [15]. Unlike cyanide poisoning, sodium thiosulfate has no role in the detoxification of H_2S [12]. Delayed anoxic brain injury may be observed in victims, depending on the duration of unconsciousness. Also, irritant properties of H_2S may produce delayed pulmonary injury, which may warrant hospitalization and symptomatic treatment [78]. Protective efficacy of bicarbonate and glucose against hydrogen sulfide-induced coma and lethality has also been shown in rat model [94]. Hydroxocobalamin has also been shown as an effective antidote against H_2S poisoning. Hydroxocobalamin cobalt can complex H_2S and also catalyze its autoxidation to form thiosulfate and sulfate, which are renally eliminated [95].

Role of OP Compounds in Induction of Hypoxia

OP compounds are widely used as pesticides, petroleum additives, and chemical warfare agents. Several occupational, accidental, and intentional morbidity and mortality have been reported due to OP poisoning [96]. Nerve agents are OP compounds when referred in context of war and terrorism [97]. Toxicity of neurotoxic OP compounds is attributed to inhibition of acetyl cholinesterase (AChE) resulting in accumulation of acetylcholine (ACh) in the neural synaptic junctions, which leads to enhancement of nerve impulses. This leads to a wide variety of hypercholinergic effects that ultimately results in muscular dysfunction and nerve damage [98]. There are three types of effects after OP poisonings. They are the following: muscarinic (e.g., miosis, excessive salivation, rhinorrhea, bronchorrhea, bronchospasm, bradycardia, incontinence of micturition and defecation, nausea, vomiting, etc.), nicotinic (e.g., muscle weakness, respiratory paralysis, miosis, sweating, tachycardia, hypertension, etc.), and central effects (e.g., anxiety, headache, excitement, ataxia, disorientation, coma, seizures, etc.). All these effects have been associated with AChE inhibition [99]. However, there are sufficient evidences on non-cholinergic effects of OP as well. They include changes in other enzymes, neurotransmitters, immune changes, anaphylactoid reaction, behavior, etc. [100]. Non-cholinergic toxicity of OP is ascribed to generation of NO and other free radicals and induction of ROS, leading to oxidative damage [101]. Literature shows that OP compounds are neither categorized as chemical asphyxiants nor known to cause hypoxia [12]. However, possible role of hypoxia in mediating OP-induced cellular and molecular aberrations has been documented [21–23]. A recent review discussed possible implications of OP compounds in electron supply pathways, respiratory chain enzymes (complex I, II, III, and IV), mitochondrial respiration, ATP synthesis and hydrolysis, oxidative stress, mitochondrial membrane changes, calcium uptake, and mitochondrial-dependent apoptotic pathways [24]. Effect of OP compounds on respiratory chain enzymes severely affects the energy metabolism leading to hypoxia [102]. Current therapy for OP poisoning relies on the use of atropine, a cholinergic muscarinic antagonist, and an oxime, such as pralidoxime (2-PAM), to reactivate phosphorylated AChE. Anticonvulsant like diazepam serves as a good adjunct to challenge OP-induced convulsions [97]. However, in view of several non-cholinergic effects of OP compounds, beneficial role of other agents like antioxidants, alkalizers, calcium-channel blockers, oxygen therapy, etc. has also been considered [97, 103]. Possibly, OP-induced hypoxia-mediated cellular and molecular events could be important sites for various pharmacological interventions.

Conclusion

Chemical asphyxiants are gases that reduce the utilization of dissolved oxygen in the blood by body tissues. This results in acute hypoxia, more particularly histotoxic hypoxia. Such hypoxia leads to several alterations at biochemical, cellular, and molecular levels, possibly resulting in tissue pathology and death. If confronted during any chemical emergency, such hypoxia should be immediately, appropriately, and vigorously treated for better prognosis. In all cases of exposure to chemical asphyxiants, a successful outcome relies on the evacuation of casualties, decontamination, resuscitation, oxygen therapy followed by institution of specific antidotes, and follow-up with good supportive care. Community preparedness and emergency disaster response should be given priority.

Acknowledgements Authors acknowledge Dr. B.K. Bhattacharya, Head, Division of Biochemistry and Biotechnology, and Dr. Pravin Kumar, Head, Division of Pharmacology and Toxicology, DRDE, Gwalior, for their valuable suggestions. Authors also thank Mr. Shiv Kumar Yadav and Ms. Poonam Singh, Research Scholars, Division of Pharmacology and Toxicology, DRDE, Gwalior, for their excellent technical support in preparation of this manuscript.

References

1. Moody EJ, Simon BA, Johns RA (2001) Therapeutic gases: oxygen, carbon dioxide, nitric oxide and helium. In: Hardman G, Limbird LE (eds) Goodman & Gilman's the pharmacological basis of therapeutics, 10th edn. McGraw-Hill, New York, pp 385–397
2. Patel PM, Patel HH, Roth DM (2011) General anesthetics and therapeutic gases. In: Bruton LL (ed) Goodman & Gilman's the pharmacological basis of therapeutics, 12th edn. McGraw-Hill Companies Inc., New York, pp 528–564
3. Kiang J, Tsen KT (2006) Biology of hypoxia. Chin J Physiol 49:223–233
4. Coricor G (2012) Study of cellular responses under chemically induced hypoxia. PhD dissertation, Seton Hall University, South Orange, NJ, USA, pp 1–239. http://schoraship.shu.edu/theses. Accessed 12 July 2013
5. Baue AE, Durham R, Faist E (1998) Systemic inflammatory response syndrome (SIRS), multiple organ dysfunction syndrome (MODS), multiple organ failure (MOF): are we winning the battle? Shock 10:79–89
6. Fawcett TA, Moon RE, Fracica PJ, Mebane GY, Theil DR, Piantadosi CA (1992) Warehouse workers' headache: carbon monoxide poisoning from propane-fueled forklifts. J Occup Med 34:12–15
7. Melius JM (1998) The Bhopal disaster. In: Rom WN (ed) Environmental and occupational medicine, 3rd edn. Lippincott–Raven, Philadelphia, pp 1239–1244
8. Okumura T, Suzuki K, Fukuda A, Kohama A, Takasu N, Ishimatsu S, Hinohara S (1998) The Tokyo subway sarin attack: disaster management, part 2: hospital response. Acad Emerg Med 5:618–624
9. Fuller DC, Suruda AJ (2000) Occupationally related hydrogen sulfide deaths in the United States from 1984 to 1994. J Occup Environ Med 42:939–942
10. Okudera H (2002) Clinical features on nerve gas terrorism in Matsumoto. J Clin Neurosci 9:17–21
11. Spanjaard H, Khabib O (2003) Chemical weapons. In: Levy BS, Sidel VW (eds) Terrorism and public health. Oxford University Press, New York, pp 199–219
12. Kales SN, Christiani DC (2004) Acute chemical emergencies. N Engl J Med 350:800–806
13. Eckstein M (2008) Enhancing public health preparedness for a terrorist attack involving cyanide. J Emerg Med 35:59–65
14. CDC (2013) Chemical emergencies overview. In: Emergency preparedness and response. Centers for Disease Control and Prevention, Atlanta, USA. http://emergency.cdc.gov/chemical/overview.asp. Accessed 16 July 2013
15. Tan K-H, Wang T-L (2005) Asphyxiants: simple and chemical. Ann Disaster Med 4:S35–S40
16. Chang S, Lamm SH (2003) Human health effects of sodium azide exposure: a literature review and analysis. Int J Toxicol 22:175–186
17. Díaz-Castro B, Pintado CO, Piruat JI (2012) Differential impairment of catecholaminergic cell maturation and survival by genetic mitochondrial complex II dysfunction. Mol Cell Biol 32:3347–3357
18. Delivoria-Papadopoulosm M, Mishra OP (2004) Nuclear mechanisms of hypoxic cerebral injury in the newborn. Clin Perinatol 31:91–105
19. Mishra OP, Randis T, Ashraf QM, Delivoria-Papadopoulos M (2006) Hypoxia-induced Bax and Bcl-2 protein expression, caspase-9 activation, DNA fragmentation, and lipid peroxidation in mitochondria of the cerebral cortex of newborn piglets: the role of nitric oxide. Neuroscience 141:1339–1349
20. Mishra OP, Zubrow AB, Ashraf QM, Delivoria-Papadopoulos M (2006) Effect of nitric oxide synthase inhibition during post-hypoxic reoxygenation on Bax and Bcl-2 protein expression and DNA fragmentation in neuronal nuclei of newborn piglets. Brain Res 1101:20–28
21. Millis RM, Archer PW, Whittaker JA, Trouth CO (1988) The role of hypoxia in organophosphorus nerve agent intoxication. Neurotoxicology 9:273–285
22. Binukumar BK, Bal A, Kandimalla R, Sunkaria A, Gill KD (2010) Mitochondrial energy metabolism impairment and liver dysfunction following chronic exposure to dichlorvos. Toxicology 270:77–84
23. Karami-Mohajeri S, Abdollahi M (2012) A comprehensive review on experimental and clinical findings in intermediate syndrome caused by organophosphate poisoning. Toxicol Appl Pharmacol 258:309–314
24. Karami-Mohajeri S, Abdollahi M (2013) Mitochondrial dysfunction and organophosphorus compounds. Toxicol Appl Pharmacol 270:39–44
25. Loiacono LA, Shapiro DS (2010) Detection of hypoxia at the cellular level. Crit Care Clin 26:409–421
26. Giaccia AJ, Simon MC, Johnson R (2004) The biology of hypoxia: the role of oxygen sensing in development, normal function, and disease. Genes Dev 18:2183–2194
27. Krohn KA, Link JM, Mason RP (2008) Molecular imaging of hypoxia. J Nucl Med 49:129S–148S
28. Eltzschig HK, Carmeliet P (2011) Hypoxia and inflammation. N Engl Med 364:656–665
29. Connett RJ, Honig CR, Gayeski EJ, Brooks GA (1990) Defining hypoxia: a systems view of V02, glycolysis, energetics, and intracellular P02. J Appl Physiol 68:833–842
30. Johnson JD, Conroy WG, Isom GE (1987) Alteration of cytosolic calcium levels in PC12 cells by potassium cyanide. Toxicol Appl Pharmacol 88:217–224
31. Cook R, Karch C, Nahar P, Rivera A, Ko J (2010) Effects of deferoxamine-induced hypoxia on neuronal human mu-opioid receptor gene expression. Biochem Biophys Res Commun 398:56–61
32. Dendorfer A, Heidbreder M, Hellwig-Burgel T, Johren O, Qadri F, Dominiak P (2005) Deferoxamine induces prolonged cardiac preconditioning via accumulation of oxygen radicals. Free Radic Biol Med 38:117–124

33. Sato M, Ikeda K, Yoshizaki K, Koyano H (1991) Responses of cytosolic calcium to anoxia and cyanide in cultured glomus cells of newborn rabbit carotid body. Brain Res 551:327–330
34. Kiang JG, Smallridge RC (1994) Sodium cyanide increases cytosolic free calcium: evidence for activation of the reversed mode of the Na^{+}/Ca^{2+} exchanger and Ca^{2+} mobilization from inositol trisphosphate-insensitive pools. Toxicol Appl Pharmacol 127:173–181
35. Kiang JG, Tsokos GC (1998) Heat shock protein 70 kDa: molecular biology, biochemistry, and physiology. Pharmacol Ther 80:183–201
36. Kiang JG, Ding XZ, McClain DE (1998) Over expression of HSP-70 attenuates increases in $[Ca^{2+}]i$ and protects human epidermoid A-431 cells after chemical hypoxia. Toxicol Appl Pharmacol 149:185–194
37. Hierholzer C, Billiar TR (2001) Molecular mechanism in the early phase of hemorrhagic shock. Langenbecks Arch Surg 386:302–308
38. Kiang JG, Krishnan S, Lu X, Li Y (2008) Inhibition of inducible nitric-oxide synthase protects human T cells from hypoxia-induced apoptosis. Mol Pharmacol 73:3738–3747
39. Gurova KV, Kwek SS, Koman IE, Komarov AP, Kandel E, Nikiforov MA, Gudkov AV (2002) Apoptosis inhibitor as a suppressor of tumor progression: expression of Bcl-2 eliminates selective advantages for p53-deficient cells in the tumor. Cancer Biol Ther 1:39–44
40. Kiang JG, Varke VG, Tsokos GC (2003) NaCN-induced chemical hypoxia is associated with altered gene expression. Mol Cell Biochem 254:211–216
41. Jeong JI, Lee YW, Kim YK (2003) Chemical hypoxia-induced cell death in human glioma cells: role of reactive oxygen species, ATP depletion, mitochondrial damage and Ca^{2+}. Neurochem Res 28:1201–1211
42. Semenza GL (2000) HIF-1: mediator of physiological and pathophysiological responses to hypoxia. J Appl Physiol 88:1474–1480
43. Lee J, Bae S, Jeong J, Kim S, Kim K (2003) Hypoxia-inducible factor (HIF-l) α: its protein stability and biological functions. Exp Mol Med 36:1–12
44. Yan J, Zhou B, Taheri S, Shi H (2011) Differential effects of HIF-l inhibition by YC-l on the overall outcome and blood-brain barrier damage in a rat model of ischemic stroke. PLoS One 6:e27798
45. Ke Q, Costa M (2006) Hypoxia-inducible factor-l (HIF-l). Mol Pharm 70:1469–1480
46. Okumura T, Ninomiya N, Ohta M (2003) The chemical disaster response system in Japan. Prehosp Disaster Med 18:189–192
47. Rotenberg JS (2003) Cyanide as a weapon of terror. Pediatr Ann 32:236–240
48. Bhattacharya R (2000) Antidotes to cyanide poisoning: present status. Ind J Pharmacol 32:94–101
49. Megarbane B, Delahaye A, Goldgran-Toledano D, Baud FJ (2003) Antidotal treatment of cyanide poisoning. J Chin Med Assoc 66:193–203
50. Bhattacharya R, Flora SJS (2009) Cyanide toxicity and its treatment. In: Gupta RC (ed) Handbook of toxicology of chemical warfare agents. Elsevier Inc., New York, pp 255–270
51. Cummings TF (2004) The treatment of cyanide poisoning. Occup Med 54:82–85
52. Baud FJ (2007) Cyanide: critical issues in diagnosis and treatment. Hum Exp Toxicol 26:191–201
53. Vesey CJ, Cole PV (1985) Blood cyanide and thiocyanate concentrations produced by long-term therapy with sodium nitroprusside. Br J Anaesth 57:148–155
54. Ryan JG (1998) Cyanide. In: Viccellio P (ed) Emergency toxicology, 2nd edn. Lippincott-Raven Publishers, Philadelphia, pp 969–978
55. Bromley J, Hughes BG, Leong DC, Buckley NA (2005) Life-threatening interaction between complementary medicines: cyanide toxicity following ingestion of amygdalin and vitamin C. Ann Pharmacol 39:1566–1569
56. Ellenhorn MJ, Schonwald S, Ordog G, Wasserberger J (1997) Cyanide poisoning. In: Ellenhorn MJ, Schonwald S, Ordog G, Wasserberger J (eds) Ellenhorn's medical toxicology: diagnosis and treatment of human poisoning, 2nd edn. Williams and Wilkins, Baltimore, pp 1476–1484
57. Kamalu BP (1995) The adverse effects of long-term cassava (Manihot esculenta Crantz) consumption. Int J Food Sci Nutr 46:65–93
58. Pettersen JC, Cohen SD (1993) The effects of cyanide on brain mitochondrial cytochrome oxidase and respiratory activities. J Appl Toxicol 13:9–14
59. Maduh EU (1989) Mechanism of cyanide neurotoxicity. PhD dissertation, Purdue University, West Lafayette, IN, USA, pp 1–199
60. Ardelt BK, Borowitz JL, Maduh EU, Swain SL, Isom GE (1994) Cyanide-induced lipid peroxidation in different organs: subcellular distribution and hydroperoxide generation in neuronal cells. Toxicology 89:127–137
61. Kanthasamy AG, Ardelt B, Malave A, Mills EM, Powley TL, Borowitz JL, Isom GE (1997) Reactive oxygen species generated by cyanide mediate toxicity in rat pheochromocytoma cells. Toxicol Lett 93:47–54
62. Hariharakrishnan J, Satpute RM, Prasad GBKS, Bhattacharya R (2008) Oxidative stress mediated cytotoxicity of cyanide in LLC-MK2 cells and its attenuation by Alpha-ketoglutarate and N-acetyl cysteine. Toxicol Lett 185:132–141
63. Hariharakrishnan J, Anand T, Satpute RM, Jayaraj R, Prasad GBKS, Bhattacharya R (2009) Activity and gene expression profile of certain antioxidant enzymes in different organs of rats after subacute cyanide exposure: effect of alpha-ketoglutarate. Drug Chem Toxicol 32:268–276
64. Sun P, Rane SG, Gunasekar PG, Borowitz JL, Isom GE (1997) Modulation of NMDA receptor by cyanide: enhancement of receptor-mediated responses. J Pharmacol Exp Ther 280:1341–1348

65. Kanthasamy AG, Borowitz JL, Pavlakovic G, Isom GE (1994) Dopaminergic neurotoxicity of cyanide: neurochemical, histological, and behavioural characterization. Toxicol Appl Pharmacol 126:156–163
66. Yamamoto H-A (1995) A hypothesis for cyanide-induced tonic seizures with supporting evidence. Toxicology 95:19–26
67. Satpute RM, Hariharakrishnan J, Bhattacharya R (2008) Alpha ketoglutarate and N-acetyl cysteine protect PC12 cells from cyanide-induced cytotoxicity and altered energy metabolism. NeuroToxicol 29:170–178
68. Bhattacharya R (2004) α-ketoglutarate: a promising antidote to cyanide poisoning. In: Flora SJS, Romano JA, Baskin SI, Sekhar K (eds) Pharmacological perspectives of toxic chemicals and their antidotes. Narosa Publishing House, New Delhi, pp 411–430
69. Mittal G, Singh T, Kumar N, Bhatnagar A, Tripathi RP, Tulsawani R, Vijayaraghavan R, Bhattacharya R (2010) Radiolabeling and dose fixation study of oral alpha-ketoglutarate as a cyanide antidote in healthy human volunteers. Clin Toxicol 48:509–515
70. Omaye ST (2002) Metabolic modulation of carbon monoxide toxicity. Toxicology 180:139–150
71. Valent F, McGwin G Jr, Bovenzi M, Barbone F (2002) fatal work related inhalation of harmful substances in the United States. Chest 121:969–975
72. Mcdow S, Tollerud DJ (2003) In: Grrenberg MI, Hamilton RJ, Phillips SD, McCluskey GJ (eds) Occupational, industrial and environmental toxicology, 2nd edn. Mosby Inc., Pennsylvania, p 628
73. Prockop LD, Chichkova RI (2007) Carbon monoxide intoxication: an updated review. J Neurol Sci 262:122–130
74. Raub JA, Mathieu-Nolf M, Hampson NB, Thom SR (2000) Carbon monoxide poisoning- a public health perspective. Toxicology 145:1–14
75. Hardy KR, Thom SR (1994) Pathophysiology and treatment of carbon monoxide poisoning. J Toxicol Clin Toxicol 32:613–629
76. Weaver LK (2009) Clinical practice. Carbon monoxide poisoning. N Engl J Med 360:1217–1225
77. Seger DL, Welch LW (2001) Carbon monoxide. In: Sullivan JB Jr, Krieger GR (eds) Clinical environmental health and toxic exposures, 2nd edn. Lippincott Williams & Wilkins, Philadelphia, pp 722–727
78. Holland M (2003) Chemical manufacturing. In: Grrenberg MI, Hamilton RJ, Phillips SD, McCluskey GJ (eds) Occupational, industrial and environmental toxicology, 2nd edn. Mosby Inc., Philadelphia, pp 489–491
79. Gorman D, Drewry A, Huang YL, Sames C (2003) The clinical toxicology of carbon monoxide. Toxicology 187:25–38
80. Bateman DN (2003) Carbon monoxide. Medicine 31:233
81. Alonso JR, Cardellach F, Lopez S, Casademont J, Miro O (2003) Carbon monoxide specifically inhibits cytochrome c oxidase of human mitochondrial respiratory chain. Pharmacol Toxicol 93:142–146
82. Blumenthal I (2001) Carbon monoxide poisoning. J R Soc Med 94:270–272
83. Varon J, Marik PE, Fromm RE Jr, Gueler A (1999) Carbon monoxide poisoning: a review for clinicians. J Emerg Med 17:87–93
84. Guidotti TL (1996) Hydrogen sulphide. Occup Med 46:367–371
85. Gerasimon G, Bennett S, Musser J, Rinard J (2007) Acute hydrogen sulphide poisoning in a dairy farmer. Clin Toxicol 45:420–423
86. Costigan MG (2003) Hydrogen sulphide: UK occupational exposure limits. Occup Environ Med 60:308–312
87. Khan AA, Schuler MM, Prior MG, Yong S, Coppock RW, Florence LZ, Lillie LE (1990) Effects of hydrogen sulphide exposure on lung mitochondrial respiratory chain enzymes in rats. Toxicol Appl Pharmacol 103:482–490
88. Truong DH, Eghbal M, Hindmarsh W, Roth SH, O'Brien PJ (2006) Molecular mechanisms of hydrogen sulphide toxicity. Drug Metab Rev 38:733–744
89. Eghbal MA, Pennefather PS, O'Brien PJ (2004) H_2S cytotoxicity mechanism involves reactive oxygen species formation and mitochondrial depolarization. Toxicology 203:69–76
90. Thompson RW, Valentine HL, Valentine WM (2003) Cytotoxic mechanisms of hydrosulfide anion and cyanide anion in primary rat hepatocyte culture. Toxicology 188:149–159
91. Szabó C (2007) Hydrogen sulphide and its therapeutic potential. Nat Rev Drug Discov 6:917–935
92. Roberts ES, Thomas RS, Dorman DC (2008) Gene expression changes following acute hydrogen sulphide (H_2S)-induced nasal respiratory epithelial injury. Toxicol Pathol 36:560–567
93. Kimura H, Nagai Y, Umemura K, Kimura Y (2005) Physiological roles of hydrogen sulphide: synaptic modulation, neuroprotection, and smooth muscle relaxation. Antioxid Redox Signal 7:795–803
94. Almeida AF, Nation PN, Guidotti TL (2008) Mechanism and treatment of sulphide-induced coma: a rat model. Int J Toxicol 27:287–293
95. Truong DH, Mihajlovic A, Gunness P, Hindmarsh W, O'Brien PJ (2007) Prevention of hydrogen sulphide (H_2S)-induced mouse lethality and cytotoxicity by hydroxocobalamin (vitamin B_{12a}). Toxicology 242:16–22
96. Afshari R, Majdzadeh R, Balali-Mood M (2004) Pattern of acute poisonings in Mashhad, Iran 1993–2000. J Toxicol Clin Toxicol 42:965–967
97. Balali-Mood M, Saber H (2012) Recent advances in the treatment of organophosphorus poisonings. Iran J Med Sci 37:74–91

98. Qian Y, Venkatraj V, Barhoumi R, Pal R, Datta A, Wild JR, Tiffany-Castiglioni E (2006) The non-cholinergic mechanisms of organophosphorus toxicity in neural cell cultures. Toxicol Appl Pharmacol 219:162–171
99. Okumura T, Taki K, Suzuki K, Satoh T (2009) The Tokyo subway sarin attack: a toxicological whole truth. In: Gupta RC (ed) Handbook of toxicology of chemical warfare agents. Elsevier Inc., Los Angeles, pp 25–32
100. Jun D, Bajgar J, Kuca K, Kassa J (2009) Monitoring of blood cholinesterase activity in workers exposed to nerve agents. In: Gupta RC (ed) Handbook of toxicology of chemical warfare agents. Elsevier Inc., Los Angeles, pp 877–886
101. Ranjbar A, Solhi H, Mashayekhi FJ, Susanabdi A, Rezaie A, Abdollahi M (2005) Oxidative stress in acute human poisoning with organophosphorus insecticides; a case control study. Environ Toxicol Pharmacol 20:88–91
102. Rousseau JM, Rüttimann M, Brinquin L (2000) Acute neurotoxic organophosphate poisoning: insecticides and chemical weapons. Ann Fr Anesth Reanim 19:588–598
103. Nurulain SM, Szegi P, Tekes K, Naqvi SN (2013) Antioxidants in organophosphorus compounds poisoning. Arh Hig Rada Toksikol 64:169–177

Hypertension at High Altitude

Akpay Sarybaev, Deepak Mulajkar, Gaurav Sikri, Tsering Norboo, Amit Badhya, Shashi Bala Singh, and Maramreddy Prasanna Reddy

Abstract

Systemic hypoxia is associated with conditions like anemia, lung diseases, sleep disorders, breathing, and exposure to high altitude. The physiological responses including blood pressure regulation to acute, chronic, intermittent normobaric, and hypobaric hypoxia have been under intense investigation for the past many years; however, the regulatory mechanisms are incompletely understood. The available literature indicates a differential response to hypoxia in pulmonary versus systemic circulation. Multiple physiological and metabolic changes contribute towards high-altitude acclimatization; yet, in some individuals, exposure to high altitude could be life threatening due to maladaptation. There are a few studies on the prevalence of hypertension in high-altitude natives and sea-level dwellers exposed to high altitude (acute and chronic). Elevated blood pressure is an established risk factor for different cardiovascular disease and the evidence suggests that the blood pressure rises to a modest extent in patients with mild to moderate hypertension upon acute ascent to high altitude (Luks et al., Congenit Heart Dis 5:220–232, 2010), but there is no clear evidence on increased risk of complications due to increased systemic blood pressures. This book chapter reviews available literature on systemic blood pressure responses to high altitude.

A. Sarybaev (✉)
National Centre of Cardiology & Internal Medicine (NCCIM), Togolok Moldo Str. 3, 720040 Bishkek, Kyrgyz Republic
e-mail: ak_sar777@mail.ru

D. Mulajkar • G. Sikri
High Altitude Medical Research Center, Leh, Ladakh, India

T. Norboo
Ladakh Institute of Prevention, Leh, Ladakh, India

A. Badhya • M.P. Reddy
Cardiorespiratory Division, Defense Institute of Physiology & Allied Sciences, Defense Research & Development Organization, Lucknow Road, Timarpur, Delhi, India

S.B. Singh
Defence Institute of Physiology and Allied Sciences (DIPAS), New Delhi, India
e-mail: drshashisingh@gmail.com

S.B. Singh et al. (eds.), *Translational Research in Environmental and Occupational Stress*, DOI 10.1007/978-81-322-1928-6_20,

Introduction

Elevated blood pressure is an established risk factor for different cardiovascular disease and the evidence suggests that the blood pressure rises to a modest extent in patients with mild to moderate hypertension upon acute ascent to high altitude (HA) [1]. And also it is well known that exposure to high altitude (HA) is associated with significant changes in several cardiopulmonary functions [2]. It has been observed that ventilation, heart rate (HR), arterial blood pressure (ABP), pulmonary arterial pressure (PAP), and sympathetic nerve activity (SNA) are increased during exposure to high altitude (HA) in order to maintain adequate arterial partial pressure of oxygen (Pao_2) and tissue oxygenation [3–6]. Many of these physiological parameters gradually come back to near normal values within 2 weeks of acclimatization [2]. But in some individuals, PAP, ABP, and SNA remain significantly elevated throughout their stay at high altitude and sustained even after returning to the plains [7].

It is well known that hypoxia causes vasodilation in the systemic circulation, which can lead to systemic hypotension. Hypoxia by direct stimulatory effect on adrenal medulla causes a rapid increase in epinephrine levels [8]. Epinephrine causes dilation of systemic arteries, which helps to improve the blood supply of tissues by increasing the capacity of the vascular bed. It has been shown that hypoxic dilation of systemic vessels is not only due to the stimulation of β2-adrenergic receptors [9], but also the action of vasoactive substances, such as adenosine [10], prostaglandins [11], and nitric oxide [12]. Indeed, recently the increase of circulating nitric oxide levels was shown in lowlanders during high-altitude acclimatization [13].

Hypoxia influences peripheral chemoreceptors to increase the systemic blood pressure by activation of sympathetic nervous system [10, 14]. It should be noted, however, that despite the vasoconstrictor effects of the sympathetic nervous system, the selective dilatation of the coronary [15] and cerebral blood vessels [16] occurs. Moreover, the reduction of coronary artery constriction happens in response to angiotensin and norepinephrine [17]. All of these events contribute to the redistribution of blood flow to vital organs. There are several studies demonstrating the effects of hypoxia on pulmonary circulation. However, there are a very few studies that have explored the effects of hypobaric hypoxia on systemic circulation and understanding the possible factors responsible for a sustained rise in blood pressure upon acute and chronic exposure of humans to high altitude. Unlike pulmonary hypertension that occurs shortly after the onset of hypoxia, sustained systemic hypertension seems to be set in several days after the onset of hypoxia [18]. The possibility of chronic hypoxia causing systemic hypertension is not entirely unexpected, since hypoxia is known to cause acute and chronic changes in the pulmonary circulation [19]. Substantiating this assumption, several recent studies on intermittent and/or chronic hypoxia clearly demonstrate that the effect of hypoxia is not only confined to the pulmonary circulation and can, in fact, involve the systemic circulation as well [20–22]. However, the effects of prolonged exposure to hypobaric hypoxia on systemic circulation continue to be a subject of debate.

Elevated blood pressure is reported to be an establish risk factor for several cardiopulmonary and vascular diseases including dementia [20, 23]. It is possible that systemic hypertension at HA may influence and/or aggravate the HA induced maladies like thrombosis, stroke, pulmonary edema, pulmonary hypertension, cerebral edema, and cognitive impairment. Nevertheless, several unanswered questions remain, like the cause for sustained elevation of blood pressure during exposure to high altitude and the factors associated with interindividual variation. Further, it remains to be addressed whether the high-altitude-induced systemic hypertension is a clinical entity or not.

Blood Pressure During the Ascent to High Altitude in Lowlanders

Ascent to high altitude causes some changes in the blood pressure (BP). The main determinants of BP at high altitude, as well as at sea level, are cardiac output, which depends on the heart rate and stroke volume; systemic peripheral resistance; and central venous pressure, which reflects the volume of circulating blood. High-altitude hypoxia, which causes vasoconstriction in pulmonary circulation, has direct vasodilating effects on systemic arteries as mentioned above, thereby reducing the systemic arterial pressure. However, almost immediately, due to significant activation of the sympathetic system, the increased cardiac output and peripheral vasoconstriction are observed that overcome the hypoxia-induced vasodilation within a few hours and increases BP [8, 24]. In other words, there is an interaction between the direct influence of hypoxia on the vascular wall and processes, mediated by chemoreceptors [24, 25].

Recognition of underlying molecular processes in hypoxic vasodilation of systemic arteries is the subject of active research. Possibly several mechanisms are involved, such as the release of ATP from red blood cells and the generation of nitric oxide by the endothelium of blood vessels [26, 27]. The data of BP changes in normal (nonhypertensive) people during induction at high altitude are ambiguous. Most studies have reported an increase in systemic arterial pressure during acute hypoxia [28–33], while others report that BP can remain unchanged [34–37] or even decrease [38]. Some studies reported that the long-term sojourning (for a year or longer) leads to decrease in both systolic and diastolic blood pressure [39, 40]. Kamat and Banerji [29] observed an increase in BP in 31 of 32 subjects brought to 3,500–4,000 m above sea level. The systolic blood pressure increased from 115 to 125 mmHg and diastolic blood pressure from 78 to 93 mmHg. Wolfel and his associates (1991) reported a significant increase in BP and peripheral vascular resistance in 7 healthy young people, who stayed at 4,300 m for 21 days.

Mean arterial pressure increased from 124/71 to 145/88 mmHg and peripheral vascular resistance from 17 to 28 Wood units (mmHg/L/min). Pressure during exercise was higher than at sea level. Norepinephrine levels were also higher at high altitude. In another study, 4 plain residents climbed to high altitude (4,350 m), where their mean arterial pressure increased from 100 to 128 mmHg upon arrival and maintained elevated for 10 days [41]. A researcher from Japan reported about excessive BP increase in some normotensive subjects when shifting from the 1,336 to 5,400 m during the Japanese expedition to Mount Everest [42]. Examination of workers at the telescopic station on Mauna Kea in Hawaii revealed a significant increase (on 10 %) in diastolic BP after arrival compared to the baseline level, which persisted for 40 days of staying at high altitude (Forster, 1986). Wolfel et al. [7] also reported about the significant increase of systemic BP in 11 subjects who had ascended to 4,300 m and a lowering in BP after propranolol intake, suggesting that increased sympathetic activity was a causal factor in increasing BP. In another study, a group of 24 trekkers climbed to Everest Base Camp, where the average BP rose to 140/90 mmHg at an altitude of 5,300 m from 130/80 mmHg at sea level [43]. Taken together, the available studies suggested that probably some additional mechanisms besides sympathetic system activation may play a role in the sustained rise of BP and peripheral vascular resistance during high-altitude exposure, since the oxygen administration, as well as alpha- or beta-blockers, do not return the system pressure to baseline levels [44, 45].

Very interesting data were obtained by Kumar et al. [46]. The authors examined 46 normotensive subjects, residents of low altitude, in the course of their 30-day stay at 3,500 m. After acute induction with altitude in 18 (32.4 %) of them, the altitude-induced systemic arterial hypertension (BP $\geq$ 140/90 mmHg) developed, which persisted for 30 days. Mean arterial pressure at low altitude had been 121.8 $\pm$ 1.9/74.6 $\pm$ 1.9 mmHg, and on the 1st, 10th, and 30th days of stay at high altitude, the mean blood pressure were 141.6 $\pm$ 2.6/92.8 mmHg, 139.7 $\pm$ 2.3/91.0 $\pm$ 1.3 mmHg, and 140.1so $\pm$ 1.9/96.3 $\pm$ 1.5 mmHg ($p < 0.05$), respectively. The remaining 28 subjects had no significant changes during the stay at high altitude. It is important to note that the frequency of D allele of angiotensin-converting enzyme was significantly higher in hypertensive subjects than in normotensive subjects (0.67 vs. 0.32 $\chi^2{}_1 = 10.6$, $P < 0.05$). Similarly data from different high-altitude medical centers under Northern, Eastern, and Western Command hospitals revealed an increase in preponderance of high-altitude-induced systemic hypertension among sea-level native soldiers deployed at high and extreme altitude (unpublished report). Recently, we carried out a cross-sectional study on 132 Indian army soldiers (sea-level natives) who were admitted with elevated blood pressure ($\geq$140/90 mmHg) at Army hospital, Leh, Ladakh.

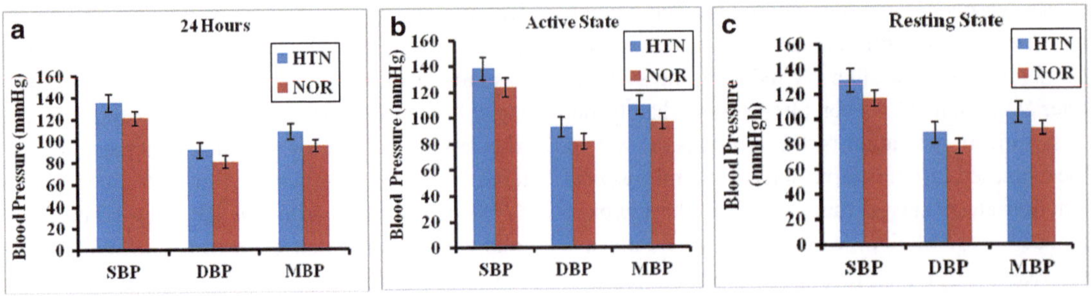

Fig. 1 Shows ambulatory BP values in hypertensive and normotensive subjects; (**a**) 24 h mean BP ≥ 130/80 mmHg, (**b**) daytime (active state) mean BP ≥ 135/85 mmHg, and (**c**) nighttime (resting state) mean BP ≥ 120/75 mmHg

All the patients were sea-level natives, reported to be normotensive before inducted to high altitude and had been living in high altitudes (11,500–20,000 ft) for the past 1–12 months or more. To confirm the diagnosis of hypertension, all the admitted patients were subjected to 24 h ambulatory BP monitoring (ABPM, daytime, every 15 min and, nighttime, every 30 min). The following clinical and physiological parameters like anthropometry, electrocardiogram (ECG), chest X-Ray, fundus examination, and ultrasound (kidney) were examined along with ABPM. Blood and serum samples were collected for blood sugar (fasting), biochemical and hematological tests. Serum was used to evaluate the levels of pro-inflammatory marker, high-sensitive C-reactive protein (hs-CRP), monocyte chemoattractant protein (MCP-1), and anti-inflammatory marker, interleukin 10 (IL-10) in healthy control versus hypertensive patients. Our results showed that around 50 % of admitted patients (68 out of 132) with elevated blood pressure were diagnosed with hypertension. ABPM data of these patients is as follows: 24 h mean BP ≥ 130/80 mmHg, daytime mean BP ≥ 135/85 mmHg, and nighttime mean BP ≥ 120/75 mmHg, and the remaining 64 subjects were found to be normotensive (Fig. 1).

In hypertensive patients, the ECG, chest X-Ray, and fundus examination revealed no abnormal changes. Similarly, ultrasound examination of kidney showed normal status. The lipid profile (total cholesterol, triglycerides, LDL, and HDL), fasting blood sugar, blood urea, creatinine, and sodium and potassium levels were found to be within the normal limits. Further, no significant difference in hemoglobin and hematocrit values was observed between normotensive and hypertensive subjects (Table 1).

Interestingly, the markers of inflammation (hs-CRP; 1.37 ± 0.2 vs. 0.64 ± 0.25 ng/ml, $P < 0.01$ and MCP-1; 37.2 ± 3.46 vs. 32.1 ± 5 pg/ml, $P < 0.05$) were found to be significantly higher in hypertensive group compared to normotensive. On other hand, anti-inflammatory marker (IL-10) was found to be significantly lower in hypertensive patients compare to normotensive group ($202.2 \pm 25.4 \pm 314 \pm 41$ pg/ml, $P < 0.01$) (Fig. 2).

Our preliminary results indicate that the inflammation and/or altered immune function may also play an important role in the pathophysiology of high-altitude hypertension besides proposed established mechanism like exacerbated sympathetic activity; however, further research is needed in this direction in larger cohort of subjects to prove the same. The magnitude of BP response during high-altitude exposure may be determined by a balance between the hypoxia-mediated carotid body chemoreceptors-induced sympatho-vasoconstriction and hypoxia-induced systemic vasodilatation at vascular level, as well as, possibly, by altitude level, ascent rate, ethnicity, and other factors like cold, humidity, radiation, dehydration, sleep disturbance, and psychological stress either alone or in combination. Therefore, it should be noted that the systemic BP response to high-altitude exposure is a result of the interaction of multiple factors like direct hypoxic effects on blood vessels, autonomic nerves system, and immunomodulation. Nevertheless, the prevalence of high-altitude-induced systemic hypertension and its clinical complication are not clearly

Table 1 Shows physiological parameters between hypertensive and normotensive subjects

	Age (year)	Body wt. (kg)	BMI	Hb (g/dl)	TCol (mg/dl)	TG (mg/dl)	LDL (mg/dl)	HDL (mg/dl)	VLDL (mg/dl)	Na (mmol/L)	K (mmol/L)	Urea	Creatinine (mg/dl)
HTN	36.6 ± 1	70.5 ± 1	24.3 ± 0.3	14.75 ± 0.7	190 ± 27	149 ± 15	109 ± 28	42 ± 6	30 ± 3	139 ± 3	3.89 ± 0.2	30 ± 4	0.9 ± 0.1
NOR	36.2 ± 1	69.3 ± 1	24.1 ± 0.31	149 ± 0.5	179 ± 26	153 ± 25	104 ± 28	42 ± 11	31 ± 5	140 ± 3	3.94 ± 0.18	27 ± 5	1.02 ± 0.6

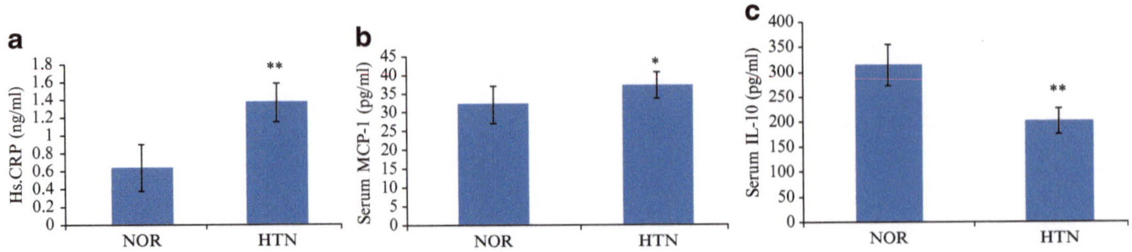

Fig. 2 Graphs (**a**, **b** & **c**) showing comparison of serum high-sensitive C-reactive protein (Hs-CRP), monocyte chemoattractant protein-1 (MCP-1), and anti-inflammatory cytokine interleukin 10 (IL-10) levels between normotensive and hypertensive subjects. Statistical significance between the means was determined with the Mann-Whitney test. $*P < 0.05$; $**P < 0.01$

understood. One of the main difficulties in the analysis of these studies is their heterogeneity, as they differ in ascent rate, achieved height, length of staying, physical activity, nutritional restrictions and differences, and other methodological features. It should be noted, however, that large-scale longitudinal human studies are needed to address the above issues with certainty.

Hypertension at High Altitude

Hypertension is defined as values ≥140 mmHg systolic blood pressure (SPB) and/or ≥90 mmHg diastolic blood pressure (DPB) [47]. Overall, the prevalence of hypertension appears to be around 30–45 % of the general population, with a steep increase with aging. Precise data on the prevalence of hypertension among high-altitude travelers is lacking, but survey studies suggest that 6–14 % of travelers to altitudes between 1,900 and 2,900 m are hypertensive [48, 49] and therefore this requires special attention.

Studies on BP in hypertensive subjects during short-term visits to high altitudes show wide variability. In hypertensive patients, pressure elevation is often more expressed because of endothelial dysfunction associated with hypertension which may restrict hypoxic vasodilation and strengthen sympathetic vasoconstriction. On the other hand, the ambient low humidity, increased ventilation, and physical activity may lead to dehydration, which could reduce the blood pressure. In hypertensive patients taking diuretics or other medications, this effect may be increased.

However, most studies report that, for hypertensive subjects ascending high altitude moderate increase in BP is more common [32, 50, 51]. However, the individual differences still have importance [51, 52]. Furthermore, the changes in BP may depend on the severity of hypoxia. At lower height, the average increase of BP is minimal. Palatini et al. [32] examined 12 normotensive subjects and 12 patients with mild hypertension using 24 h ambulatory BP monitoring, at sea level, 12 h after arrival at 1,210 m and 1.5–3 h after arrival at 3,000 m. BP was higher at 1,210 m in all subjects during the day but not at night, and BP elevation was similar in both groups compared with sea-level figures (6.1 mmHg for systolic/1.5 mmHg for diastolic BP in normotensive subjects and 5.5 mmHg for systolic/4.3 mmHg for diastolic blood pressure in hypertensive ones). However, the marked interindividual variability was observed. Maximum increase in BP was 17.4 mmHg for systolic and 16.3 mmHg for diastolic pressure, while other subjects did not have significant changes in systolic and diastolic blood pressure. Short-term exposure to 3,000 m caused further, although nonsignificant, increase in BP in both groups. Average levels of serum catecholamines also increased stepwise, but due to the wide interindividual differences, these changes were not statistically significant.

Roach et al. [33] described 4 subjects with hypertension whose average blood pressure during the first day at an altitude of 2,500 m rose up to 197 ± 7/101 ± 9 mmHg, and other 5 hypertensive subjects whose BP values maintained in the normal ranges during their stay at high altitude (5 days). There are reports about differences between races [53]. Several authors found only slight changes in BP, which did not reach statistical significance [32, 37, 54]. In studies with significant BP increase, the average increase of systolic BP usually did not exceed 15 mmHg. So, according to Savonito et al. [52], the systolic blood pressure rose from 154 ± 18 mmHg at 1,370 m to 166 ± 15 mmHg at 3,460 m, while Wu et al. (2007) reported that the systolic BP rose from 154 ± 18 to 168 ± 17 mmHg at altitudes between 3,486 and 5,072 m. Wu et al. [55] and Roach et al. [33] noted that after the initial rise, the BP reduced within a few days or weeks, but they did not perform any statistical comparison of these data with baseline values at sea level. Nevertheless, in some studies, elevated systemic pressure after ascent to high altitude has been reported to be sustained for months [56].

The abovementioned studies examined BP changes at rest. However, it is obvious that many of the people visiting mountains will not remain at rest; on the contrary, they will engage in a variety of physical activity, such as skiing, climbing, etc. Such situations, certainly, are followed by increase of BP. Very few researchers have paid attention for this issue. In a study by Savonitto et al. [52], hypertensive patients performed bicycle stress test with incremental load from 50 to 150 W at 1,370 and 3,460 m. Systolic BP increased from 154 ± 18 to 207 ± 3 mmHg, while at 3,460 m, pressure rose from 166 ± 15 to 223 ± 24 mmHg. Changes in diastolic BP were not given and were not compared with the control group. D'este et al. [54] compared changes in the systemic pressure during exercise in healthy subjects and hypertensive patients at sea level and at 2,572 m. They found a slightly greater increase in systolic pressure during submaximal load in hypertensive patients at high altitude at the same time, the maximum load did not show any differences. In well-controlled hypertensive patients, blood pressure elevation was small both at rest and at exercise [38, 57, 58]. It is necessary to note that studies of arterial hypertension at high altitude included only patients with mild to moderate hypertension. However, there is no information about the BP response at high altitude in people with severe or unstable hypertension.

Summing up the available data, it must be emphasized that there is a wide interindividual variability in hypertensive patients during ascent to high altitude, and BP response to high-altitude hypoxia is often unpredictable and variable. Nowadays, we do not have adequate markers that could predict which patients will experience a significant increase in BP at high altitude. Nevertheless, it is reasonable to let a person with well-controlled hypertension go to high altitude without any serious risk of complications. However, all hypertensive patients should be trained to control their BP and change their antihypertensive medication [59] in case of symptomatic hypotension or increase in BP, especially during the first days at high altitude.

Hypertension, Risk of Acute Mountain Sickness, and Other Adverse Events

The prevalence of AMS is about the same among healthy individuals and hypertensive patients who have risen to high altitude [49]. Due to the fact that hypertensive patients have an increased hypoxic pulmonary vasoconstriction compared with healthy individuals [60], the increased probability of high-altitude pulmonary edema can be expected, but this assumption was never tested and needs clarification. Also, there are no data on the relationship of hypertension and AMS.

It is difficult to draw definite conclusions about the risk of short- and long-term complications of arterial hypertension, because information on this issue is very scanty. Wu et al. [55] did not report about complications, such as retinopathy, hypertension, intracranial hemorrhage, myocardial infarction, or stroke, despite the fact that some subjects' systolic pressure exceeded 190 mmHg or the diastolic pressure reached 125 mmHg during their stay at high altitude for 4–6 weeks. Roach et al. [33] also did not report about similar complications in patients with high values of systolic and/or diastolic BP, but the length of their stay at high altitude was much shorter – 5 days. There were no complications in other studies investigating the response of blood pressure during exercise at high altitude [52, 54]. Only one study [61] refers to the increase in the odds ratio (1.5 for mountain hiking and 9.0 for skiing) for sudden death from hypertension during hiking or skiing in the ski resorts.

Despite the fact that the process of continuous adaptation of high populations of people in different parts of the world had some differences [62], most previous studies indicated that levels of systemic BP in adult high-altitude residents were lower than in populations observed at sea level [40, 63–65]. For example, scientists from Kyrgyzstan surveyed 774 male Kyrgyz aged from 30 to 59 years, permanent residents of the high-altitude regions of Pamir and Tien Shan (3,600–4,200 m). As a control group, 817 male Kyrgyz of same age, living at 700–900 m, were examined. Systolic BP at high-altitude residents was 116.08 $\pm$ 0.62 mmHg and diastolic BP was 75.09 $\pm$ 0.39 mmHg, while in lowlanders, pressure was 129.43 $\pm$ 0.72 and 81.88 $\pm$ 0.49 mmHg, respectively. The prevalence of hypertension in this high-altitude Kyrgyz group was 13 $\pm$ 1 %, while in lowlanders, this value was 26 $\pm$ 1 %. According to Sun [66], the prevalence of hypertension among the Tibetans of China was 11.02 %.

In other words the frequency of essential hypertension among high-altitude residents is significantly lower than among people living at sea level. Moreover, some authors believe that long-term residence at high altitude (up to 15 years) decreases the systolic and diastolic pressure. This was explained by the fact that chronic hypoxia had relaxing effect on smooth muscle cells, followed by vasodilation [67]; furthermore the prolonged hypoxia led to increased vascularization [68, 69], followed by decrease in peripheral vascular resistance. That is, the lower systemic pressure could be regarded as a "by-product" of tissue adaptation to high-altitude hypoxia.

This standpoint has long prevailed in the scientific medical world. Occasionally, this position is supported by the results of studies in recent years [70–72]. But over the past decade, a number of studies received the data which are contrary to this popular point of view [73–76]. So, Otsuka K et al. [75] showed higher diastolic pressure and thicker walls of arteries in high-altitude residents (Leh, Ladakh) compared with sea-level residents. Moreover, in another study [76] which examined a large sample size (more than 1,000 subjects) of Tibetans at different altitudes, significantly lower BPs were revealed only in children and adolescents living at high altitude, while there were no differences between adult residents living at low and high altitude. Tripathy et al. [77] reported that the prevalence of hypertension among adult Tibetans in India was 37.9 %. More recent studies conducted by Chinese authors [78] showed the greater prevalence of hypertension among the Tibetans in China, which in people older than 40 years reached 55.9 %.

Previously, many researchers believed that long-term residence or stay at high altitude had so-called "protective" effect on the systemic pressure [79–81]. Now, there is evidence that systemic hypertension at high altitude can have a more unfavorable course [82, 83], due to similar processes occurring both in pulmonary and systemic hypertension at the molecular level [84, 85]. There are some changes with high-altitude residents despite the successful adaptation. It was reported that there was carotid body hyperplasia and blunted response to hypoxia in older high-altitude native population [86]. Besides that, these residents demonstrated earlier cardiovascular degenerative changes with aging [87].

A population-based epidemiological study was conducted (Norboo T; Stobdan T et al. unpublished data) to determine the prevalence of hypertension and its relation to altitude, occupation, dietary habits, and ethnicity in a widely dispersed (45,110 km^2) representative group of Ladakhis in Northern India (altitude 2,600–4,900 m). 2,910 subjects (aged 20–94 years) were enrolled to complete a questionnaire including demographics, core behavior (tobacco, alcohol, dietary habits, and physical activities); personal and family history of hypertension, diabetes, coronary artery disease, and stroke; and modified Rose questionnaire. Measurements included height and weight for body mass index, waist circumference, ECG, Cardio-Ankle-Vascular Index (CAVI), Ankle-Brachial Index (ABI), and SpO_2. BP was recorded with home BP machine (Omron Dalian co., Ltd). Systolic blood pressure ≥140 mmHg and/or diastolic BP of ≥90 mmHg was taken as the criteria for labelling the subject as hypertensive. Following overnight fasting, blood samples were taken for fasting sugar and lipid profile estimation. The rural population comprised of six subdivisions with a distinct altitude and dietary and occupational pattern. Two special groups of monks and service personnel were analyzed separately because of their distinct diet and lifestyle pattern.

The prevalence of hypertension was found to be high in all the rural and urban subgroups; however, it was highest in Tibetan migrants settled in Leh town (47 %) followed by Ladakhi rural to urban migrants (43 %) and Tibetans born in Leh (41 %). The lowest prevalence in service personnel can be explained by their younger age, healthy diet, and regular physical activity. Relatively lower prevalence of hypertension in Ladakhi (31 %) and Tibetan natives (21 %) living at higher altitude (4,000–4,900 m) clearly indicates that altitude does not play a major role in causation of hypertension. However, there is a definite relationship of BP with age with all the subgroups showing a gradual increase in BP with age with a significant steep increase in both systolic and diastolic BP after the age of late 60s in the subgroup of highest altitude region (4,000–4,900 m). Like everywhere else in the world, hypertension prevalence in high-altitude population has multifactorial etiology. The Ladakh study shows that age, gender, socioeconomic factors, diet, culture, race, and changing lifestyle play major roles than altitude in prevalence of hypertension. However, at very high altitude (over 4,000 m), chronic hypoxia leads to accelerated steep in both systolic and diastolic hypertension after the age of 69 years. The pathogenesis may have some resemblance to chronic mountain sickness, which is an accelerated aging process occurring at high altitude.

Despite increase in prevalence of arterial hypertension and the number of people going or living at high altitude, there are very few studies to our knowledge addressing hypertension course and its management at high altitude. In order to clarify this issue, we need, firstly, the large-scale clinical studies, which we will take into account, and the different confounding factors, for example, the impact of urbanization, ethnic heterogeneity, and lifestyle. Secondly, it is very important to take up the molecular genetic studies that will determine the effects of chronic hypoxia (favorable or negative) on systemic BP regulation.

References

1. Luks AM, Stout K, Swenson ER (2010) Evaluating the safety of high-altitude travel in patients with adult congenital heart disease. Congenit Heart Dis 5:220–232
2. Higgins JP, Tuttle T, Higgins JA (2010) Altitude and the heart: is going high safe for your cardiac patient? Am Heart J 159:25–32
3. Calbet JA, Boushel R, Radegran G, Sondergaard H, Wagner PD, Saltin B (2003) Why is VO2 max after altitude acclimatization still reduced despite normalization of arterial O2 content? Am J Physiol Regul Integr Comp Physiol 284:R304–R316
4. Curran LS, Zhuang J, Sun SF, Moore LG (1997) Ventilation and hypoxic ventilatory responsiveness in Chinese-Tibetan residents at 3,658 m. J Appl Physiol 83:2098–2104
5. Penaloza D, Arias-Stella J (2007) The heart and pulmonary circulation at high altitudes: healthy highlanders and chronic mountain sickness. Circulation 115:1132–1146
6. Zhuang J, Droma T, Sun S, Janes C, McCullough RE, McCullough RG, Cymerman A, Huang SY, Reeves JT, Moore LG (1993) Hypoxic ventilatory responsiveness in Tibetan compared with Han residents of 3,658 m. J Appl Physiol 74:303–311
7. Wolfel EE, Selland MA, Mazzeo RS, Reeves JT (1994) Systemic hypertension at 4,300 m is related to sympathoadrenal activity. J Appl Physiol 76:1643–1650
8. Mazzeo RS, Reeves JT (2003) Adrenergic contribution during acclimatization to high altitude: perspectives from Pikes Peak. Exerc Sport Sci Rev 31:13–18
9. Weisbrod CJ, Minson CT, Joyner MJ, Halliwill JR (2001) Effects of regional phentolamine on hypoxic vasodilatation in healthy humans. J Physiol 537:613–621
10. Marshall JM (2000) Adenosine and muscle vasodilatation in acute systemic hypoxia. Acta Physiol Scand 168:561–573
11. Markwald RR, Kirby BS, Crecelius AR, Carlson RE, Voyles WF, Dinenno FA (2011) Combined inhibition of nitric oxide and vasodilating prostaglandins

abolishes forearm vasodilatation to systemic hypoxia in healthy humans. J Physiol 589:1979–1990
12. Blitzer ML, Loh E, Roddy MA, Stamler JS, Creager MA (1996) Endothelium-derived nitric oxide regulates systemic and pulmonary vascular resistance during acute hypoxia in humans. J Am Coll Cardiol 28:591–596
13. Janocha AJ, Koch CD, Tiso M, Ponchia A, Doctor A, Gibbons L, Gaston B, Beall CM, Erzurum SC (2011) Nitric oxide during altitude acclimatization. N Engl J Med 365:1942–1944
14. Lugliani R, Whipp BJ, Wasserman K (1973) A role for the carotid body in cardiovascular control in man. Chest 63:744–750
15. Kaufmann PA, Schirlo C, Pavlicek V, Berthold T, Burger C, von Schulthess GK, Koller EA, Buck A (2001) Increased myocardial blood flow during acute exposure to simulated altitudes. J Nucl Cardiol 8:158–164
16. Daugherty RM Jr, Scott JB, Dabney JM, Haddy FJ (1967) Local effects of O2 and CO2 on limb, renal, and coronary vascular resistances. Am J Physiol 213:1102–1110
17. Heistad DD, Abboud FM, Mark AL, Schmid PG (1975) Effect of hypoxemia on responses to norepinephrine and angiotensin in coronary and muscular vessels. J Pharmacol Exp Ther 193:941–950
18. Rimoldi SF, Sartori C, Seiler C, Delacretaz E, Mattle HP, Scherrer U, Allemann Y (2010) High-altitude exposure in patients with cardiovascular disease: risk assessment and practical recommendations. Prog Cardiovasc Dis 52:512–524
19. Bosc LV, Resta T, Walker B, Kanagy NL (2010) Mechanisms of intermittent hypoxia induced hypertension. J Cell Mol Med 14:3–17
20. Dempsey JA, Veasey SC, Morgan BJ, O'Donnell CP (2010) Pathophysiology of sleep apnea. Physiol Rev 90:47–112
21. Kuo TB, Yuan ZF, Lin YS, Lin YN, Li WS, Yang CC, Lai CJ (2011) Reactive oxygen species are the cause of the enhanced cardiorespiratory response induced by intermittent hypoxia in conscious rats. Respir Physiol Neurobiol 175(1):70–79
22. Prabhakar NR, Kumar GK (2010) Mechanisms of sympathetic activation and blood pressure elevation by intermittent hypoxia. Respir Physiol Neurobiol 174 (1–2):156–161
23. Nagai M, Hoshide S, Kario K (2010) Hypertension and dementia. Am J Hypertens 23:116–124
24. Bartsch P, Gibbs JS (2007) Effect of altitude on the heart and the lungs. Circulation 116:2191–2202
25. Heistad DD, Abboud FM, Dickinson W (1980) Richards lecture: circulatory adjustments to hypoxia. Circulation 61:463–470
26. Lundberg JO, Weitzberg E (2005) NO generation from nitrite and its role in vascular control. Arterioscler Thromb Vasc Biol 25:915–922
27. Singel DJ, Stamler JS (2005) Chemical physiology of blood flow regulation by red blood cells: the role of nitric oxide and S-nitrosohemoglobin. Annu Rev Physiol 67:99–145
28. Calbet JA (2003) Chronic hypoxia increases blood pressure and noradrenaline spillover in healthy humans. J Physiol 551:379–386
29. Kamat SR, Banerji BC (1972) Study of cardiopulmonary function on exposure to high altitude. I. Acute acclimatization to an altitude of 3,5000 to 4,000 meters in relation to altitude sickness and cardiopulmonary function. Am Rev Respir Dis 106:404–413
30. Kanstrup IL, Poulsen TD, Hansen JH, Andersen LJ, Bestle MH, Christensen NJ, Olsen NV (1999) Blood pressure and plasma catecholamines in acute and prolonged hypoxia: effects of local hypothermia. J Appl Physiol 87(6):2053–2058
31. Malhotra MS, Selvamurthy W, Purkayastha SS, Mukherjee AK, Mathew L, Dua GL (1976) Responses of the autonomic nervous system during acclimatization to high altitude in man. Aviat Space Environ Med 47:1076–1079
32. Palatini P, Businaro R, Berton G, Mormino P, Rossi GP, Racioppa A, Pessina AC, Dal Palu C (1989) Effects of low altitude exposure on 24-hour blood pressure and adrenergic activity. Am J Cardiol 64:1379–1382
33. Roach RC, Houston CS, Honigman B, Nicholas RA, Yaron M, Grissom CK, Alexander JK, Hultgren HN (1995) How well do older persons tolerate moderate altitude? West J Med 162:32–36
34. Kriemler S, Jansen C, Linka A, Kessel-Schaefer A, Zehnder M, Schurmann T, Kohler M, Bloch K, Brunner-La Rocca HP (2008) Higher pulmonary artery pressure in children than in adults upon fast ascent to high altitude. Eur Respir J 32:664–669
35. MacLean DA, Sinoway LI, Leuenberger U (1998) Systemic hypoxia elevates skeletal muscle interstitial adenosine levels in humans. Circulation 98:1990–1992
36. Richalet JP, Larmignat P, Rathat C, Keromes A, Baud P, Lhoste F (1988) Decreased cardiac response to isoproterenol infusion in acute and chronic hypoxia. J Appl Physiol (1985) 65:1957–1961
37. Somers VK, Mark AL, Abboud FM (1988) Potentiation of sympathetic nerve responses to hypoxia in borderline hypertensive subjects. Hypertension 11:608–612
38. Levine BD, Zuckerman JH, deFilippi CR (1997) Effect of high-altitude exposure in the elderly: the Tenth Mountain Division study. Circulation 96:1224–1232
39. Canepa A, Chavez R, Hurtado A, Rotta A, Velasquez T (1956) Pulmonary circulation at sea level and at high altitudes. J Appl Physiol 9:328–336
40. Mirrakhimov MM, Meymanaliev TS (1984) High altitude cardiology. Frunze, Kyrgyzstan, 316 p (in Russian)
41. Vogel JA, Hartley LH, Cruz JC (1974) Cardiac output during exercise in altitude natives at sea level and high altitude. J Appl Physiol 36:173–176
42. Nakashima M (1983) High altitude medical research in Japan. Prog Clin Biol Res 136:173–182
43. Levett DZ, Fernandez BO, Riley HL, Martin DS, Mitchell K, Leckstrom CA, Ince C, Whipp BJ,

Mythen MG, Montgomery HE, Grocott MP, Feelisch M (2011) The role of nitrogen oxides in human adaptation to hypoxia. Sci Rep 1:109

44. Bender PR, Groves BM, McCullough RE, McCullough RG, Huang SY, Hamilton AJ, Wagner PD, Cymerman A, Reeves JT (1988) Oxygen transport to exercising leg in chronic hypoxia. J Appl Physiol (1985) 65:2592–2597
45. Hansen J, Sander M (2003) Sympathetic neural overactivity in healthy humans after prolonged exposure to hypobaric hypoxia. J Physiol 546:921–929
46. Kumar R, Qadar Pasha MA, Khan AP, Gupta V, Grover SK, Norboo T, Srivastava KK, Selvamurthy W, Brahamchari SK (2003) Association of high-altitude systemic hypertension with the deletion allele-of the angiotensin-converting enzyme (ACE) gene. Int J Biometeorol 48:10–14
47. Mancia G, Fagard R, Narkiewicz K, Redon J, Zanchetti A, Bohm M, Christiaens T, Cifkova R, De Backer G, Dominiczak A, Galderisi M, Grobbee DE, Jaarsma T, Kirchhof P, Kjeldsen SE, Laurent S, Manolis AJ, Nilsson PM, Ruilope LM, Schmieder RE, Sirnes PA, Sleight P, Viigimaa M, Waeber B, Zannad F, Redon J, Dominiczak A, Narkiewicz K, Nilsson PM, Burnier M, Viigimaa M, Ambrosioni E, Caufield M, Coca A, Olsen MH, Schmieder RE, Tsioufis C, van de Borne P, Zamorano JL, Achenbach S, Baumgartner H, Bax JJ, Bueno H, Dean V, Deaton C, Erol C, Fagard R, Ferrari R, Hasdai D, Hoes AW, Kirchhof P, Knuuti J, Kolh P, Lancellotti P, Linhart A, Nihoyannopoulos P, Piepoli MF, Ponikowski P, Sirnes PA, Tamargo JL, Tendera M, Torbicki A, Wijns W, Windecker S, Clement DL, Coca A, Gillebert TC, Tendera M, Rosei EA, Ambrosioni E, Anker SD, Bauersachs J, Hitij JB, Caulfield M, De Buyzere M, De Geest S, Derumeaux GA, Erdine S, Farsang C, Funck-Brentano C, Gerc V, Germano G, Gielen S, Haller H, Hoes AW, Jordan J, Kahan T, Komajda M, Lovic D, Mahrholdt H, Olsen MH, Ostergren J, Parati G, Perk J, Polonia J, Popescu BA, Reiner Z, Ryden L, Sirenko Y, Stanton A et al (2013) 2013 ESH/ESC guidelines for the management of arterial hypertension: the Task Force for the Management of Arterial Hypertension of the European Society of Hypertension (ESH) and of the European Society of Cardiology (ESC). Eur Heart J 34:2159–2219
48. Faulhaber M, Flatz M, Gatterer H, Schobersberger W, Burtscher M (2007) Prevalence of cardiovascular diseases among alpine skiers and hikers in the Austrian Alps. High Alt Med Biol 8:245–252
49. Honigman B, Theis MK, Koziol-McLain J, Roach R, Yip R, Houston C, Moore LG, Pearce P (1993) Acute mountain sickness in a general tourist population at moderate altitudes. Ann Intern Med 118:587–592
50. Nault P, Halman S, Paradis J (2009) Ankle-brachial index on Kilimanjaro: lessons from high altitude. Wilderness Environ Med 20:72–76
51. Ponchia A, Noventa D, Bertaglia M, Carretta R, Zaccaria M, Miraglia G, Pascotto P, Buja G (1994) Cardiovascular neural regulation during and after prolonged high altitude exposure. Eur Heart J 15:1463–1469
52. Savonitto S, Cardellino G, Doveri G, Pernpruner S, Bronzini R, Milloz N, Colombo MD, Sardina M, Nassi G, Marraccini P (1992) Effects of acute exposure to altitude (3,460 m) on blood pressure response to dynamic and isometric exercise in men with systemic hypertension. Am J Cardiol 70:1493–1497
53. Hasler E, Suter PM, Vetter W (1997) Race specific altitude effects on blood pressure. J Hum Hypertens 11:435–438
54. D'Este D, Mantovan R, Martino A, D'Este F, Artusi L, Allibardi P, Franceschi M, Zerio C, Pascotto P (1991) The behavior of the arterial pressure at rest and under exertion in normotensive and hypertensive subjects exposed to acute hypoxia at a median altitude. G Ital Cardiol 21:643–649
55. Wu TY, Ding SQ, Liu JL, Yu MT, Jia JH, Chai ZC, Dai RC, Zhang SL, Li BY, Pan L, Liang BZ, Zhao JZ, de Qi T, Sun YF, Kayser B (2007) Who should not go high: chronic disease and work at altitude during construction of the Qinghai-Tibet railroad. High Alt Med Biol 8:88–107
56. Sizlan A, Ogur R, Ozer M, Irmak MK (2008) Blood pressure changes in young male subjects exposed to a median altitude. Clin Auton Res 18:84–89
57. Erdmann J, Sun KT, Masar P, Niederhauser H (1998) Effects of exposure to altitude on men with coronary artery disease and impaired left ventricular function. Am J Cardiol 81:266–270
58. Schmid JP, Noveanu M, Gaillet R, Hellige G, Wahl A, Saner H (2006) Safety and exercise tolerance of acute high altitude exposure (3454 m) among patients with coronary artery disease. Heart 92:921–925
59. Mancia G, De Backer G, Dominiczak A, Cifkova R, Fagard R, Germano G, Grassi G, Heagerty AM, Kjeldsen SE, Laurent S, Narkiewicz K, Ruilope L, Rynkiewicz A, Schmieder RE, Boudier HA, Zanchetti A (2007) 2007 ESH-ESC practice guidelines for the management of arterial hypertension: ESH-ESC task force on the management of arterial hypertension. J Hypertens 25:1751–1762
60. Guazzi MD, Berti M, Doria E, Fiorentini C, Galli C, Pepi M, Tamborini G (1989) Enhancement of the pulmonary vasoconstriction reaction to alveolar hypoxia in systemic high blood pressure. Clin Sci (Lond) 76:589–594
61. Burtscher M (2007) Risk of cardiovascular events during mountain activities. Adv Exp Med Biol 618:1–11
62. Beall CM (2003) High-altitude adaptations. Lancet 362(Suppl):s14–s15
63. Frisancho AR (1993) Human adaptation and accommodation. University of Michigan Press, Ann Arbor
64. Hultgren HN (1992) Effect of high altitude on cardiovascular diseases. J Wilderness Med 3:301–308
65. Ruiz L, Penaloza D (1977) Altitude and hypertension. Mayo Clin Proc 52:442–445
66. Sun SF (1986) Epidemiology of hypertension on the Tibetan Plateau. Hum Biol 58:507–515
67. Monge MC, Monge M (1966) High altitude diseases: mechanism and management. Thomas, Springfield

68. Heath D, Williams RR (1977) Man at high altitude. The pathophysiology of acclimatization and adaptation. Churchill Livingstone, Edinburg
69. Ward M (1975) Mountain medicine a clinical study of cold and high altitude. Van Nostrand Reinhold Co., New York
70. Faeh D, Gutzwiller F, Bopp M (2009) Lower mortality from coronary heart disease and stroke at higher altitudes in Switzerland. Circulation 120:495–501
71. Hanna JM (1999) Climate, altitude, and blood pressure. Hum Biol 71:553–582
72. Shrestha S, Shrestha A, Shrestha S, Bhattarai D (2012) Blood pressure in inhabitants of high altitude of Western Nepal. JNMA J Nepal Med Assoc 52:154–158
73. Fiori G, Facchini F, Pettener D, Rimondi A, Battistini N, Bedogni G (2000) Relationships between blood pressure, anthropometric characteristics and blood lipids in high- and low-altitude populations from Central Asia. Ann Hum Biol 27:19–28
74. Gamboa A, Gamboa JL, Holmes C, Sharabi Y, Leon-Velarde F, Fischman GJ, Appenzeller O, Goldstein DS (2006) Plasma catecholamines and blood volume in native Andeans during hypoxia and normoxia. Clin Auton Res 16:40–45
75. Otsuka K, Norboo T, Otsuka Y, Higuchi H, Hayajiri M, Narushima C, Sato Y, Tsugoshi T, Murakami S, Wada T, Ishine M, Okumiya K, Matsubayashi K, Yano S, Chogyal T, Angchuk D, Ichihara K, Cornelissen G, Halberg F (2005) Chronoecological health watch of arterial stiffness and neuro-cardio-pulmonary function in elderly community at high altitude (3524 m), compared with Japanese town. Biomed Pharmacother 59 (Suppl 1):S58–S67
76. Tripathy V, Gupta R (2007) Blood pressure variation among Tibetans at different altitudes. Ann Hum Biol 34:470–483
77. Tripathy V, Charan Satapathy K, Gupta R (2006) Nutritional status and hypertension among Tibetan adults in India. Hum Ecol 14:77–82
78. Zhao X, Li S, Ba S, He F, Li N, Ke L, Li X, Lam C, Yan L, Zhou Y, Wu Y (2012) Prevalence, awareness, treatment and control of hypertension among herdsmen living at 4300 in Tibet. Am J Hypertens 7:583–589
79. Henley WN, Bellush LL, Notestine MA (1992) Reemergence of spontaneous hypertension in hypoxia-protected rats returned to normoxia as adults. Brain Res 579:211–218
80. Kolar F, Ostadal B, Prochazka J, Pelouch V, Widimsky J (1989) Comparison of cardiopulmonary response to intermittent high-altitude hypoxia in young and adult rats. Respiration 56:57–62
81. Milano G, Corno AF, Lippa S, Von Segesser LK, Samaja M (2002) Chronic and intermittent hypoxia induce different degrees of myocardial tolerance to hypoxia-induced dysfunction. Exp Biol Med (Maywood) 227:389–397
82. Gupte SA, Wolin MS (2012) Relationships between vascular oxygen sensing mechanisms and hypertensive disease processes. Hypertension 60:269–275
83. Semenza GL, Prabhakar NR (2012) The role of hypoxia-inducible factors in oxygen sensing by the carotid body. Adv Exp Med Biol 758:1–5
84. Harrison DG, Guzik TJ, Lob HE, Madhur MS, Marvar PJ, Thabet SR, Vinh A, Weyand CM (2011) Inflammation, immunity, and hypertension. Hypertension 57:132–140
85. Paravicini TM, Touyz RM (2006) Redox signaling in hypertension. Cardiovasc Res 71:247–258
86. Severinghaus JW, Bainton CR, Carcelen A (1966) Respiratory insensitivity to hypoxia in chronically hypoxic man. Respir Physiol 1:308–334
87. Otsuka K, Norboo T, Otsuka Y, Higuchi H, Hayajiri M, Narushima C, Sato Y, Tsugoshi T, Murakami S, Wada T, Ishine M, Okumiya K, Matsubayashi K, Yano S, Choygal T, Angchuk D, Ichihara K, Cornelissen G, Halberg F (2005) Effect of aging on blood pressure in Leh, Ladakh, a high-altitude (3524 m) community, by comparison with a Japanese town. Biomed Pharmacother 59(Suppl 1): S54–S57

Herbs for Mitigating High Altitude Maladies

Richa Rathor, Priyanka Sharma, Geetha Suryakumar, Asheesh Gupta, Gidugu Himashree, and Gopinath Bhaumik

Abstract

A large number of armed forces are deployed at mountainous regions of the country. Nevertheless, a significant fall in work efficiency and physical performance is observed at high altitude. In view of the above, there is an essential requisite for identification of pharmacotherapeutic interventions (herbal, dietary, or chemical agents) to boost endurance and performance and to mitigate cold injuries and promote wound healing under stressful conditions. In the last few decades, DIPAS has extensively worked on this area, and this article summarizes the studies conducted under the aegis of DIPAS to ascertain the role of herbs in mitigating high-altitude maladies. Extensive scientific studies at DIPAS have focused on the development of adaptogens, herbal formulations which help to enhance nonspecific resistance to stress and improve physical performance. In this context, three herbal preparations such as Composite Indian Herbal Preparation-I (CIHP-I), Composite Indian Herbal Preparation II (CIHP-II), and "DIP-91" have been developed with significant antistress as well as adaptogenic activity. A herbal wound healer named "HERBO HEALER" is a product which possesses antioxidant, antibacterial activity and wound-healing activity in both acute as well as chronic wounds. DIPAS has also formulated a cream named "ALOCAL," which is having a promising role against cold injury. A hypolipidemic agent named "DIP-LIP" was also developed that possessed antiatherogenic, cardioprotective activities and vasorelaxation activity.

Conclusively, herbs possess a wide range of therapeutic effects, and through systematic scientific investigations using various animal models and clinical trials, research at DIPAS has translated this knowledge into products for protection against high-altitude maladies and also for enhancing performance under stressful conditions.

R. Rathor • P. Sharma • G. Suryakumar (✉)
A. Gupta • G. Bhaumik
Defence Institute of Physiology and Allied Sciences (DIPAS), DRDO, Delhi, India
e-mail: geethasuryakumar@yahoo.com

G. Himashree
High Altitude Physiology Group, Defence Institute of Physiology and Allied Sciences (DIPAS), Timarpur, Delhi, India

S.B. Singh et al. (eds.), *Translational Research in Environmental and Occupational Stress*, DOI 10.1007/978-81-322-1928-6_21,

High terrestrial altitudes have since long attracted man, both for leisure and resources. The populace at high altitude includes the indigenous natives (Tibetans and Sherpas), armed forces, as well as sojourners like trekkers, skiers, etc. High-altitude region may lead to medical problems from the mild symptoms of acute mountain sickness (AMS) to the potentially fatal high-altitude pulmonary edema (HAPE) and high-altitude cerebral edema (HACE). The most common ailment of high altitude is AMS and the symptoms are headache, nausea, anorexia, insomnia, lassitude, vomiting, and dizziness. Additionally, cold injuries and frostbites also develop at high altitude due to low temperature. The other significant factors which affect life at high altitude include hypoxia (a result of the fall in the atmospheric pressure of oxygen, required for all mammalian life), cold, wind, and isolation from society and civilization. Although individually all these factors form potent psychophysiological stressors; when present together they evoke a formidable challenge to the human adaptability and survival under such environmental vagaries.

Evidence shows a significant fall in work efficiency and physical performance under high-altitude conditions. Cardiopulmonary changes in cardiac output and blood pressure, a marked decrease in body mass, and loss of muscle function are a known response to the hypobaric hypoxia. The impairment in the physical performance capacity is further exacerbated by increased oxidative stress, polycythemia, and decreased lactate production.

In view of the deployment of a large percentage of armed forces to the mountainous regions of India, there is an essential requisite for identification of pharmacotherapeutic intervention (herbal, dietary, or chemical agents) to boost endurance and performance under strenuous conditions. In the last century, an immense interest has been shown towards Ayurveda. Ayurveda is one of the traditional systems of medicine practiced in India and can be traced back to 6000 BC [10]. Ayurveda (ayus, life; veda, knowledge, meaning the science of life) is the oldest medical system in the world, which exploits the potential of various herbs generally in polyherbal formulations. Due to fewer side effects, ease in availability, and cost effectiveness, there is an ever-increasing interest in research on different plant species to find out their therapeutic applications.

Plants have acquired an essential system to reduce and scavenge active oxygen species which are naturally generated during photosynthesis and respiration [2]. Plants, as a source of medicines and human sustenance, have been in vogue since antiquity. Ancient Indian literature like Rigveda, Atharvaveda, Upanishads, Charaka Samhita, and Sushruta Samhita cites the use of medicinal plants as drugs, essences, tools of worship, food, fuels, poisons, and agricultural tools.

The World Health Organization (WHO) has listed about 20,000 plant species in the world yielding drugs. In India, over 2,500 species are credited with medicinal values. Plants produce a number of phytonutrients which when present in an appropriate concentration act in synergistic fashion having therapeutic activity. Plants are rich source of vitamins, minerals, and other bioactive components. Major classes of phytonutrients include carotenoids, polyphenols, anthocyanins, flavonoids, isothiocyanates, sulfides, and phytosterols [5, 17, 18]. This broad range of natural compounds appears to have numerous biological functions which may be overlapping and complementary. Interest in phytonutrients has resulted in identifying mechanism of action at the cellular or molecular level for many of these compounds. Several studies have revealed that plants produce potent antioxidants to control the oxidative stress caused by excessive sun rays and oxygen and represent a source of new compounds with antioxidant activity [28].

DIPAS in the last few decades has extensively worked on developing herbal products for amelioration of high-altitude stress. This article summarizes the studies conducted under the aegis of DIPAS to ascertain the role of herbs in mitigating high-altitude maladies.

Herbal Adaptogens

DIPAS has focused on the development of adaptogens, herbal formulations which help to enhance nonspecific resistance to stress as well as improving physical performance.

The term adaptogen was first proposed by Brekhman and Dardymov [6]. These herbal derivatives improve the human system against stress and help in restoration of normal homeostasis of the body and can be viewed as general health tonics prescribed to enhance vitality (Fig. 1). Adaptogen has a large range of therapeutic activity and does not cause any metabolic alterations in the body during resting state.

Multiple Stress Animal Model for Evaluation of Adaptogenic Activity

To study the adaptogenic potential of herbal formulations, DIPAS has developed a passive multiple stress animal model way back in the 1990s, called as cold-hypoxia-restraint (C-H-R) animal model. In this model, overnight fasted rats are exposed to decompression chamber which maintained cold temperature (5 °C), a low atmospheric pressure of 428 mmHg pressure equivalent to an altitude of 4,572 m, and restraint stress simultaneously. Under such stressful conditions, a rectal probe is inserted 2 cm past the rectum and the rectal temperature (T_{rec}) is monitored once per minute by using a 16-channel Iso-Thermex temperature recorder (Columbus Instrument, Columbus, OH) (Fig. 2). The gradual fall in rectal temperature is recorded from 37 to 23 °C. Attainment of a colonic temperature of 23 °C is taken as the termination point of the exposure, as any further fall in colonic temperature resulted in a high incidence of mortality. After exposure, rat is kept inside a conditioning

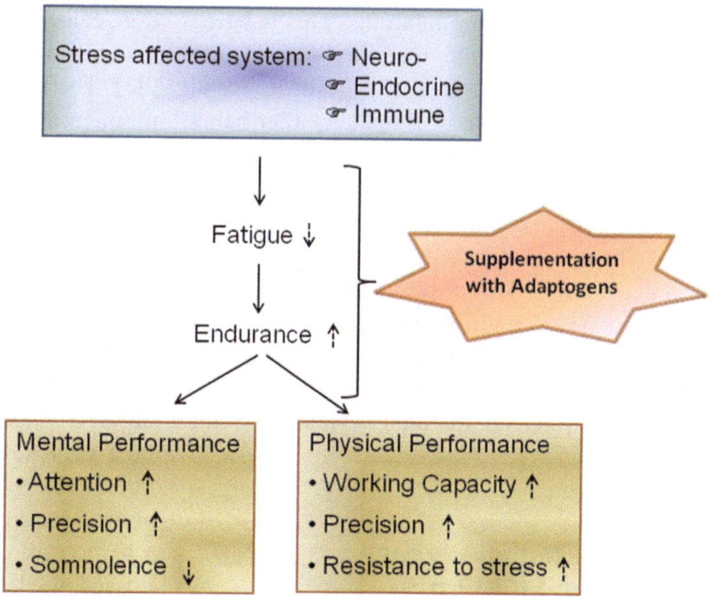

Fig. 1 Effects of adaptogen on stress-induced symptoms

Fig. 2 Animal decompression chamber

chamber for recovering temperature from 23 to 37 °C at normal atmospheric pressure and ambient temperature of 32 ± 1 °C. The rat continues to be restrained during recovery period. The time taken to attain T_{rec} of 23 °C and its recovery to T_{rec} of 37 °C is used as a measure of endurance [27].

This C-H-R model has been exploited for evaluating the adaptogenic and anti-stress efficacy of many herbal formulations. This is based on the simple principle that any formulation which is capable to resist a change in its rectal temperature (a physiological marker) after being exposed to simulated high-altitude conditions developed by C-H-R model could be considered as adaptogens for cold, hypoxia, and restraint stress. The formulations are usually administered 30 min prior to C-H-R exposure.

Using C-H-R multiple stress animal model, numerous plant extracts and composite (polyherbal, multicomponent) formulations have been evaluated for their adaptogenic and anti-stress activity.

Composite Indian Herbal Preparation-I (CIHP-I)

Basically, composite Indian herbal preparation-I (CIHP-I) is a mixture of number of herbs. The accumulative effect of these herbals is able to increase physical and mental performance. Singh et al. [32] reported that oral administration of CIHP-I to mice for 4 days at the dose of 50–150 mg/kg body weight increased survival time of swimming and also prevented stress-induced changes in the adrenals. CIHP-I was also able to increase oxygen demand at high terrestrial altitude [33] and also improved nutrition, cell-mediated immunity, and blood proteins without affecting glucose and cholesterol level [23]. The LD_{50} of the CIHP-I is 56 g/kg (orally) in mice, while it is effective at the dose of 50–150 mg/kg body weight.

Studies using CIHP-1 against cold-hypoxia-restraint stress (C-H-R) found that a 3-week oral administration of CIHP-I at the dose of 15 mg/kg resulted in low body weight gain when compared to control rats. The time taken to attain T_{rec} of 23 °C increased by 30 % at the dose of 7.5 mg. At a dosage of 150 mg/kg, the cooling time was observed to be further increased (45 %). Recovery time of T_{rec} to 37 °C was also observed to be decreased (~40 %) at a dose of 75 mg/kg and above. The time required for T_{rec} to fall to 23 °C was increased by 9 % after administration of single dosage of CIHP-I. On the other hand, five consecutive doses of CIHP-I, once a day, increased the time for T_{rec} to fall 23 °C to 29 % and recovery time of T_{rec} of 37 °C decreased by 18 % [24].

Composite Indian Herbal Preparation-II (CIHP-II)

A composite Indian herbal preparation-II (CIHP-II) is a preparation of 39 plant components and 6 minerals [15]. CIHP-II is also safe and nontoxic on prolonged use. Singh et al. (1978) also reported that the LD_{50} value of CIHP-II is 5–6 g/kg in mice while the beneficial effective dose is 100–200 mg/kg body that is quite less than LD_{50}. Administration of CIHP-II at a dose of 1 mg/kg body weight significantly delayed the time for attaining T_{rec} of 23 °C and recovery to 37 °C was significantly reduced.

DIPAS also conducted large-scale field trials of CIHP-II for introduction into the armed forces operating under extreme climatic conditions. In this study, a double-blind placebo-controlled randomized field trial was carried out on age- and body weight-matched volunteers to ascertain efficacy of CIHP in curtailing altitude-induced maladies during acute induction to an altitude of 3,500 m and following prolonged residence at extreme altitude of 4,800–6,000 m. A significant decrease in AMS symptom score was observed in CIHP-II-treated volunteers as compared to placebo-treated group. The arterial oxygen saturation (SaO_2) in CIHP-II-administered volunteers was higher than the placebo-administered group. The pulse rate in CIHP-II-administered volunteers at high altitude was lower than the placebo-administered group. These results suggested that CIHP-II administration is able to curtail significantly the altitude-induced deterioration of physical and mental performance of the soldiers during prolonged residency at high and extreme altitude.

DIP-91

"DIP-91" is a herbal anti-stress adaptogen product developed by DIPAS. Studies conducted by us show that "DIP-91" has significant anti-stress and adaptogenic activity. "DIP-91" is prepared from a single herbal component. It possesses a noncumulative potent adaptogenic activity and acts as anxiolytic, endurance promoter, revitalizer, and rejuvenator. Besides this, "DIP-91" has excellent antioxidant potential that helps to reduce oxidative stress. Interestingly, "DIP-91" does not cause any metabolic alternations in the body during resting state, and it is far better than available products as stand-alone preparation (*Withania somnifera*/*Ginkgo biloba*/*Panax ginseng*) and polyherbal and multicomponent preparations in the market. "DIP-91" is also cost-effective, nontoxic, safe, and free of heavy metal toxicity. "DIP-91" is effective as health food supplement/nutraceutical and anti-stress agent to manage day-to-day stress and increases stamina and immunity; therefore, it is useful in adapting to adverse environmental conditions.

Herbal Wound Healer

Wound repair and regeneration are essential processes in maintaining tissue homeostasis in response to injury. Wound healing is a complex process of well-orchestrated and predictable events and comprises of three overlapping phases: inflammation, granulation tissue formation, and remodeling [31]. Optimum healing of a wound requires an integration of the complex biological and molecular events of cell migration and proliferation and of extracellular matrix deposition. Interplay between blood cells, endothelial cells, fibroblasts, and keratinocytes and the local release of growth factors and cytokines influences the rate of wound repair [14, 25].

The normal healing process can be impeded at any step along its path by a variety of factors. Delayed wound healing is one of the major therapeutic and economic issues in medicine today. The World Health Organization (WHO) estimates that 6.5 million individuals suffer from chronic skin ulcers caused by prolonged pressure, venous stasis, or diabetes mellitus and burn injuries [36]. In contrast, military combat injuries are almost exclusive to penetrating wounds such as gunshot wounds, explosions, splinters, and bruises. Combat-injured patients are highly susceptible to microbial infections. Multiple pathogenic abnormalities, ranging from disease-specific intrinsic flaws in blood supply, angiogenesis, phenotypic changes in resident wound cells, altered matrix metalloproteinases profiles, and extracellular matrix turnover to extrinsic factors due to infection and the presence of biofilms, bacterial colonization, and excessive exudate, contribute to failure to heal [29]. Wounds and wound healing abnormalities cause a great deal of physical and psychological discomfort and morbidity to affected patients. Therefore, safe and effective wound care modalities are required, which are nontoxic, minimally invasive, and economically feasible for improving wound healing.

Recent technological advances have led to very promising breakthroughs in the treatment of delayed wound healing. Research on pharmacological and non-pharmacological modalities to augment wound healing is a developing area in modern biomedical sciences. Recent advances in cellular and molecular biology have greatly expanded our understanding of the biological processes involved in wound repair and tissue regeneration and have led to improvements in wound care management. The objective in wound management is to heal the wound in the shortest time possible, with minimal pain, discomfort, and scarring to the patients.

The process of wound healing is promoted by natural (honey, chitosan) and plant products which are composed of bioactive phytoconstituents, viz. alkaloids, glycosides, triterpenes, and flavonoids. These agents usually influence one or more phases of the healing process. The presence of various life-sustaining constituents in plants has urged scientist to examine these plants with a view to determine potential wound healing properties [16, 22].

Herbo Healer

Recently, a potent herbal wound healer in the form of a topical ointment has been developed for wound care applications. The preparation has been evaluated for its bioefficacy on experimental full-thickness dermal acute and chronic (burns and diabetic) animal wound models, and its possible mechanisms of action were also investigated. The developed “Herbo Healer” preparation contains a single herbal component, nontoxic, economical, and highly rich in bioactive phytochemicals (polyphenols and flavonoids).

Herbo Healer preparation showed strong antioxidant and antibacterial properties. In vitro studies revealed that Herbo Healer possessed antibacterial activity against tested wound pathogens (*Pseudomonas aeruginosa* and *Staphylococcus aureus*) and enhanced angiogenic activity observed in chick chorioallantoic membrane (CAM) model. In vivo studies demonstrated that the formulation possessed potent wound healing activity as assessed by biophysical, biochemical, molecular, and histological parameters. The herbal preparation increased cellular proliferation, collagen biosynthesis, neovascularization, and wound area contraction. The healing activity of Herbo Healer was found to be far better than silver sulfadiazine and povidone-iodine ointments (standard care). The treatment also reduced scar and upregulated the expression of growth factors (VEGF, TGF-β), extracellular matrix (ECM) protein (collagen type III), and matrix metalloproteinases (MMP-2 and MMP-9), which help in tissue regeneration and remodeling phases of the wound healing. These results were further supported by histopathological examinations. The treatment also increased the endogenous enzymatic (superoxide dismutase, catalase, glutathione peroxidases) and nonenzymatic (reduced glutathione, vitamin C) antioxidants and decreased lipid peroxide levels and reactive oxygen species in wound granulation tissue.

The studies suggested that Herbo Healer promotes wound healing in experimental acute and chronic (burns, diabetic) wounds. This was indicated by a significant increase in re-epithelialization and wound closure. The increased wound contraction in treated rats might be due to enhanced activity of myofibroblasts in regenerated wound tissue, which exert tension on the surrounding ECM and secreting ECM proteins such as collagen to stabilize the contraction. Collagen is a major

Fig. 3 "Herbo Healer" ointment for topical wound healing application

expression of VEGF in regenerated tissue. VEGF is an important proangiogenic cytokine and improves angiogenesis during wound healing by stimulating the migration and proliferation of endothelial cells through the ECM [14]. In conclusion, as illustrated in Fig. 3, Herbo Healer possesses a significant wound healing activity in both acute and chronic wounds (Fig. 4).

The detailed study for safety and toxicity evaluation (oral and dermal) of the Herbo Healer ointment formulation has been carried out by Shriram Institute for Industrial Research, Delhi (ISO and NABL accredited laboratory), and reported that herbal wound healer is completely safe for dermal applications. Furthermore, Herbo Healer ointment dermatologically tested under limited field trial studies conducted on human subjects found that it protects skin from dryness and heals skin's moisture barrier providing soothing effects at harsh climatic conditions. The developed Herbo Healer has immense wound healing potentials and wide applications in skin tears, abrasions, incision, excision injuries, superficial/deep burns, scalds, bruises, and diabetic wounds.

protein of ECM and component that ultimately contributes to wound strength. The enhanced levels of hydroxyproline and hexosamine in Herbo Healer-treated wounds probably provided the strength to the regenerated wound tissue. MMPs are key players in every phase of the repair process, like eliminate damaged protein, destroy the provisional ECM, facilitate migration to the center of the wound, remodel the granulation tissue, and probably control angiogenesis and also regulate the activity of some growth factors [16]. Increased expression of MMP-2 and MMP-9 in Herbo Healer-treated experimental rats suggested that it played an important role in remodeling of the ECM. Angiogenesis is a critical component of wound healing. The Herbo Healer treatment promoted angiogenesis in both in vivo and in vitro models as indicated by histological studies and new vessel formation in CAM model, which might be due to enhanced

Alocal (Cream for Cold Injury)

Armed forces operating at snowbound cold areas of high altitude are facing the problem of cold injuries. Hypoxia has been shown to play an important role in the reduction of blood flow to the extremities and therefore may modulate the cold injury [13]. Cold injuries can be divided into two general categories: nonfreezing cold injuries (chilblains and trench foot) and freezing cold injuries (frostbite). Chilblains are itchy, painful, reddish, or purplish areas of swelling that usually affect the fingers, toes, nose, or ears, though other areas of the body may also be involved. On the other hand, frostbite is a major health and serious medical problem for persons who are exposed to cold climatic conditions like military personnel operating in snowbound areas, mountaineers, trekkers, skiers, pilgrims, and civilian population engaged in construction work at high altitude [9].

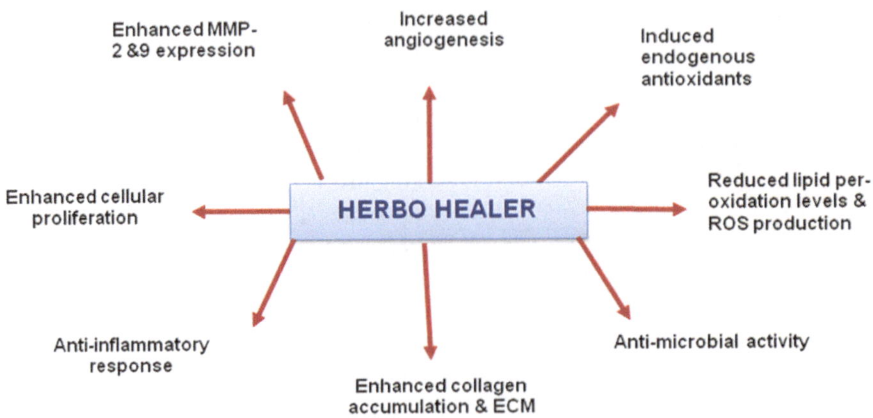

Fig. 4 Possible mechanisms by which "Herbo Healer" enhances wound healing process

Fig. 5 Trench foot

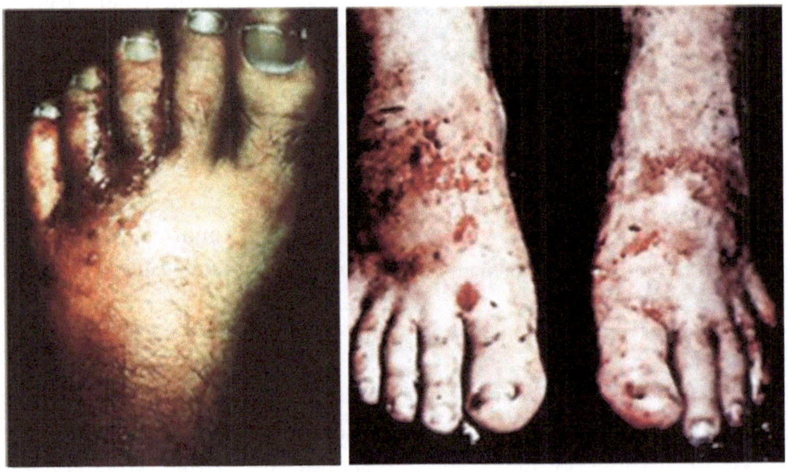

It is responsible for the loss of fingers and toes in large number of affected cases.

Trench foot (immersion injury) was named after the condition suffered by many soldiers in the trenches during World War I. Trench foot develops after prolonged exposure to a wet, cold environment and is typically a more serious condition than chilblains. The symptoms of tench foot indicate pain, numbness, and swelling. Severe trench foot leads to the development of gangrene and amputation of injured body part may also occur (Fig. 5). Frostbite occurs when there is freezing of the affected tissue and it is the most serious form of the cold injuries. Frostbite usually affects the exposed body parts, i.e., hands, feet, nose, or ears, though other parts of the body may also be affected (Fig. 6). This type of injury results from decreased blood flow and heat delivery to body tissues resulting in damaging ice crystal formation.

Freezing Simulation Machine for Inducing Frostbite Injury in Animals

Frostbite injury was induced experimentally by exposing the one of the hind limb of the harnessed rat in freezing simulation machine (Fig. 7) (designed and fabricated at DIPAS workshop) using the method described by Purkayastha et al. and Himashree et al. [26, 21, 20]. A stream of compressed air from an air

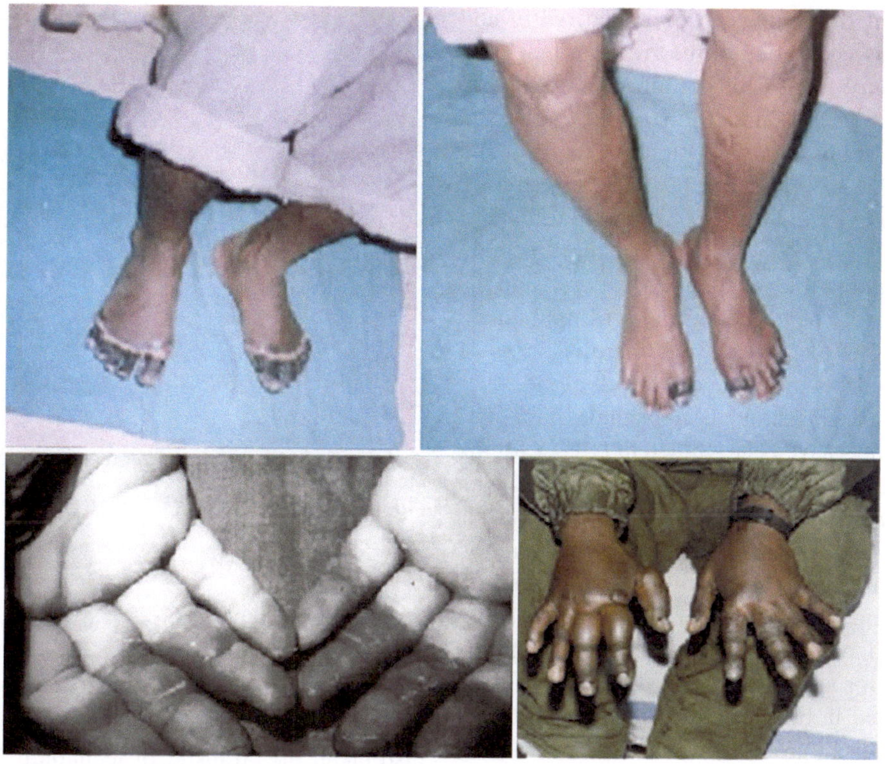

Fig. 6 Cold injuries in the glacier

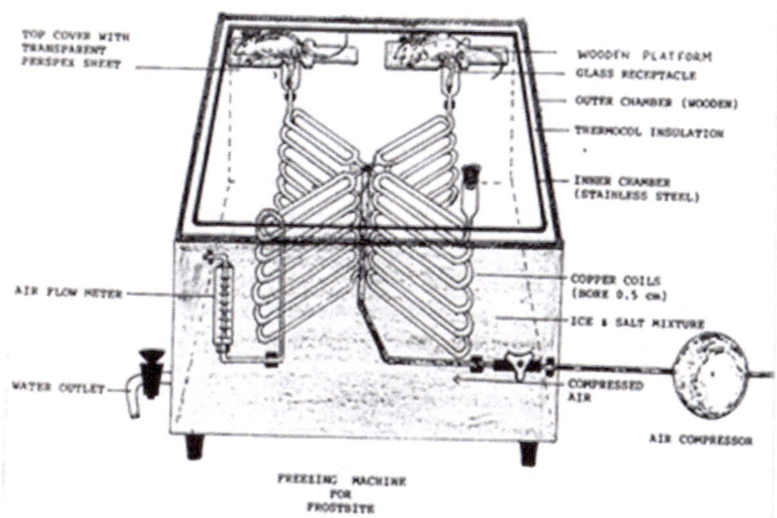

Fig. 7 Frostbite induction assembly developed at DIPAS

compressor, which has a pressure regulator, was passed through a series of copper coils submerged in a freezing mixture of ice and salt. The cold air then enters the glass receptacle in which the limb of the rat is exposed. This effectively simulates the conditions for frostbite. Consequential factors affecting frostbite, viz. temperature, duration of exposure, and velocity of the wind, can be concurrently and exclusively controlled in the freezing machine; the degree of frostbite is conditional to all these factors [35]. Left hind limb of the harnessed rats was exposed at −20 ± 1 °C with wind flow of 25–30 l/min, for 15 min; temperature recording was done by YSI telethermometer (Model 46 TUC, USA). During exposure, only one of the hind limb of the rat is exposed and the temperature around the animal was maintained at 26–28 °C to prevent hypothermia so as to study the effect of local cold injury alone.

Using freezing simulation machine, numerous plant extracts have been studied in DIPAS for the protective efficacy against cold injury in which *Aloe vera* is found to be promising. *Aloe vera* is known to have tissue regeneration, vasodilatory, anti-inflammatory, and antimicrobial properties [11, 12] and have been extensively used in traditional Indian, Tibetan, Chinese, and Mongolian medicine for wounds and burns [7, 19, 30, 34]. It is also used in burns as the pathophysiology of frostbite is similar to burns. Hence the efficacy of *Aloe vera* in frostbite was tested at DIPAS. *Aloe vera* contains different compounds such as polysaccharides, steroids, organic acid, and enzymes. *Aloe vera* can penetrate through the injured tissue, dilates the blood capillaries, and increases blood circulation. It also inhibits platelet aggregation and has anti-inflammatory properties. *Aloe vera* also prevents tissue necrosis by wound healing.

Aloe vera was developed by DIPAS as a topical cream, named "Alocal." The cream is having 50 % *Aloe vera* with aqueous base and moisturizer. It is not oily or greasy and importantly it does not freeze or flake at subzero temperature even at −30 °C. The cream has also been found to be effective in offering protection from sunburn and chafing of cheeks, lips, and fingers. Prophylactic/therapeutic effect of "Alocal" is due to its anti-inflammatory, vasodilatory, wound healing, and tissue-regeneratin properties. The cream needs to be applied on all parts of the exposed body three times a day for its prophylactic use against frostbite/cold injury.

Fig. 8 Alocal cream

With these specifications, we approached the industry and Ms Fem Pharma (now Dabur India) accepted the proposal and made the "Alocal cream" (Fig. 8). Presently other companies have also taken the "transfer of technology" to make and market the same.

DIP-LIP Hypolipidemic Agent

High-altitude exposure from one to several days or weeks results in hypoxic condition, emphasized by diminished ambient oxygen pressure, decreased temperature, lower humidity, and increased ultraviolet radiation. These hypoxic conditions cause changes in the myocardium with profound effect on the morphology and function of the cardiopulmonary system [4]. Prolonged exposure to high altitude leads to an increase in heart rate, myocardial contractility, and cardiac output [8]. Atland and Highman [1] found severe atherosclerotic lesions in the pulmonary vessels with calcification in rabbits when the animals are exposed to 11–17 weeks at 16,000 ft simulated height.

Keeping complications in mind, DIPAS has developed a hypolipidemic agent named "DIP-

LIP." Besides hypolipidemic property, it also possesses antiatherogenic and cardioprotective activities. It is basically synthesized by super-critical CO_2 extraction. In a study conducted on rabbit model, administration of DIP-LIP along with high-cholesterol diet restricted further rise of total cholesterol and caused a significant decline of triglyceride and LDL cholesterol as compared to animals fed on high-cholesterol diet only. The rise in HDL cholesterol over the basal values in DIP-LIP-treated animals was significantly higher than the nontreated animals. Besides this, DIP-LIP also caused a significant vasorelaxation activity [3].

Conclusion

This series of studies highlights the translation of traditional ethnopharmacological wisdom into a process and product with a scientific rationale and development of products for acclimatization to high altitude. Herbs have wide range of therapeutic effects, and through systematic scientific investigations using various animal models and clinical trials, research at DIPAS has translated this knowledge into products for protection against high-altitude maladies and also for enhancing performance under stressful conditions.

References

1. Atland PD, Highman B (1960) Effects of high altitude on cholesterol-fed rabbits: production of severe pulmonary atherosclerosis with calcification. Arch Pathol 70:349–357
2. Asada K (1994) Production and action of active oxygen species in photosynthetic tissues. In: Foyer CH, Mullineaux PM (eds) Causes of photooxidative stress and amelioration of defense systems in plants. CRC Press, Boca Raton
3. Basu M, Prasad R, Jayamurthy P, Pal K, Arumughan C, Sawhney RC (2007) Anti-atherogenic effects of Sea buckthorn (*Hippophae rhamnoides*) seed oil. Phytomedicine 14:770–777
4. Bohuslavova R, Kolar F, Kuthanova L, Neckar J, Tichopad A, Pavlinkova G (2010) Gene expression profiling of sex differences in HIF1-dependent adaptive cardiac responses to chronic hypoxia. J Appl Physiol 109:1195–1202
5. Bravo L (1998) Polyphenols: chemistry, dietary sources, metabolism and nutritional significance. Nutr Rev 56:317–333
6. Brekhman II, Dardymov IV (1968) New substances of plant origin which increase nonspecific resistance. Annu Rev Pharmacol 8:419–430
7. Briggs C (1995) Herbal medicine: Aloe. Can Pharm J 128:48–50
8. Calbet J (2003) Chronic hypoxia increases blood pressure and noradrenaline spillover in healthy humans. J Physiol 551:379–386
9. Cattermole TJ (1999) The epidemiology of cold injury in Antarctica. Aviat Space Environ Med 70 (2):135–140
10. Charak Samhita (1949) Translator: Shree Gulabkunvrba Ayurvedic Society, Jamnagar
11. Coats BC (1979) The silent healer, a modern study of Aloe Vera. B. C Coats, Garland
12. Danof IE (1987) Remarkable Aloe. Omnimedicus Press, Grand Prairie
13. Durand J, Martineud JP (1971) High altitude physiology in cardiac and respiratory aspects. In: Porter R, Knight J (ed) Resistance and capacitance vessels of the skin in permanent and temporary residents at high altitude. Churchill Livingstone, Edinburg, London, pp 159–186
14. Falanga V (2005) Wound healing and its impairment in the diabetic foot. Lancet 366:1736–1743
15. Grover SK, Divekar IM, Kumar R, Pahw MI, Bharadwaj SK, Gupta AK, Srivastava KK (1995) Experimental evaluation of a Composite Indian Herbal Preparation-I (CIHP-I) as an adaptogen and its mechanism of action. Int J Pharmacogn 33:148–154
16. Gupta A, Upadhyay NK, Sawhney RC, Kumar R (2008) A poly-herbal formulation accelerates normal and impaired diabetic wound healing. Wound Rep Reg 16:784–790
17. Harborne JB, Williams CA (2000) Advances in flavonoid research since 1992. Phytochemistry 55(6): 481–504
18. Harborne JB, Williams CA (2001) Anthocyanins and other flavonoids. Nat Prod Rep 18:310–333
19. Heck E, Head M, Nowak D (1981) *Aloe vera* (gel) cream as a topical treatment for outpatient burns. Burns 7(4):291–294
20. Himashree G, Dass D, Singh S, Lama H, Chauhan SKS, Baumik G, Banerjee PK (2007) Thermographic evaluation of experimentally induced superficial and deep frostbite in rats. Thermol Int 17(2):51–58
21. Himashree G, Purkayastha SS, Dass D, Singh S, Lama H, Chauhan SKS, Baumik G, Banerjee PK (2005) Clinical studies on the treatment of cold injury: using therapeutic rewarming in decoction of tea leaves and combined therapy of pentoxifylline, aspirin and vitamin C with prophylactic application of Aloe Vera cream. Technical report, DIPAS, DRDO, report no DIPAS/224/2005
22. Kumar B, Vijayakumar M, Govindarajan R, Pushpangadan P (2007) Ethnopharmacological

approaches to wound healing-exploring medicinal plants of India. J Ethnopharmacol 114:103–113

23. Kumar P, Singh SP, Doval DC, Singh N, Kohli RP (1982) A clinical assessment of changes in cell mediated immune responses induced by 'Geriforte' (An Ayurvedic herbal rejuvenative drug). Antiseptic 10:560–563
24. Kumar R, Grover SK, Shyam R, Divaker HM, Gupta AK, Srivastava KK (1999) Enhanced thermogenesis in rats by a Composite Indian Herbal Preparation-I and its mechanism of action. J Altern Complement Med 5:245–251
25. Martin P, Parkhurst SM (2004) Parallels between tissue repair and embryo morphogenesis. Development 131:3021–3034
26. Purkayastha SS, Bhaumik G, Chauhan SKS et al (2002) Immediate treatment of frostbite using rapid rewarming in tea decoction followed by combined therapy of pentoxifylline, aspirin and vitamin C. Indian J Med Res 116:29–34
27. Ramachandran U, Divekar HM, Grover SK, Srivastava K (1990) New experimental model for the evaluation of adaptogenic products. J Ethanopharmacol 29:275–281
28. Scartezzini P, Speroni E (2000) Review on some plants of Indian traditional medicine with antioxidant activity. J Ethnopharmacol 71:23–43
29. Schreml S, Szeimies RM, Prantl L, Landthaler M, Babilas P (2010) Wound healing in the 21st century. J Am Acad Dermatol 63:866–881
30. Ship AG (1977) Is topical *Aloe vera* plant mucus helpful in burn treatment? JAMA 238:1770
31. Singer AJ, Clark RAF (1999) Cutaneous wound healing. N Engl J Med 341:738–746
32. Singh N, Nath N, Misra N, Kohli RP (1978) An experimental evaluation of anti stress effects of Gariforti (An Ayurvedic drug). J Crude Drug Res 16:125–136
33. Srivastava KK, Grover SK, Ramachandran U (1990) Membrane integrity changes at high terrestrial altitudes. Probe 29:112–117
34. t'Hart LA, Nibbering PH, van den Barselaar MT, van Dijk H, van den Berg AJ, Labadie RP (1990) Effects of low molecular constituents from *Aloe vera* gel on oxidative metabolism and cytotoxic and bactericidal activities of human neutrophils. Int J Immunopharmacol 12(4):427–434
35. Weatherly-White RC, Sjostrom B, Paton BC (1964) Experimental studies in cold injury. II. The pathogenesis of frostbite. J Surg Res 4:17–22
36. Yildirimer L, Thanh NTK, Seifalian AM (2012) Skin regeneration scaffolds: a multimodal bottom-up approach. Trends Biotechnol 30:638–648

Lessons from a 20-Year Investigation of Intermittent Hypoxia: Principles and Practices

T.V. Serebrovskaya

Abstract

Widespread use of the intermittent hypoxic training/treatment (IHT) methods in sports, military, and medical practice during recent decades has provoked a discussion: "What is 'intermittent hypoxia'?" In contrast to studies from the former Soviet Union countries that emphasized mainly the beneficial effects of IHT on an organism, intermittent hypoxia research in Western Europe and North America was primarily focused on the detrimental effects associated with sleep apnea. However, during the past decade, such a gap of division between East and West is progressively shrinking, and mutual understanding on what "intermittent hypoxia" means becomes clearer. Potential mechanisms underlying both beneficial and adverse effects of IHT have been described. Basic investigations led to the proliferation of various methods of IHT exposure and the development of different medical equipment – hypoxicators – for its implementation in sport practice and military operations and also for clinical application. However, wide array of different protocols and measurements makes the results difficult to harmonize. Meanwhile, the mode of hypoxic influence (depth, duration, and intermittence) appeared to be critical for the determination of healing or harmful result. Therefore, special purposeful investigations are needed to elucidate basic mechanisms of different IHT effects depending on the modality of hypoxic stimuli and elaborate the most effective and safe regimen for the introduction in human practice.

Introduction

Intermittent hypoxia (periodic hypoxia, interval hypoxia, hypoxic preconditioning, etc.) became today "the talk of the town" among physiologists and clinicians who deal with hypoxic problems. Although the roots of this topic go deep into Middle Ages, sharply intensifying in the 30th year of the twentieth century in Soviet Union due to military needs, the most fundamental investigations were made during the last two decades. The number of publications indexed in PubMed under the keyword "intermittent hypoxia" increased from 49 in 1993 to 520 during

T.V. Serebrovskaya (✉)
Department of Hypoxic States, Bogomoletz Institute of Physiology NASU, Kiev, Ukraine
e-mail: sereb@biph.kiev.ua

S.B. Singh et al. (eds.), *Translational Research in Environmental and Occupational Stress*,
DOI 10.1007/978-81-322-1928-6_22, © Springer India 2014

the first half-year of 2013. Several monographs have been published [1–4].

Many types of protocol with different numbers of hypoxia episodes, severity, and total exposure duration have been used by investigators, and these combinations may have resulted in various physiological responses. Principles of IHT application for cell cultures and animal experiments (mice, dogs, cats, rabbits, pigs, horses, and even insects) have been elaborated. A variety of technical implementations for treatment of animals and humans have been tested.

Widespread use of the intermittent hypoxic training/treatment (IHT) methods in sports, military, and medical practice during recent decades has provoked a discussion: "What is 'intermittent hypoxia'?" [5]. All papers using this term should be divided into four main classes: (1) hypoxic hypoxia (intermittent hypoxic training using gas mixtures or barochambers, recurrent sojourn at high altitudes, hypoxic preconditioning in stem cell transplantation therapy), (2) ischemic preconditioning (cardiac, cerebral, etc.), (3) hypoxia induced by breath holding (divers, yogic technique *pranayama*, training with extra dead space), and (4) obstructive sleep apnea syndrome (OSAS) and other diseases associated with brainstem disorders.

The three first classes are generally considered as beneficially influencing on an organism, whereas the fourth one (which is characterized by the similar pattern of hypoxic and normoxic episodes) is an example of the pathological process. Rats exposed to chronic intermittent hypoxia (CIH) simulating recurrent apnea in OSAS patients demonstrate autonomic morbidities and hypertension similar to those described in recurrent apneic patients [6, 7 and many others]. Meanwhile, such comparison seems to be rather mechanistic because it does not take into account several significant differences between other factors accompanying hypoxia in these four paradigms.

For example, most researchers do not take into account that IHT methods in the vast majority of cases use eucapnic hypoxia which results in hyperventilation and hence hypocapnia. At the same time, ischemic preconditioning which was proved to activate endogenous defense mechanisms and shows marked protective effects is accompanied by hypercapnia, acidosis, and the accumulation of metabolites absent during IHT. In experiments on rats, only hypoxic component is modulated, whereas inspired CO_2 is maintained at normal level. Meanwhile, pCO_2 and pH play one of the main regulative roles in respiration and metabolism and could affect the organism very differently from hypoxia per se. Intracellular acidosis due to hypercapnia raises concerns about potential harmful effects. In contrast to intermittent hypoxia, the effects of intermittent hypercapnia and its cohabitation with hypoxia are the areas of research that remain to be explored. Therefore, a direct comparison of IHT, ischemia, and sleep apnea effects seems inconsistent.

Although intermittent hypoxia research in Western Europe and North America was primarily focused on the detrimental effects of chronic intermittent hypoxia associated with sleep-disordered breathing, during the last decade such a gap of division is progressively shrinking, and mutual understanding on what "intermittent hypoxia" means becomes clearer.

In this mini-review we will just outline the main recent achievements in the field of intermittent hypoxia focusing on recent advances in the mechanisms of IH investigation.

Mechanisms

An impressive amount of scientific information has been gathered with regard to the responses to hypoxia, from the integrative systems level to the molecular and genomic level, such as (1) regulation of respiration and circulation, (2) free radical production, (3) mitochondrial respiration, (4) role of genetic factors (HIF, MTF-1, NF-βκ, c-Fos, c-Jun, etc.), and (5) epigenetic mechanisms of adaptation to IH. Repeated exposures to hypoxia have been examined for both their beneficial and adverse effects. The following questions arise: what are the key mechanisms determining the adaptive versus maladaptive nature of different paradigms of intermittent hypoxia and what molecular pathways are mediating the observed pathological or physiological response? Until now there is no exact evidence about the precise mechanism for switching adaptive or maladaptive responses to hypoxic impact. The most important arguments are presented in recent papers [8, 9].

Many discoveries demonstrated that intermittent hypoxia leads to remodeling of the carotid body function manifested by augmented sensory response to hypoxia and induction of sensory long-term facilitation (LTF). More than 20 years ago we have shown that intermittent normobaric hypoxia augments hypoxic ventilatory response (HVR) and does not substantially influence hypercapnic ventilatory sensitivity (HCVR) [10]. Later on John Weil and his co-workers [11] described variations in the HVR in human subjects. There are many reviews that reflected further investigations in this field [12–14 and oth.]. Recent studies strongly indicate that endothelin-1 takes part in this process resulting from reactive oxygen species-dependent activation of endothelin-converting enzyme [15]. The role of such gasotransmitters as nitric oxide, carbon dioxide, and hydrogen sulfide (H2S) in the regulation of respiration under intermittent hypoxia was excellently described by N. Prabhakar, 2013 [16].

It is widely known that during acute episodes of hypoxia, chemoreceptor-mediated sympathetic activity increases heart rate, cardiac output, peripheral resistance, and systemic arterial pressure. Tyrosine hydroxylase (TH) is the rate-limiting enzyme for catecholamine synthesis. Several mechanisms contribute to the short- and long-term regulation of TH which are well established. IH-mediated activation of TH leads to the increase in catecholamine level in the brainstem and adrenal medulla [9]. In our lab, it was shown that a 2-week IHT course increased dopamine synthesis in adult and old rats and the animals with experimental Parkinson's disease (PD), especially in the right striatum, restoring partially the skewness of DA distribution between brain hemispheres which has been lost during aging [17].

However, different IH paradigms produce remarkably divergent effects on systemic arterial pressure in the posthypoxic steady state [18]. The hypertensive effects of OSA versus the depressor effects of therapeutic hypoxia exemplify this divergence. Why do OSA and IHT produce such disparate effects on blood pressure? It is useful to consider the fundamental differences between the two phenomena: duration of hypoxic periods, hypercapnia and acidemia versus hypocapnia and alkalemia, hypoxic episodes occurring at day- or nighttime, etc. As a result, OSA ignites a crescendo of factors which activate the sympathetic nervous system and systemic inflammation, culminating in maladaptive, persistent hypertension. In contrast, therapeutic IHT activates the parasympathetic system and dampens other factors.

Another IH effects on the cardiorespiratory system should be only mentioned here. There are increased alveolar ventilation and lung diffusion capacity, increased hematopoiesis, increased capillary density and tissue perfusion, suppressed function of mitochondrial enzyme complex I (MEC I), and the alternative activation of MEC II (see reviews [8, 13, 19–22]). Some authors [23] consider intermittent hypoxia as a multifunctional tool of a natural mitochondria-rejuvenative strategy.

Besides, hypoxic exposure significantly increases the tolerance and regenerative properties of stem cells and progenitor cells. During the last decade it was shown that short-term hypoxic exposures can mobilize hematopoietic stem cells (HSC) and increase their presence in peripheral circulation [24–27]. Different intensities and durations of hypoxia could have important and diverse effects on stem cell development. Special study was designed to compare the effects of intermittent versus acute hypoxia on human HSCs and some immune parameters [28]. The effect of a 2-week program of cyclic 5 min exposures to 10 % O_2 were (1) decrease in circulating hematopoietic stem cells, (2) complement activation, and (3) phagocytic and bactericidal activities of neutrophil stimulation while suppressing proinflammatory cytokines. In contrast to the 14d program, a single IHT session provoked appreciable yet transitory increase in circulating HSC which quickly subsided after hypoxic exposures. Results raise the possibility that IH induces HSC emigration from niches into the circulation, followed by homing and sequestration in target tissues during posthypoxic recovery. The IH-induced decrease in blood TNF-α content with simultaneous increase in IFN-γ could contribute to the moderation of infectious inflammatory processes.

One of the key mechanisms of cell damage during hypoxia and reoxygenation is an excessive production of reactive oxygen and nitrogen species (ROS and RNS) in mitochondria. ROS and RNS generation leads to mitochondrial protein, lipid, and DNA oxidation which impedes normal mitochondrial physiology and initiates cellular death pathways [29]. On the other hand, ROS function as signaling molecules in a variety of physiological systems [30, 31]. Several attempts were undertaken to analyze this question [20, 32, 33]. It was shown that low levels of ROS production are protective and may serve as a trigger for hypoxic adaptations. At the cellular level, intermittent hypoxia leads to reprogramming of mitochondrial metabolism that ensures adequate ATP generation and prevents adverse consequences of excess mitochondrial ROS generation. These metabolic adaptations are due to hypoxia-inducible factors 1 and 2 (HIF-1 and HIF-2) transcriptional regulation of glycolytic enzymes, mitochondrial electron transport chain components, and other metabolic enzymes [8, 34]. Recent studies have shown that HIF-1 and HIF-2-regulate the expression of gene products with opposing functions that regulate the redox state [16]. For instance, HIF-1 regulates the expression of prooxidant enzymes, including NADPH oxidases, whereas HIF-2 regulates the expression of antioxidant enzymes.

In our lab, Drevytska et al. [35] investigated the role of another subunit – HIF-3α – in adaptation to IH and physical load. It was shown that this subunit plays a negative role in the adaptation to hypoxia. HIF-3α mRNA expression increased sharply under acute hypoxia in the heart, lung, and kidney but did not changed after a 5-week IHT. Inhibition of HIF-3α expression led to an increase in physical endurance. Thus, every HIF subunits plays different role in response to hypoxic load. It seems that the investigation of their ensemble functioning under different IH modes (depth, duration, and intermittence) could explain the mechanism for switching adaptive or maladaptive cellular and systemic responses to hypoxic impact.

One of the new directions in the investigation of hypoxic adaptations is epigenetics – heritable modifications of DNA that do not involve changes in the DNA primary sequence [16, 36, 37]. Epigenetic mechanisms can determine whether a gene is activated or silenced. These studies seem to be very promising in this rapidly emerging area.

While all the abovementioned fundamental studies provided important insights into mechanisms of HIF activation by hypoxia, they cannot answer as yet practical question on what dose and regimen of hypoxic impact could be mostly beneficial for animals and humans.

Use in Clinical Practice

To the present days, intermittent hypoxic training (IHT) has been used extensively for altitude preacclimatization, for treatment of a variety of clinical disorders, and in sports. Wide spectrum of protocols for IHT is represented now in literature showing both beneficial and detrimental effects. Beneficial results were shown for treatment and prophylaxis of numerous disorders in pulmonology (chronic obstructive diseases, bronchial asthma, chronic rhinitis, etc.), cardiology (ischemic heart disease, hypertension, cardiosclerosis, etc.), hematology (hypoplastic and iron deficiency anemia, postradiation hematological disturbances, etc.), neurology (functional neurological disorders, Parkinson's and Alzheimer's diseases, neurosis, syndrome of autonomic dystonia, diabetic neuropathy, psychosomatic disorders), diabetes mellitus, obstetrics and gynecology (juvenile bleedings, toxicosis of expectant mothers, pathology of climacteric period, etc.), gastrointestinal diseases (gastroduodenitis, peptic ulcer), professional diseases (pneumoconiosis, vibration- and dust-induced pathology, acute and chronic intoxication, etc.), postradiation disorders of the immune system and male reproductive system, and many others. In this mini-review we cannot mention all spectrum of papers devoted to this problem. The interested reader is referred to several reviews and monographs [3, 4, 37–39 and many others]. Much literature may be found on the websites www.go2altitude.com and www.bionova.ru. Here we mention just some last publications.

IHT clinical applications are clearly presented by S. Basovich in his last review, 2013 [40]. Among others, he described beneficial results of IHT application for treatment of bronchial asthma, chronic obstructive pulmonary disease, and hypertension; to correct abnormalities during pregnancy; in epilepsy treatment; for preparation of patients to surgery to increase nonspecific resistance, etc. The efficacy of IHT was demonstrated for improving male subfertility and other andrological disorders [41]. Intermittent hypoxia protocols may be developed for treatment and prevention of osteopenia and osteoporosis [42, 43].

Recently, a new mode of adaptive training was explored, which combines periods of hypoxia and hyperoxia [44–46]. A novel principle of short-term periodic adaptive training by varying the oxygen level from hypoxia to hyperoxia is substantiated both theoretically and experimentally. Studies support the viewpoint that moderate periodic generation of free radical signal during hypoxic/hyperoxic bouts causes better induction of antioxidant enzyme protein synthesis than hypoxic/normoxic exposures that may be an important trigger for specific adaptations.

Another new direction in IHT application is developing during the last years: hypoxic postconditioning [47–51]. While preconditioning is induced before stroke onset, experiments on animals have shown that ischemic postconditioning performed after reperfusion attenuates brain injury. Clinical investigations testify on cardioprotective impact of postconditioning in patients with acute myocardial infarction and cardiac surgery patients.

Some works are devoted to the application of hypoxic-hypercapnic or intermittent hypercapnic treatment to clinical practice. This question is elucidated in the review of Pokorski and Serebrovskaya [52]. The effects of hypercapnia are somewhat surprising. CO_2 is a recognized vasodilator of myocardial blood vessels; it is capable to substantially increase cerebral blood flow leading to increased tissue oxygenation. Hypercapnic acidosis may have a beneficial effect in its own right in severe respiratory conditions and may, paradoxically, be helpful in patients with organ failure due to ischemia-reperfusion-related cellular injury. That brings us to the use of "therapeutic hypercapnia," a purposefully increased inspired CO_2 concentration to achieve some beneficial health effects. Hypoxia and hypercapnia, used in tandem, may strengthen the curative effects of either. So, intermittent hypercapnia seems an obvious area of future research focusing not only on the mechanisms of long-term potentiation and synaptic plasticity in the brainstem respiratory network but also on the health-related applicability of this kind of respiratory strategy. The controversies that surround the use of therapeutic hypercapnia uphold research interest. The potential of intermittent hypercapnia is just starting to be realized and hopefully will be further explored.

During the past few years, numerous debates about the ethical evaluation of diagnostic and therapeutic use of hypoxia in humans are raised. Although the works devoted to this problem obtained the approval from the Human Research Ethics Committees, there is the lack of evidences about strong evaluation of risk/benefit ratio. The analysis of such ratio and the creation of standardized guidelines for hypoxic treatment/training application are complicated due to the differences in criteria for individual dosage and utilized methods. One of the attempts to solve this problem was made by applying a new mathematical method – "Method of Expert Assessing Scales" (MEAS) – for the estimation of IHT application safety in human practice [53]. MEAS dilates capabilities of traditional probabilistic safety assessment and allows determining the danger degree at the most early stage of its development and fulfilling well-timed actions for danger prevention. It includes the description of (a) hazard causal factors, (b) situations as a set of values of causal factors, (c) influences of separate factors on the origin of basic events, and (d) joint influence of factors on basic event probability. The methodology provides the forming of the system of indexes characterizing the risk of IHT-negative effects and determination of legitimate value scopes for basic physiological parameters, creation of the classification system allowing to set human individual cardiorespiratory reactivity, and development of proper IHT regimen for every class of reactivity.

But this is just one of the first steps which is far from the elaboration of concrete methodic recommendations. Mode of hypoxic influence (depth, duration, and intermittence) appeared to be critical for the determination of beneficial or detrimental effects of IHT. Low doses of hypoxia might not be sufficient stimuli to mobilize adaptive mechanisms, while severe or prolonged hypoxia may provoke dangerous pathological processes. Meanwhile, in practice hypoxic regimens which are used for the study of hypoxic adaptations vary broadly from 3 to 12 short hypoxic sessions (2–10 min) with 2–20 min normoxic breaks during 7–30 days to hypoxic influences lasting from 1 to 12 h during 2–90 days. In our lab, we compared the effects of the five most spread modes of IHT on rat gastrocnemius muscle PO_2 and heart and liver mitochondrial respiration [54]. Minutes of hypoxia, % O_2, and recovery minutes on air in each mode were (1) 5, 12 %, 5; (2) 15, 12 %, 15; (3) 5, 12 %, 15; (4) 5, 7 %, 5; and (5) 5, 7 %, 15. Our experimental data indicated that among 5 tested modes of IHT, optimal hypoxic dose for muscle oxygen supply is 5-min breathing with 12 % O_2 gas mixture and 5-min breaks (Mode 1), 5–6 times a day during 2 or 3 weeks. Under such mode, PmO_2 dropped minimally to the end of every hypoxic period and recovered quickly after every hypoxic set to initial level or even exceeded it. A 2-week training with this mode raised basal tissue oxygenation during normoxia and provided higher PmO_2 level during acute hypoxia. Such mode caused the substrate-dependent reorganization of liver and heart mitochondrial energy metabolism favoring NADH-dependent oxidation and improving the efficiency of oxidative phosphorylation.

However, we must take into account that all these beneficial results were obtained on rat models. Are we ready to propose this as a clinical therapeutic method? More rigorous studies need to be provided in the near future on patients with several diseases. Besides, in actual human practice including sports and military applications of hypoxic training [55], the IHT regimen (the degree of hypoxia, exposure duration, and number of sessions) could be also titrated to the mission requirements, such as the operational target altitude, risk of developing acute mountain sickness, or anticipated physical activity levels.

Basic investigations led to the proliferation of various methods of IHT exposure and the development of different medical equipment – hypoxicators – for its implementation in sport practice and military operations and also for clinical application [56].

In conclusion, intermittent hypoxic treatment/training represents a promising field of study in prevention and treatment of many diseases. The proper choice of the hypoxic dosage depending on individual's reactivity must be titrated for each patient to avoid negative effects of hypoxia and to augment the favorable properties. We can envisage a bright future for individualized IHT, which may play a significant role in the fast-developing field of personalized preventive medicine against various human diseases.

References

1. Kolb JC (2004) Physiological responses to intermittent hypoxia in humans. German Sport University, Boca Raton, 196pp
2. Gozal D, Kheirandish-Gozal L, Yang Wang, Zhang XL (2011) In: Pack AI (ed) Sleep apnea: pathogenesis, diagnosis and treatment. CRC Press, 570pp
3. Xi L, Serebrovskaya T (eds) (2009) Intermittent hypoxia: from molecular mechanisms to clinical applications. Nova Science, New York, p 615
4. Xi L, Serebrovskaya T (eds) (2012) Intermittent hypoxia and human diseases. Springer, London, p 316
5. Semenza GL (2012) Foreword. In: Xi L, Serebrovskaya TV (eds) Intermittent hypoxia and human diseases. Springer, London

6. Fletcher EC (2001) Invited review: physiological consequences of intermittent hypoxia: systemic blood pressure. J Appl Physiol 90(4):1600–1605
7. Prabhakar NR, Peng YJ, Jacono FJ, Kumar GK, Dick TE (2005) Cardiovascular alterations by chronic intermittent hypoxia: importance of carotid body chemoreflexes. Clin Exp Pharmacol Physiol 32:447–449
8. Prabhakar NR, Semenza GL (2012) Adaptive and maladaptive cardiorespiratory responses to continuous and intermittent hypoxia mediated by hypoxia-inducible factors 1 and 2. Physiol Rev 92:967–1003
9. Raghuraman G, Prabhakar NR, Kumar GK (2012) Differential regulation of tyrosine hydroxylase by continuous and intermittent hypoxia. Adv Exp Med Biol 758:381–385
10. Serebrovskaya TV (1992) Comparison of respiratory and circulatory human responses to progressive hypoxia and hypercapnia. Respiration 59(1):35–41
11. Weil JV (2003) Variation in human ventilatory control – genetic influence on the hypoxic ventilatory response. Respir Physiol Neurobiol 135:239–246
12. Prabhakar NR, Kline DD (2002) Ventilatory changes during intermittent hypoxia: importance of pattern and duration. High Alt Med Biol 3(2):195–204
13. Serebrovskaya TV (2002) Intermittent hypoxia research in the former Soviet Union and the Commonwealth of Independent States (CIS): history and review of the concept and selected applications. High Alt Med Biol 3(2):205–221
14. Teppema LJ, Dahan A (2010) The ventilatory response to hypoxia in mammals: mechanisms, measurement, and analysis. Physiol Rev 90(2):675–754
15. Peng YJ, Nanduri J, Raghuraman G, Wang N, Kumar GK, Prabhakar NR (2013) Role of oxidative stress induced endothelin converting enzyme activity in the alteration of carotid body function by chronic intermittent hypoxia. Exp Physiol 98(11):1620–1630
16. Prabhakar NR (2013) Sensing hypoxia: physiology, genetics and epigenetics. J Physiol 591(9):2245–2257
17. Belikova MV, Kolesnikova EE, Serebrovskaya TV (2012) Intermittent hypoxia and experimental Parkinson's disease (Chapter 12). In: Xi L, Serebrovskaya TV (eds) Intermittent hypoxia and human diseases. Springer, London, pp 147–153
18. Serebrovskaya TV, Manukhina EB, Smith ML, Downey HF, Mallet RT (2008) Intermittent hypoxia: cause of or therapy for systemic hypertension? Exp Biol Med (Maywood) 233(6):627–650
19. Schmidt W (2002) Effects of intermittent exposure to high altitude on blood volume and erythropoietic activity. High Alt Med Biol 3(2):167–176
20. Lukyanova LD, Germanova EL, Kopaladze RA (2009) Development of resistance of an organism under various conditions of hypoxic preconditioning: role of the hypoxic period and reoxygenation. Bull Exp Biol Med 147(4):400–404 [In Russian]
21. Faiss R, Léger B, Vesin JM, Fournier PE, Eggel Y, Dériaz O, Millet GP (2013) Significant molecular and systemic adaptations after repeated sprint training in hypoxia. PLoS One 8(2):e56522
22. Mankovska IM, Gavenauskas BL, Hosar VI, Nazarenko AI, Rozova KV, Bratus LV (2009) Mechanisms of muscular tissue adaptation to intermittent hypoxia. Sport Med (Ukr) 1:3–11 [In Ukrainian]
23. Prokopov AF (2012) Intermittent hypoxia and health: from evolutionary aspects to mitochondria rejuvenation (Chapter 21). In: Xi L, Serebrovskaya TV (eds) Intermittent hypoxia and human diseases. Springer, London, pp 253–269
24. Viscor G, Javierre C, Pagès T, Ventura JL, Ricart A, Martin-Henao G, Azqueta C, Segura R (2009) Combined intermittent hypoxia and surface muscle electrostimulation as a method to increase peripheral blood progenitor cell concentration. J Transl Med 29:7–91
25. Serebrovskaya TV, Nikolsky IS, Nikolska VV, Mallet RT, Ishchuk VA (2011) Intermittent hypoxia mobilizes hematopoietic progenitors and augments cellular and humoral elements of innate immunity in adult men. High Alt Med Biol 12(3):243–252
26. Tang YL, Zhu W, Cheng M, Chen L, Zhang J, Sun T, Kishore R, Phillips MI, Losordo DW, Qin G (2009) Hypoxic preconditioning enhances the benefit of cardiac progenitor cell therapy for treatment of myocardial infarction by inducing CXCR4 expression. Circ Res 104(10):1209–1216
27. Ranera B, Remacha AR, Álvarez-Arguedas S, Romero A, Vázquez FJ, Zaragoza P, Martín-Burriel I, Rodellar C (2012) Effect of hypoxia on equine mesenchymal stem cells derived from bone marrow and adipose tissue. BMC Vet Res 8:142
28. Serebrovska TV, Nikolsky I, Ishchuk V (2010) Human adaptation to intermittent hypoxia: effects on hematopoietic stem cells and immune function. In: Wang P, Kuo C-H, Takeda N, Singal PK (eds) Cell adaptations and challenges, vol 6, Adaptation biology and medicine. Narosa Publisher, Canada, pp 181–191
29. Thompson JW, Dave KR, Young JI, Perez-Pinzon MA (2013) Ischemic preconditioning alters the epigenetic profile of the brain from ischemic intolerance to ischemic tolerance. Neurotherapeutics 10(4):789–797
30. Prabhakar NR (2011) Sensory plasticity of the carotid body: role of reactive oxygen species and physiological significance. Respir Physiol Neurobiol 178(3):375–380
31. Wang ZH, Chen YX, Zhang CM, Wu L, Yu Z, Cai XL, Guan Y, Zhou ZN, Yang HT (2011) Intermittent hypobaric hypoxia improves postischemic recovery of myocardial contractile function via redox signaling during early reperfusion. Am J Physiol Heart Circ Physiol 301(4):H1695–H1705
32. Sazontova TG, Arkhipenko YuV (2009) Intermittent hypoxia in resistance of cardiac membrane structures: role of reactive oxygen species and redox signaling (Chapter 5). In: Xi L, Serebrovskaya TV (eds) Intermittent hypoxia: from molecular mechanisms to clinical applications. Nova Science, New York, pp 113–150
33. Lukyanova LD, Kirova YuI, Germanova EL (2012) Energotropic effects of intermittent hypoxia: role of succinate-dependent signaling (Chapter 20). In: Xi L, Serebrovskaya TV (eds) Intermittent hypoxia and human diseases. Springer, London, pp 239–252

34. Semenza GL, Prabhakar NR (2007) HIF-1-dependent respiratory, cardiovascular, and redox responses to chronic intermittent hypoxia. Antioxid Redox Signal 9(9):1391–1396
35. Drevytska T, Gavenauskas B, Drozdovska S, Nosar V, Dosenko V, Mankovska I (2012) HIF-3α mRNA expression changes in different tissues and their role in adaptation to intermittent hypoxia and physical exercise. Pathophysiology 19(3):205–214
36. Gluckman PD (2011) Epigenetics and metabolism in 2011: epigenetics, the life-course and metabolic disease. Nat Rev Endocrinol 8(2):74–76
37. Nanduri J, Makarenko V, Reddy VD, Yuan G, Pawar A, Wang N, Khan SA, Zhang X, Kinsman B, Peng YJ, Kumar GK, Fox AP, Godley LA, Semenza GL, Prabhakar NR (2012) Epigenetic regulation of hypoxic sensing disrupts cardiorespiratory homeostasis. Proc Natl Acad Sci U S A 109(7):2515–2520
38. Karash YM, Strelkov RB, Chizhov AY (1988) Normobaric hypoxia in treatment, prophylaxis and rehabilitation. Meditsina, Moscow, p 352
39. Berezovskii VA, Levashov MI (1992) Physiological premises and mechanisms of normalizing effect of normobaric hypoxia and inhalation therapy. Fiziol Zh 38(5):3–12
40. Fesenko ME, Lisyana TO (1992) Approach to employment of hypoxic stimulation for treatment of lingering and relapsing bronchitis in children of early age. Fiziol Zh 38(5):31–33
41. Basovich SN (2013) Trends in the use of preconditioning to hypoxia for early prevention of future life diseases. Biosci Trends 7(1):23–32
42. Swanson RJ, Serebrovska ZA (2012) Intermittent hypoxia remedies male subfertility. In: Xi L, Serebrovskaya TV (eds) Intermittent hypoxia and human diseases (Chapter 18). Springer, London, pp 221–227
43. Berezovskii VA, Litovka IG, Kostyuchenko AS (2004) Low oxygen tension may defence the bone tissue from unloading simulated osteopenia. J Gravit Physiol 11(2):153–154
44. Guner I, Uzun DD, Yaman MO, Genc H, Gelisgen R, Korkmaz GG, Hallac M, Yelmen N, Sahin G, Karter Y, Simsek G (2013) The effect of chronic long-term intermittent hypobaric hypoxia on bone mineral density in rats: role of nitric oxide. Biol Trace Elem Res 154(2):262–267
45. Arkhipenko YV, Sazontova TG, Zhukova AG (2005) Adaptation to periodic hypoxia and hyperoxia improves resistance of membrane structures in heart, liver, and brain. Bull Exp Biol Med 140(3):278–281
46. Glazachev OS, Zvenigorodskaia LA, Dudnik EN, Iartseva LA, Mishchenkova TV, Platonenko AV, Spirina GK (2010) Interval hypoxic-hyperoxic training in the treatment of the metabolic syndrome. Eksp Klin Gastroenterol 7:51–56
47. Gonchar O, Mankovska I (2012) Moderate hypoxia/hyperoxia attenuates acute hypoxia-induced oxidative damage and improves antioxidant defense in lung mitochondria. Acta Physiol Hung 99(4):436–446
48. Leconte C, Tixier E, Freret T, Toutain J, Saulnier R, Boulouard M, Roussel S, Schumann-Bard P, Bernaudin M (2009) Delayed hypoxic postconditioning protects against cerebral ischemia in the mouse. Stroke 40(10):3349–3355
49. Maslov LN, Mrochek AG, Hanus L, Pei J, Zhang Y, Wang H, Naryzhnaia NV (2012) Phenomenon of heart ischemic postconditioning. Ross Fiziol Zh Im I M Sechenova 98(8):943–961
50. Joo SP, Xie W, Xiong X, Xu B, Zhao H (2013) Ischemic postconditioning protects against focal cerebral ischemia by inhibiting brain inflammation while attenuating peripheral lymphopenia in mice. Neuroscience 243:149–157
51. Xie R, Wang P, Ji X, Zhao H (2013) Ischemic post-conditioning facilitates brain recovery after stroke by promoting Akt/mTOR activity in nude rats. J Neurochem [Epub ahead of print]
52. Pokorski M, Serebrovskaya T (2009) Intermittent hypercapnia. In: Xi L, Serebrovskaya TV (eds) Intermittent hypoxia: from molecular mechanisms to clinical applications. Nova Science, New York, pp 297–310
53. Serebrovsky A, Serebrovska T (2009) Models and algorithms for the assessment of intermittent hypoxia application safety and efficacy in medical practice. In: Hypoxia and consequences: from molecule to malady. New York, Book of Abstracts, Session II, abstract # 25
54. Serebrovskaya TV, Nosar VI, Bratus LV, Gavenauskas BL, Mankovska IM (2013) Tissue oxygenation and mitochondrial respiration under different modes of intermittent hypoxia. High Alt Med Biol 14(3):280–288
55. Muza SR (2007) Military applications of hypoxic training for high-altitude operations. Med Sci Sports Exerc 39(9):1625–1631
56. Lopata VA, Serebrovskaya TV (2012) Hypoxicators: review of the operating principles and constructions (Chapter 24). In: Xi L, Serebrovskaya TV (eds) Intermittent hypoxia and human diseases. Springer, London, pp 291–302